POWER QUALITY

VAR Compensation in Power Systems

POWER QUALITY

VAR Compensation in Power Systems

R. Sastry Vedam
Mulukutla S. Sarma

CRC Press is an imprint of the
Taylor & Francis Group, an **informa** business

CRC Press
Taylor & Francis Group
6000 Broken Sound Parkway NW, Suite 300
Boca Raton, FL 33487-2742

Printed in the United States of America on acid-free paper
10 9 8 7 6 5 4 3 2 1

International Standard Book Number-13: 978-1-4200-6480-3 (Hardcover)

Visit the Taylor & Francis Web site at
http://www.taylorandfrancis.com

and the CRC Press Web site at
http://www.crcpress.com

Dedication

This book is dedicated to all the members of my family:
To my parents, Narayaniah and Kamakshamma, who are
no more to see the achievements of their son
To my wife, Amruta Lakshmi
To my children, Rama, Uma, and Hima
To David, who has become a new member of this family
To all the other extended members of the family,
too numerous to list here
and the most important member of the family, who is also
the future of this family, granddaughter Manasa Veena

R. Sastry Vedam

Contents

Chapter 9

Preface

In many countries the electricity supply industry has been deregulated. Though the designs of electricity markets vary, there are some common features. Generation, transmission, and distribution functions have been split up. There is competition in generation and distribution, whereas the situation in transmission everywhere approaches a monopoly. Further, the regulatory authorities ensure that no company providing power may overcharge customers and make huge profits. There are also regulations to help ensure the reliability and quality of power supply. The utilities must ensure a reasonable quality of supply, that is, without too many sags, swells, and high harmonic voltages affecting the performance of customer equipment. These changes have put considerable pressure on the utilities to become efficient and at the same time provide reliable supply without too many interruptions.

In addition to the above effects of the deregulation and creation of today's electricity markets, utilities have to supply increased loads to the customers. There has been widespread use of power electronic equipment that generates harmonics. To meet these increased loads, utilities would like to strengthen the transmission system by building more interconnectors. It is rather difficult to obtain permission from the local authorities for new routes of transmission lines due to (a) the increase in built-up areas, (b) the need to retain good farm land, and (c) other aesthetic and environmental considerations. Hence, utilities have to find other technical solutions like static VAR compensators (SVCs), capacitors, and reactors for reactive compensation purposes. At the same time, utilities must ensure that their power quality is not affected due to harmonics generated by electronic equipment. In quite a few situations such as high-voltage DC transmission (HVDC), SVCs-suitable harmonic filters must be provided.

There is a realization by most governments that they must decrease the use of fossil fuels and increase the use of the other renewal energy sources—wind, solar, fuel cell, geothermal technologies, etc. These technologies also use power electronic devices to connect the generated power to the grid.

Use of FACTS (flexible AC transmission systems) devices, SVCs, universal power flow controllers, dynamic voltage regulators, etc., is increasing, and so are solid-state devices, mobile phones, color TV sets, personal computers, and fluorescent tubes. This, in turn, raises harmonic levels in the power system. All these create power quality problems.

Furthermore, to reduce labor costs for operation, and to increase reliability of supply, single-pole auto recloning is employed. Because 80% of the single-phase-to-ground faults are temporary due to lightning, etc., the power supply can be restored quickly by installing single-pole autoreclosing facilities.

Because transmission lines and interconnections are loaded to full capacity, any faults on the system or elsewhere cause reliability problems. In addition to these negatives, unbalanced distortion of three-phase voltages can occur if there are significant single-phase loads (e.g., traction loads) in an area. Hence, this reference book was undertaken to cover in a single volume all of these different topics: power quality; harmonics—their sources, analysis, measurement, and monitoring; harmonic filters and their design; harmonic standards; SVCs in main transmission systems, and SVCs for compensating traction loads to reduce flicker due to arc furnaces, that is, for load balancing, power factor improvement, and improving power system damping in general.

Some professional engineers may have to write specifications for the purchase of the equipment and later may be involved in their operation and maintenance. Others may be working on single-pole autoreclosing projects and the design of neutral reactors in four-reactor schemes to improve reliability of supply, transient stability, and also other traditional reactive compensation devices like shunt capacitors and shunt reactors. In a few cases, series capacitors to increase transmitted power are used. All these are reviewed.

This book is organized into 11 chapters: Chapter 1 deals with general terminology used in power quality, reliability, and the ITI (CBEMA) curve. Chapter 2 discusses SVCs used in main transmission systems, as well as those used to correct unbalances due to traction loads. As wye–delta transformers are generally used with these SVCs, expressions are derived to relate the active and reactive powers from the delta side to the wye side and vice versa to enable the choice of the SVC transformer rating and associated harmonic filters.

Chapter 3 deals with the control of the SVCs both for main transmission systems and those used to compensate unsymmetric traction single-phase loads—noting the difference that whereas the transmission SVC has equal rating across the phases AB, BC, and CA, the SVCs used in the traction system have rather unequal ratings across phases AB, BC, and CA to reduce the cost of capacitors used in the system.

Chapter 4 considers harmonics and their sources. Chapter 5 describes the utility harmonic regulations and standards. Internationally, IEC standards are used, and in the United States IEEE 519-1992 is the standard. The main philosophical differences between these two sets of standards are discussed. Application of the IEEE 519-1992 to a utility system involves many practical issues. How these issues are handled in the British Columbia hydro system is also covered in this chapter.

Chapter 6 focuses on the undesirable effects of harmonics and mitigation techniques. The design details of the tuned filters, filter components, their tolerance variation, and the relative merits of the tuned and damped filters are included. To enable readers to perform a harmonic scan, models of the common power system elements like generators, transformers, and transmission lines loads, etc., are described.

Chapter 7 presents the computation tools that are available for the design and analysis of SVCs and harmonic filters. The relative merits of the digital and analog computers and the special problems that power electronic circuits present using either digital or analog computers are discussed. Brief details of programs like EMTP, ATP, and commercial programs such as PSCAD/EMTDC, PSS TME version 30, DADiSP, and SuperHarm are described. The three-phase power flow program developed at Powerlink Queensland is detailed. The main differences are in the representation of loads and transformer taps. In the positive sequence, power flow loads are represented as constant P and Q loads, whereas in three-phase power flow programs they are represented as constant impedances to ensure convergence. In the three-phase transformer models, the magnitude of one line-voltage controls the taps on the other two-line voltages; also taps move uniformly on all the three phases. In the positive-sequence power flow, only one primary (say, HV) and secondary voltage (say, LV) needs to be considered, and the taps control only one of these voltages.

Chapter 8 is devoted to the different aspects of monitoring power quality. Commonly available commercial instruments are briefly described. The method for the measurement of frequency response of the instrument transformers by Prof. A.P. Sakis Eliopoulos and his group at Georgia Institute of Technology is described in some detail. Also, the continuous harmonic analysis in real-time (CHART) system developed by Allan J.V. Miller and Michael B. Dewe at the University of Canterbury in New Zealand is included. Brief details of two projects needing harmonic survey in the Powerlink Queensland system in Australia are described, as well as the IEC flicker meter specification to measure flicker due to the arc furnaces.

Chapter 9 presents the constructional details of HV reactors and the considerations involved in the choice of their location at the end of EHV lines, tertiary windings of transformers, etc. Readers will observe how single-pole reclosing is being employed on HV and EHV lines to improve reliability, as 70–80% of the faults on these lines are single-line to ground faults; a formula was derived to enable the size of the neutral reactor in the common four-reactor scheme. Brief recommendations from the field tests of several projects regarding the secondary arc current are found in this chapter.

Chapter 10 describes the use of shunt capacitors for reactive compensation and power factor correction. An expression was derived for the percentage of overvoltage of the remaining units when some units have failed in a series group. This is helpful in the design of the capacitor banks, as it is essential to ensure that the remaining units are not unduly stressed when some units fail. Protection of the capacitor banks with different configurations grounded or ungrounded, double wyes, series capacitors, etc., are discussed. Series capacitor applications and an NGH damping scheme to avoid subsynchronous resonance are described, in addition to metal oxide arresters, their modeling, and calculation of their parameters for use in ATP program. Their typical ratings in different power systems with different voltages are included.

Chapter 11 presents fast Fourier transforms (FFT). Most older engineers experience considerable difficulty in grasping this subject. Nowadays, most harmonic instrumentation utilizes FFT techniques. Hence, a reasonable understanding of this topic is essential for intelligent choice of sampling interval and to avoid the other pitfalls while using FFT techniques. Initially, the Fourier series, discrete Fourier transform (DFT), and the Cooley–Tukey algorithm—both decimation in time and decimation in frequency (Sande's approach) versions—were described. Also, some computational details such as in-place calculations and bit reversal are explained. Initially, the results were illustrated with eight-point transforms and later extended to the general case. From the general FFT algorithm it was shown that $N \log_2 N$ complex additions and, at most, ½ N $\log_2 N$ complex multiplications, are required for the computation of the DFT of an N-point sequence, when n is a power of 2. This number is very small compared with N^2, which will be the number of operations required if the straightforward procedure is adopted for the calculation of the DFT. Brief mention is made of the other developments in this field such as the FFT of two real functions simultaneously, mixed-radix FFT, split radix FFT, and FFT pruning. Also some guidelines for using the FFT for harmonic analysis are provided.

Finally, who would benefit from reading this reference? It will be useful for most practicing engineers, whose responsibilities include power quality problems, harmonics, enforcement of the standards, analysis and measurement of harmonics, SVCs (both in main transmission and traction applications), reactive compensation using shunt reactors and capacitors, protection of capacitor banks, MOV arresters for insulation coordination, and single-pole autoreclosing. Also finding it valuable should be professional engineers who may have to write specifications for the purchase of equipment and later oversee their operation and maintenance. Some others who will find it useful will be involved in single-pole autoreclosing projects and the design of neutral reactors in four-reactor schemes to improve reliability of supply.

This book also can be used by senior students and first-year graduate students in power system courses. Although some universities might have specialized power quality courses, most will spend at least a quarter of a year dealing with power quality topics in their power system courses. The majority of the topics discussed here will need a computer to perform the involved calculations. These days many students are exposed to some basic programming, and they have some programming skills. Further, several commercial packages are available for use in the industrial environment and universities. Hence, depending upon the facilities available for a particular institution, the instructor can suggest some manageable projects to the students. Further, if there is a collaboration between a utility and a local university, then some of the problems in the industry can be taken as relevant by the concerned professor and solved with the help of graduate students.

In the initial stage of this project, the authors were considering a book much more limited in scope, more in the style of a monograph. It was Nora Konopka who encouraged the authors to embark on much broader coverage, developing it into a reference book.

Although there are books and several papers in technical journals covering power quality, harmonics, FFT, and other related subjects in depth, there is no single book with the comprehensiveness of this volume. Hence, the authors are confident that this broad coverage will appeal to many practicing engineers and students.

R. Sastry Vedam

Mulukutla S. Sarma

Acknowledgments

1. The authors thank the International Electrotechnical Commission (IEC) for permission to reproduce information from its International Standards:
   IEC 61000-4-15 ed 1.1 (2003)
   IEC 61000-3-6 ed.1.0 (1996),
   IEC 61000-4-7 (2002)
   Subclause 3.2.3
   Subclause 3.4.3
   Subclause 5.3 Table 1
   and
   IEC 61000-4-7 (1991)
   Subclause 9.3
   Subclause 10.1
   All such extracts are copyrighted by IEC, Geneva, Switzerland. All rights reserved. Further information on the IEC is available from www.iec.ch. IEC has no responsibility for the placement and context in which the extracts and contents are reproduced by the author, nor is IEC in any way responsible for the other content or accuracy therein.
2. The authors would also like to thank IEEE for permission to reproduce from several IEEE papers and IEEE standards.
3. The authors would like to thank the following organizations to reproduce the material from their Web sites:
   a. Siemens–PTI
   b. Electrotek Concepts, Inc.
   c. DSP Development Corporation
   d. Information and Technology Information Council (ITIC)
   e. EPRI from one of their journals
4. The authors would like to thank the following organizations for reports on some of the projects on which one of the authors worked while employed with Powerlink Queensland:
   a. Powerlink Queensland
   b. Queensland Railways
4. The authors thank B. Venkatesan for all the assistance he has given during the preparation of the manuscript of this book. He can justifiably call himself the third invisible author of this book. A mechanical engineer by training, he mastered the electrical engineering jargon with remarkable speed.
5. The following people have assisted in the preparation of the figures for this book: Matthew Simpson, Peter Mangan, B. Venkatesan, Bollu Mohanakrishna,

and Rama Addepalli. We very much appreciate their assistance and quality of work.

6. The printing of the two hard copies was done at the QUT printing office. Thanks are due to them for this assistance.
7. Finally, we would like to thank Nora Konopka, publisher, Engineering Book Group, and Marsha Pronin, Project Coordinator, Editorial Project Development Group, Jim McGovern, Project Editor, CRC Press, a Taylor & Francis Group company, for their interest in this project and their encouragement of the authors. We would also like to thank the other editorial staff members of CRC Press for their professional and cooperative approach in getting this manuscript of the book ready for printing and seeing that this project is successfully completed.

Finally, thanks are due to an anonymous reviewer for a painstaking thorough review and pointing out the errors and correcting punctuation, and the indexer for preparing the index.

R. Sastry Vedam

Mulukutla S. Sarma

1 Power Quality

1.1 INTRODUCTION

Power quality has assumed increasing importance in view of the widespread use of power electronic equipment. For reactive var compensation, in addition to shunt capacitors and reactors, static var compensators (SVCs) are used. They are also used to solve several power quality problems—for reducing voltage sags, overvoltages after fault clearing, voltage regulation, negative sequence voltages, etc. In some cases, harmonics can cause misoperaton of the protective equipments, contributing to a reduction in power quality. Harmonic filters are used to absorb undesirable harmonics.

Further, with the deregulation of the power industry, competitive pressures force electric utilities to cut costs, which sometimes affects power quality and reliability. Hence, it must be ensured by suitable regulations that customers do not suffer from reduced power quality and reliability.

This chapter covers issues of power quality, and later chapters deal with other topics such as SVCs, harmonics, filters and shunt capacitors, and reactors for reactive var compensation.

Already several books[1–8] and papers[9,10] have been published in the technical literature dealing with power quality. Hence, it is not necessary, nor will it be practicable, to cover the topic of power quality in detail in this chapter. The aim is to introduce the concept of power quality and its importance, and explain the common terms used in describing power quality. Any person who is interested in pursuing this subject further can find material from the references at the end of this chapter.

The term *power quality* is rather nebulous and may be associated with reliability by electric utilities. However, equipment manufacturers can interpret it rather differently, referring to those characteristics of power supply that enable the equipment to work properly. Recently, people working in the field appear to have agreed on the following definition of a power quality problem.[1,11] A power quality problem is any occurrence manifested in voltage, current, or frequency deviation that results in failure or misoperation of customer equipment. Although people talk of "power quality" quite often, they are actually referring to "voltage quality" because most of the time the controlled quantity is voltage.

Another term that is used to indicate the nonavailability of electricity supply to consumers because of sustained interruptions is *reliability*.

1.2 IMPORTANCE OF POWER QUALITY

Before the widespread use of power electronic equipment, microprocessors for industrial control, and automation in factories and offices, minor variations in power did not seriously affect the operation of conventional equipment such as lights and induction motors. If the supply voltage dipped because of a fault (i.e., a sag in voltage occurred), the lights just dimmed, and the induction motor produced a lower output. These days the effects of power interruptions are rather costly. Reference 10 lists the following cases to illustrate the cost of short-duration power interruptions:

a. One glass plant estimates that a five-cycle interruption, a momentary interruption less than a tenth of second, can cost about $200,000.
b. A major computer center reports that a 2-s interruption can cost some $600,000.
c. In some factories, following a voltage sag, the restarting of assembly lines may require clearing the lines of damaged work, restarting of boilers, and reprogramming automatic controls at a typical cost of $50,000 per incident.
d. One automaker estimated that total losses from momentary power interruptions at all its plants run to about $10 million a year.

1.3 COMMON DISTURBANCES IN POWER SYSTEMS

The common disturbances in a power system are

a. Voltage sag
b. Voltage swell
c. Momentary interruptions
d. Transients
e. Voltage unbalance
f. Harmonics
g. Voltage fluctuations

Table 1.1 (reproduced from Reference 11) describes the characteristics of these electromagnetic disturbances. The following equipment is most susceptible to these common disturbances:

Programmable logic controllers
Automated data processors
Adjustable speed drives

1.4 SHORT-DURATION VOLTAGE VARIATION

A *voltage sag* (*dip*) is defined as a decrease in the root-mean-square (rms) voltage at the power frequency for periods ranging from a half cycle to a minute.[11] It is caused by voltage drops due to fault currents or starting of large motors. Sags may trigger shutdown of process controllers or computer system crashes.

A *voltage swell* is defined as an increase up to a level between 1.1 and 1.8 pu in rms voltage at the power frequency for periods ranging from a half cycle to a minute.

An *interruption* occurs when the supply voltage decreases to less than 0.1 pu for a period of time not exceeding 1 min. Interruptions can be caused by faults, control malfunctions, or equipment failures.

TABLE 1.1
Categories and Typical Characteristics of Power System Electromagnetic Phenomena

Categories	Typical Spectral Content	Typical Duration	Typical Voltage Magnitude
1.0 Transients			
1.1 Impulsive			
1.1.1 Nanosecond	5 ns rise	<50 ns	
1.1.2 Microsecond	1 μs rise	50 ns–1 ms	
1.1.3 Millisecond	0.1 ms rise	>1 ms	
1.2 Oscillatory			
1.2.1 Low frequency	<5 kHz	0.3–50 ms	0–4 pu
1.2.2 Medium frequency	5–500 kHz	20 μs	0–8 pu
1.2.3 High frequency	0.5–5 MHz	5 μs	0–4 pu
2.0 Short-duration variations			
2.1 Instantaneous			
2.1.1 Sag		0.5–30 cycles	0.1–0.9 pu
2.1.2 Swell		0.5–30 cycles	1.1–1.8 pu
2.2 Momentary			
2.2.1 Interruption		0.5–30 cycles	<0.1 pu
2.2.2 Sag		30 cycles–3 s	0.1–0.9 pu
2.2.3 Swell		30 cycles–3 s	1.1–1.4 pu
2.3 Temporary			
2.3.1 Interruption		3 s–1 min	<0.1 pu
2.3.2 Sag		3 s–1 min	0.1–0.9 pu
2.3.3 Swell		3 s–1 min	1.1–1.2 pu
3.0 Long-duration variations			
3.1 Interruption, sustained		>1 min	0.0 pu
3.2 Undervoltages		>1 min	0.8 pu
3.3 Overvoltages		>1 min	1.1–1.2 pu
4.0 Voltage imbalance		Steady state	0.5–2%
5.0 Waveform distortion			
5.1 Dc offset		Steady state	0–0.1%
5.2 Harmonics	0–100th H	Steady state	0–20%
5.3 Interharmonics	0–6 kHz	Steady state	0–2%
5.4 Notching		Steady state	
5.5 Noise	Broadband	Steady state	0–1%
6.0 Voltage fluctuations	>25 Hz	Intermittent	0.1–7%
7.0 Power frequency variations		<10 s	

Source: From IEEE1159.3 (2003), Recommended Practice for the Transfer of Power Quality Data. With permission from IEEE.

All these types of disturbances, such as voltage sags, voltage swells, and interruptions, can be classified into three types, depending on their duration.

a. Instantaneous: 0.5–30 cycles
b. Momentary: 30 cycles–3 s
c. Temporary: 3 s–1 min

It is helpful to distinguish the term *outage* used in reliability terminology from *sustained interruption* when the supply voltage is zero for longer than 1 min. Outage refers to the state of a component in a system that has failed to function as expected and is used to quantify reliability statistics regarding continuity of service, whereas sustained interruptions as used in monitoring power quality to indicate the absence of voltage for long periods of time.

1.5 LONG-DURATION VOLTAGE VARIATIONS

An *undervoltage* is a decrease in the rms ac voltage to less than 90% at the power frequency for a duration longer than 1 min. These can be caused by switching on a large load or switching off a large capacitor bank.[1,11] Undervoltages are sometimes due to a deliberate reduction of voltage by the utility to lessen the load during periods of peak demand. These are often referred to by the nontechnical term *brownout*.

An undervoltage will lower the output from capacitor banks that a utility or customer will often install to help maintain voltage and reduce losses in the system by compensating for the inductive nature of many conductors and loads.

An *overvoltage* is an increase in the rms ac voltage to a level greater than 110% at the power frequency for a duration longer than 1 min. These are caused by switching off a large load or energizing a capacitor bank.

Incorrect tap settings on transformers can also cause undervoltages and overvoltages. As these can last several minutes, they stress computers, electronic controllers, and motors. An overvoltage may shorten the life of power system equipment and motors.

1.6 TRANSIENTS

Transients can be classified into two categories, impulsive and oscillatory, which we shall discuss in the following subsections.[1,11]

1.6.1 IMPULSIVE TRANSIENTS

An impulsive transient is a sudden nonpower frequency change in the steady-state condition of voltage or current, or both, which is unidirectional in polarity (either positive or negative). Some people use the term *surge* to describe an impulsive transient, whereas others employ it to denote *any* transient. Because this is ambiguous, it is better to avoid its use without qualification.

Impulsive transients are normally characterized by their rise and decay times. They can also be described by their spectral content. For example, a 1.2-/50-μs 4000-V

TABLE 1.2
Oscillatory Transients

	Typical Spectral Content	Typical Duration	Typical Voltage Magnitude	Examples
Low frequency	300–900 Hz	0.5–10 ms depending upon system damping	1.3–1.5 pu can reach 2.0 pu	Capacitor bank energization
Medium frequency	Tens of kHz	20 μs	0–8 pu	1. Back-to-back capacitor energization 2. Cable switching 3. System response to an impulsive transient
High frequency	>500 kHz	5 μs	0–4 pu 0.1 pu (less the 60-Hz component)	1. Switching transients 2. Commutation transients in power electronic devices

impulsive transient rises to its peak value of 4000 V in 1.2 μs, and then decays to half its peak value in 50 μs. The most common cause of impulsive transients is lightning.

1.6.2 Oscillatory Transients

An *oscillatory transient* consists of a voltage or current whose instantaneous value changes polarity rapidly. It is described by its spectral content.

Table 1.2 gives some examples of oscillatory transients.

1.6.3 Voltage Imbalance

Voltage imbalance (unbalance) is defined as the ratio of a negative- or zero-sequence component to a positive-sequence component. The voltage imbalance in a power system is due to single-phase loads. In particular, single-phase traction loads connected across different phases produce negative-phase-sequence voltages, which in many cases have to be reduced to less than 2% with the help of SVCs.

Severe voltage imbalance can lead to derating of induction motors because of excessive heating. Voltage imbalance can also occur from a blown fuse on one phase of a three-phase bank. There are occasions when a severe voltage imbalance greater than 5% can occur from single-phasing conditions.

Voltage or current imbalance is estimated sometimes (less commonly) using the following definition:

> Maximum deviation from the average of the three-phase voltages (or currents) divided by the average of the three-phase voltages (or currents)

1.7 HARMONICS

When a nonlinear load is supplied from a supply voltage of 60-Hz or 50-Hz frequency, it draws currents at more than one frequency, resulting in a distorted current waveform. Fourier analysis of this distorted current waveform resolves it into its fundamental component and different harmonics. *Harmonics* are sinusoidal voltages or currents having frequencies that are integer multiples of the fundamental frequency (usually 60 Hz or 50 Hz in power systems).

Harmonic distortion is a growing concern for many customers and the utilities because of increasing application of power electronics equipment. The nonlinear harmonic-producing devices can be modeled as current sources that inject harmonic currents into the power system.

We will discuss in detail in a later chapter the sources and disadvantages of harmonics. In this section, we will limit our discussion to some of the common terms associated with harmonic distortion.

Harmonic voltage distortion (U_n): The rms value of a harmonic voltage of order n, expressed as a percentage of the rms value of the fundamental component.
Total harmonic voltage distortion (*THD*): This is calculated from the expression THD = $\sqrt{\Sigma (U_n)^2}$ and expressed as a percentage of the fundamental component.

Similarly, harmonic current distortion and total harmonic current distortion can be defined. Harmonic distortion levels can be characterized by the complete harmonic spectrum with magnitude and phase angles of each individual harmonic component.

When dealing with harmonic current quantities, the total harmonic current distortion value can be misleading. Some harmonic-producing loads (for example, adjustable speed drives) will exhibit high total harmonic current distortion values under light load conditions. This is not a significant concern because the magnitude of the harmonic current is low even though its relative distortion is high.

To handle this concern, IEEE Std 519-1992 defines another term, *total demand distortion* (*TDD*). This is the same as total harmonic current distortion except that the distortion is expressed as a percentage of some rated load current instead of the fundamental current magnitude. Some authors use harmonic current values in amps rather than as a percentage of the fundamental current magnitude. IEEE Std 519-1992 provides guidelines for harmonic voltage and current distortion levels on transmission and distribution circuits.[12]

1.8 INTERHARMONICS

Interharmonics are defined as frequency components of voltages or currents that are not an integer multiple of the normal system frequency (e.g., 60 or 50 Hz).[1,11]

The main sources of interharmonics are static frequency converters, cycloconverters, induction motors, and arcing devices. Power line carrier signals can be considered as interharmonics. The effects of interharmonics are not well known but have been shown to affect power line carrier signaling and induce visual flicker in display devices such as cathode ray tubes (CRTs).

Two other phenomena in power electronic devices contribute to waveform distortion. These are (1) dc offset and (2) notching, which we shall explain in the following sections.

1.9 DC OFFSET

The presence of a dc voltage or current in an ac power system is termed dc offset.[1,11] This phenomenon can occur from the effect of half-wave rectification or as the result of a geomagnetic disturbance. Half-wave rectification is sometimes used in light dimmer circuits and TV power supplies. Direct current in alternating current networks can cause (1) transformer saturation with consequent increased losses, additional heating, and reduction in transformer life, and (2) the electrolytic erosion of grounding electrodes.

1.10 NOTCHING

Three-phase converters that convert ac to dc require commutation of the alternating current from one phase to another. During this period, there is a momentary short circuit between the two phases. This causes a periodic voltage disturbance, which is called *notching*.[1,11] The frequency components associated with notching can be quite high and may not be characterized with the help of measurement equipment normally used for harmonic analysis. The severity of the notch at any point in the system is determined by the source impedance and isolating inductance between the converter and the point being monitored.

1.11 NOISE

Noise refers to unwanted electrical signals (with broadband spectral content lower than 200 kHz) that produce undesirable effects in the circuits of control systems in which they occur.[1,11] (For the purpose of this definition, control systems are intended to include sensitive electronic equipment either wholly or in part.) Noise in power systems can be caused by power electronic devices, control circuits, arcing equipment, loads with solid-state rectifiers, and switching power supplies. Noise problems are often exacerbated by improper grounding. There are two types of noise voltages:

Common-mode noise voltage: A noise voltage that appears between current-carrying conductors and ground. That is, this noise voltage appears equally and in phase from each current-carrying conductor to the ground.

Normal-mode noise voltage: A noise voltage that appears between or among active circuit conductors, but not between the grounding conductor and the active circuit conductors.

Noise disturbs electronic devices such as microcomputer and programmable controllers. The problem can be mitigated by using filters, isolation transformers, and some line conditioners.

1.12 VOLTAGE FLUCTUATIONS

Loads that exhibit continuous, rapid variations in load current can cause voltage variations erroneously referred to as flicker. ANSI C84.1-1992 recommends that the system voltages should lie in the range 0.9–1.1 pu.

IEC 1000-3-3 (1994) defines various types of voltage fluctuations. We will concentrate on voltage fluctuations of the IEC 1000-3-3 (1994), Type (d). This type is characterized as systematic variations of voltage envelopes or a series of random voltages.

Arc furnaces are the most common cause of voltage fluctuations in the transmission and distribution system. Voltage fluctuations are defined by their rms magnitude expressed as a percentage of the fundamental magnitude.[1,11] They are the response of the power system to the varying load, and light flicker is the response of the lighting system as observed by the human eye. Even though flicker is caused by voltage fluctuations, some authors use the term "voltage flicker" to represent either of these terms.

Voltage fluctuations generally appear as a modulation of the fundamental frequency. Hence, the magnitude of voltage fluctuations can be obtained by demodulating the waveform to remove the fundamental frequency and then measuring the magnitude of the modulation components. Typically, magnitudes as low as 0.5% can result in perceptible light flicker if the frequencies are in the range of 6–8 Hz.

1.13 POWER FREQUENCY VARIATIONS

At any instant, the frequency depends on the balance between the load and the capacity of the available generation.[1,11] When dynamic balance changes, small changes in frequency occur. In modern interconnected power systems, frequency is controlled within a tight range as a result of good governor action. Frequency variations beyond ±0.1 Hz are likely to occur under fault conditions or from the loss of a major load or generating unit. However, in isolated systems, governor response to abrupt load changes may not be adequate to regulate them within the narrow bandwidth required by frequency-sensitive equipment.

Voltage notching can sometimes cause frequency or timing errors on power electronic machines that count zero crossings to derive frequency or time. The voltage notch may produce additional zero crossings that can cause frequency or timing errors and affect the performance of digital electric clocks.

Many other power quality terms and their definitions are listed in References 1 and 11 to promote standardization in the power quality literature.

1.14 SOLUTIONS TO POWER QUALITY PROBLEMS

Some of the solutions for power quality problems[10] adopted by the utilities and customers are shown in Table 1.3. As indicated in the table, in a given situation the ultimate solution will depend on such factors as economic issues, system configuration, the utility–customer relationship, and the customer's electrical environment.

TABLE 1.3
Solutions on Both Sides of the Meter

Solutions on Both Sides of the Meter

Disturbance	Possible Causes	Utility-Side Solutions	Customer-Side Solutions
Voltage sag	• Lightning strike • Tree or animal contact with lines	• Dynamic voltage restorer • Static condenser	• Line conditioner • Uninterruptible power supply
Overvoltage	• Fault on another phase • Load rejection	• Dynamic voltage restorer • Fault current limiter • High-energy surge arrester	• Line conditioner • Voltage regulator • Uninterruptible power supply
Interruption	• Blown fuse • Breaker operation in response to fault	• Solid-state circuit breaker • Static condenser	• Uninterruptible power supply • Motor-generator set
Transient	• Lightning strike • Utility switching	• High-energy surge arrester	• Line conditioner • Surge suppressor
Harmonic distortion	• Nonlinear loads • Ferroresonance	• Filter • Static condenser • Dynamic voltage restorer	• Line conditioner • Filter
Electrical noise	• Improper customer wiring or grounding		• Grounding and shielding • Line conditioner • Filter

Note: The solutions listed are not inclusive. In a given situation, the ultimate solution will depend on such factors as economic issues, system configuration, the utility-customer relationship, and the customer's electrical environment.

Source: Douglas, J. (December 1993). Solving problems of power quality, *EPRI Journal*, 8–15. With permission from EPRI.

1.15 AMBIGUOUS TERMS

The following terms cited below have a varied history of use in the development of the literature on power quality, and some may have specific definitions in other applications.[1,11] In view of their ambiguous and nontechnical nature, IEEE Std 1159-1995 discourages their use in describing power quality phenomena; such terms are listed below

Blink, brownout, bump, clean ground, glitch, outage, interruption, power surge, surge, wink, raw power, clean power, dirty power, dirty ground, spike.

1.16 CBEMA AND ITI CURVES

The original CBEMA curve was originally developed by the Computers Business Equipment Manufacturers Association (CBEMA) and adopted by IEEE Standard 446. CBEMA has been renamed as the Information Technology Industry (ITI) Council, and a new curve as shown in Figure 1.1 has been developed to replace the original CBEMA curve.[3,13,14] Please see the appendix to this chapter for details of the application note listed in the ITI Council Web site.

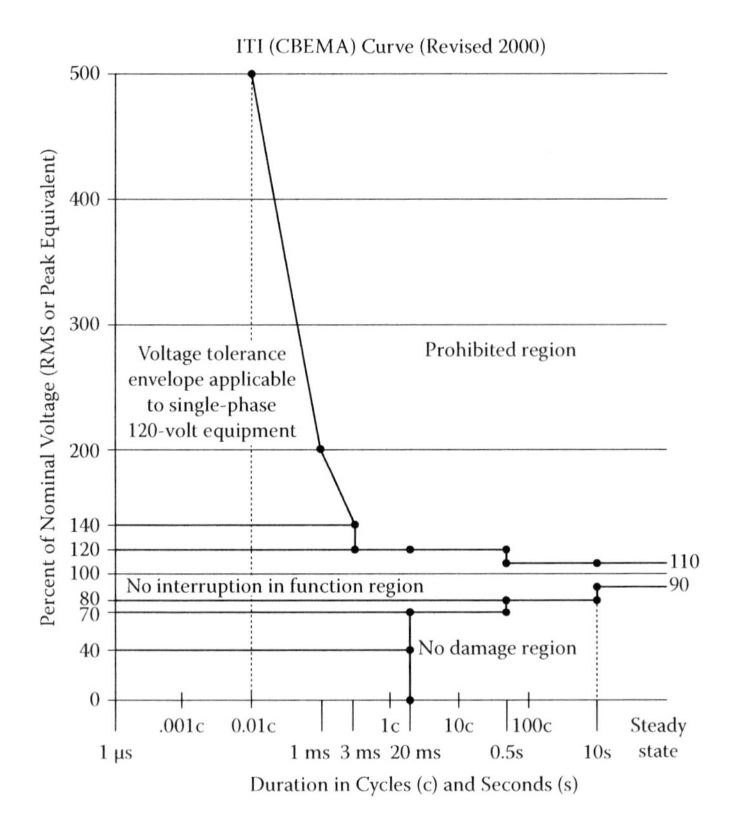

FIGURE 1.1 ITI (CBEMA) curve. (From ITI curve Web site, http://www.itic.org/iss-pol/techdocs/curve.Pdf. With permission from ITI.)

Outside the bounded tolerance region, in the no-damage region, the applied voltages are very low, and sensitive computer equipment will not function properly; however, no damage occurs to the equipment. In the prohibited region, sensitive computer equipment will be damaged due to the occurrence of severe voltage swells. Both the CBEMA and ITI curves were specially developed for use in the 60-Hz 120-V distribution voltage system. The guidelines expect users in 50-Hz 240-V distribution systems to exercise their own judgment when applying the CBEMA and ITI curves.

Although there is no legal requirement to conform to these curves, most original equipment manufacturers build equipment that meet or exceed the limits set forth by these curves, with the occasional exception.

1.17 FEATURES OF VOLTAGES IN POWER SYSTEMS

UNIPEDE (Association of the European Electricity Industry), starting in 1989, has been collecting statistics to describe the actual status in low- and medium-voltage systems.[7] On the basis of this document, CENELEC (European Committee for Electrotechnical standardization) established a European standard, EN 50160, in 1993, which described the features of voltage and frequency in public supply systems at the customer connection point (point of common coupling [PCC]). This standard has been in force since October 1995. High-voltage systems are not considered under this standard. Because this standard does not deal with electromagnetic compatibility and safety aspects, it does not have a German standard VDE (Association for Electrical, Electronic and Information Technologies founded in Germany) classification number.

Many of these limits on voltage and frequency (voltage sag limits, flicker, short- and long-time interruptions [numbers permitted, duration, etc.], limits on temporary power and frequency overvoltages and transient overvoltages, voltage unbalance limits, THD limits are stipulated in the EN 50160 standard. These quantities can be found on pages 29 and 30 of Reference 7.

1.18 GROUNDING*

Many power quality problems are related to grounding and wiring within a facility.[1,6,15–21] The National Electrical Code (NEC) and other standards provide minimum standards for grounding and wiring.

Grounding describes a conducting connection by which an electrical circuit or equipment is connected to earth or to some conducting body of relatively large extent that serves in place of earth. **Bonding** is the intentional electrical interconnecting of conductive paths to ensure common electrical potential between the bonded parts.

The primary purpose of grounding is safety of the personnel working within a facility and of the general public, and protection of all the equipment against damage,

* A substantial portion of the material in Section 1.18 is from Sankaran, C., *Power Quality*, CRC Press, 2001. With permission from CRC Press.

TABLE 1.4
Effect of Current Flow Through Human Body

Current Level	Shock Hazard
100 μA	Threshold of perception
1–5 mA	Sensation of pain
5–10 mA	Increased pain
10–20 mA	Intense pain; unable to release grip
30 mA	Breathing affected
40–60 mA	Feeling of asphyxiation
75 mA	Ventricular fibrillation, irregular heartbeat

Source: From Sankaran, C. (2001). *Power Quality*, CRC Press. With permission from CRC Press.

for example, from fires due to short circuits. Table 1.4 lists the physiological hazards associated with the passage of electrical current through an average human body.[6] The resistance of an average human under conditions when the skin is dry is about 100 kΩ or higher. When the skin is wet, the resistance drops to 10 kΩ or lower. It is not difficult to see how susceptible humans are to shock hazards.

Suppose a person touches a motor frame that is not grounded. Then the electrical circuit will be completed through the source, motor input cable stray capacitance, the human body, the ground on which the person is standing, and the ground connection of the source. If we assume a stray cable capacitive reactance of 15 kΩ, currents significant enough to cause a shock would flow through the person in contact with the motor body.

NEC Article 250–251 states that an effective grounding path (the path to ground from circuits, equipment, and conducting enclosures) shall:

- Be permanent and continuous.
- Have capacity to conduct safely any fault current likely to be imposed on it.
- Have sufficiently low impedance to limit the voltage to ground and facilitate the operation of the protective devices in the circuit.
- Besides, the earth shall not be used as the sole equipment ground conductor.

1.18.1 Ground Electrodes

The NEC states that the following elements are part of a ground electrode system in a facility:

- Metal underground water pipes
- Metal frames of buildings or structures
- Concrete-enclosed electrodes
- Ground rings
- Other electrodes such as underground structures, rod and pipe electrodes, and plate electrodes, when none of the previously listed items is available

The code prohibits the use of a metal underground gas piping system as a ground electrode. Also, aluminum electrodes are not permitted. The NEC also mentions that, when applicable, each of the items listed previously should be bonded together. The purpose of this requirement is to ensure that the ground electrode system is large enough to present low impedance to the flow of fault energy. It should be recognized that, while any one of the ground electrodes may be adequate by itself, bonding all of these together provides a superior ground grid system.

Earth-Ground Grid Systems

The requirements of the ground grids are

- To provide a metal grid of sufficient area of contact with the earth so as to derive low resistance between the ground electrode and the earth
- To be stable with time
- Not to have chemical reactions with other metal objects in the vicinity, such as buried water pipes, building reinforcement bars, etc., and cause corrosion either in the ground grid or the neighboring metal objects

Figure 1.2 shows the elements of a properly grounded scheme.

1.18.1.1 Ground Rods, Ground Rings, Plates

According to the NEC, ground rods should be not less than 8 ft long and should consist of the following:

- Electrodes of conduits or pipes that are no smaller than 3/4-in. trade size; when these are made of steel, the outer surface should be galvanized or otherwise metal-coated for corrosion protection.

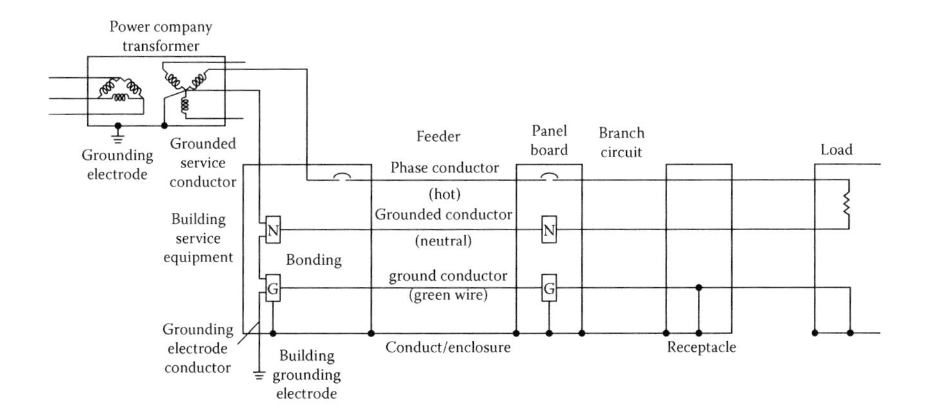

FIGURE 1.2 Basic elements of a properly grounded system. (From Dugan, R.C. et al., 1996. *Electrical Power Systems Quality*, McGraw-Hill, New York, p. 222, Figure 7.3. With permission from McGraw-Hill.)

Electrodes of rods of iron or steel that are at least 5/8 in. in diameter; the electrodes should be installed so that at least an 8-ft length is in contact with the soil. Typically, copper-clad steel rods are used for ground rods.

The basic components of resistance in a ground rod are

1. Electrode resistance due to the physical connection of the grounding wire to the grounding rod.
2. Rod–earth contact resistance, which is inversely proportional to the surface area of the grounding rod, that is, the more the area of contact, the lower the resistance.

The resistance of the ground rod connection is important for the following reasons:

1. It influences the transient voltage levels during switching events and lightning transients.
2. High-magnitude currents during lightning strokes result in a voltage across the resistance, raising the ground reference for the entire facility. The difference in voltage between the ground reference and the true earth ground will appear at grounded equipment in the facility, and this can result in dangerous touch potentials.

In addition to ground rods, rectangular and circular plates with a minimum area of 2-ft^2 exposure to the soil are used. To provide a better equipotential ground for the grounding electrode, a **ground ring** encircling the building in direct contact with the earth should be installed at a depth of not less than 2.5 ft below the surface of the earth.

1.18.1.2 Signal Reference Ground (SRG)

The power ground (Figure 1.3) and signal reference ground (Figure 1.4) have different purposes. The power ground is for personal safety and equipment protection. The signal reference ground merely provides a common-reference low-impedance plane over a wide frequency range from which sensitive loads may operate.

A signal reference ground consists of a #2 AWG or large copper conductor laid underneath the floor of the computer or communication center to form a grid of 2 × 2-ft squares (Figure 1.4). Some installations use copper strips instead of circular conductors to form the grid. Other facilities might use sheets of copper under the floor of the computer center as the SRG. The SRG is also bonded to the building steel and the stanchions that support the raised floor of the computer center. It is important that all noise-producing loads be kept away from the SRG. If such loads are present, they should be located at the outer periphery of the data center and bonded to the building steel, if possible.

1.18.2 Single-Point and Multipoint Grounding

With multipoint grounding, every piece of equipment sharing a common space or building is individually grounded (Figure 1.5), whereas with single-point grounding,

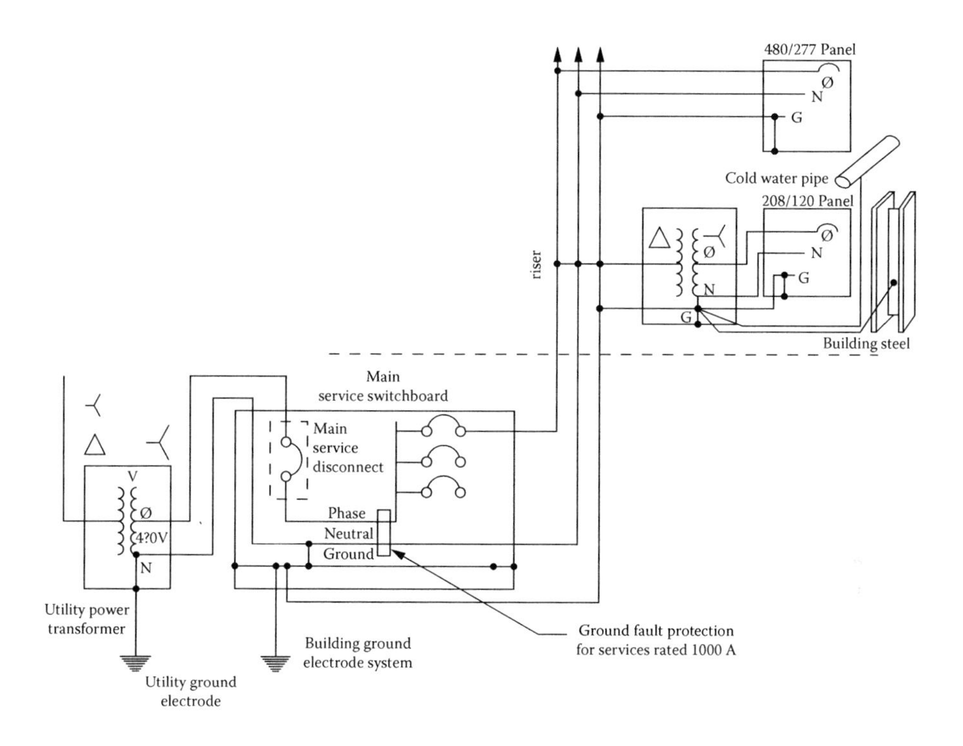

FIGURE 1.3 Typical power system grounding scheme. (From Sankaran, C., 2001. *Power Quality*, CRC Press.)

each piece of equipment is connected to a common bus or reference plane (Figure 1.6). Multipoint grounding is adequate at power frequencies. For typical power systems, various transformers, UPS systems, and emergency generators located in each area or floor of the building are grounded to the nearest building ground electrode, building steel, or cold pipe. Generally, this method is both convenient and economical but not effective, nor is it recommended for grounding

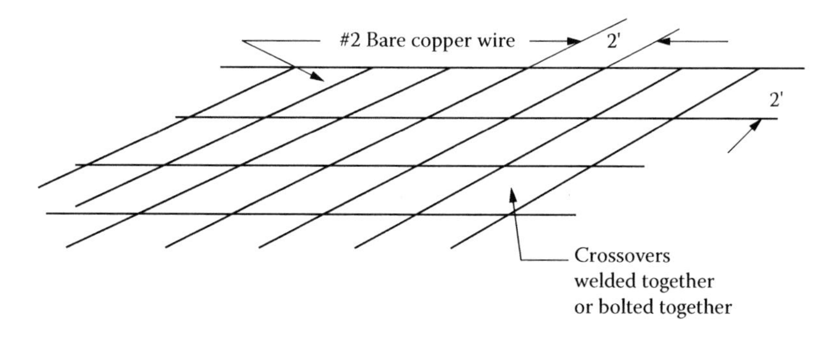

FIGURE 1.4 Typical 2 × 2-ft signal reference ground system. (From Sankaran, C., 2001. *Power Quality*, CRC Press. With permission from CRC Press.)

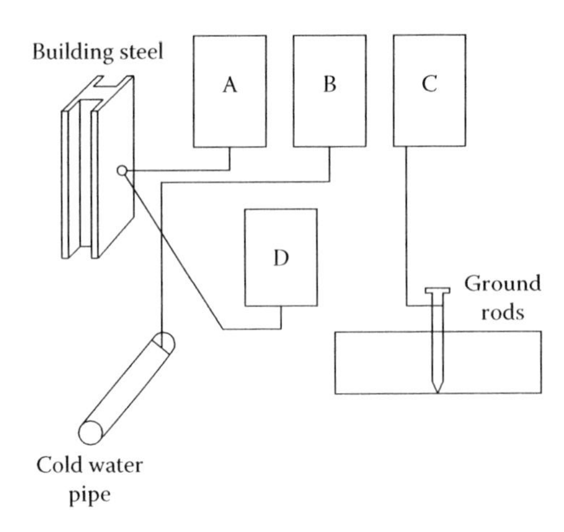

FIGURE 1.5 Multipoint ground system.

sensitive devices and circuits. In installations with SRG, the SRG must be bonded to the building ground electrode to ensure personal safety.

However, in distribution systems, the only neutral to the ground bond should be at the service entrance. The neutral and ground should be kept separate at all panel boards and junction boxes. Downline neutral to ground bonds result in parallel paths for the load return current where one of the paths becomes the ground circuit. This can cause misoperation of protective devices. Also, during a fault condition, the fault current will split

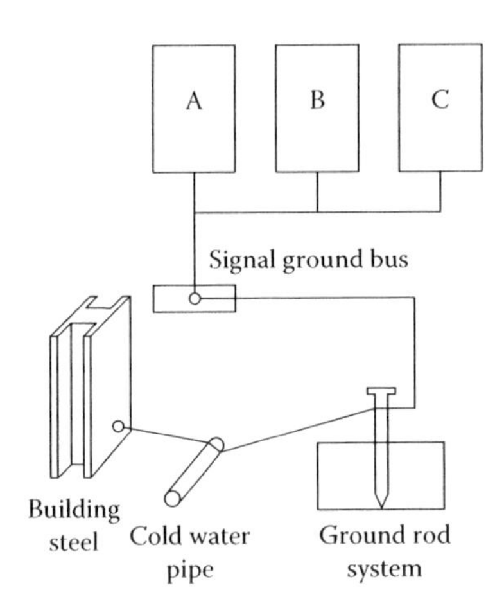

FIGURE 1.6 Single-point grounding of sensitive equipment. Typical 2 × 2 ft signal reference ground system. (From Sankaran, C., 2001. *Power Quality*, CRC Press. With permission from CRC Press.)

between the ground and the neutral, which could prevent proper operation of protective devices (a serious safety concern). This is a direct violation of the NEC regulations.

1.18.3 Ground Loops

Ground loops are the result of faulty or improper wiring practices that cause stray currents to flow in the ground path, creating a voltage difference between two points in the ground system. They may also be due to a high-resistance or high-impedance connection between a device and the ground plane. A solution to this problem is the use of optical couplers, or the grounded conductors in the signal cable may have to be supplemented with heavier conductors or better shielding.

1.18.4 Isolated Ground

The noise performance of the supply to sensitive loads can sometimes be improved by providing an isolated ground to the load. This is done using isolated ground receptacles, which are orange in color. If an isolated ground receptacle is used downline from the panel board, the isolated conductor is not connected to the conduit or enclosure in the panel board.

1.18.5 Electrochemical Reactions Due to Ground Grids

When two dissimilar metals are installed in damp or wet soil, an electrolytic cell is formed. Because copper pipe and ground ring are electronegative and structural steel is electropositive, over time, structural steel members will start to disintegrate as they are asked to supply electrons to support the current flow. By suitably coating the steel or copper, current flow is interrupted and the electrolytic action is minimized.

In some installations, to prevent corrosion of a specific metal member, sacrificial anodes are installed in the ground. They are more electropositive than the metals they protect, so they are sacrificed to protect the structural steel.

Reference 6 provides some examples of installations with grounding problems, including one in which a fatal accident occurred. References 19 and 21 provide some standard definitions regarding grounding. Further, References 15–21 provide valuable information regarding grounding and wiring practices. Also, Reference 6 has some useful data on the following aspects:

I. Resistivities of common materials
II. Effect of moisture on soil resistivity
III. Effect of temperature on earth resistivity
IV. Change in earth resistance with multiple ground rods
V. Effect of ground rod length on earth resistance
VI. Effect of ground rod diameter on earth resistance
VII. Resistance of circular plates buried 3 ft below surface
VIII. Earth resistance of buried conductors
IX. Electromotive series of metals

1.19 REACTIVE POWER IN POWER SYSTEMS WITH HARMONIC DISTORTION

In power systems with perfect sinusoidal waveforms, the definitions of active power and reactive power are universally accepted.[3,22–30] However, several definitions have been proposed by different authors for reactive power, apparent power, power factor, etc., when voltage and current waveforms are distorted as a result of harmonics. So far there is no generally accepted definition, although Budeanu's reactive power definition has been endorsed by the IEEE and IEC, and his work is described in Ref. 22.

Before the widespread use of power electronic equipment, capacitor banks were used for reactive compensation and improvement of the power factor. However, in power systems with significant harmonics, when capacitor banks are used for reactive compensation, proper design is necessary to avoid harmonic resonance problems. Hence, some harmonic filters at suitable locations may have to be installed to ensure that harmonic voltage levels are reduced to the limits recommended in the relevant standards. Further, power factor improvement may not be achieved only with the use of capacitor banks.

In power systems with less harmonic distortion, for the determination of tariffs, instruments of the Ferraris type are used in the measurement of power factor. With non-sinusoidal waveforms, the power factor must be determined using the apparent power.

1.19.1 Single-Phase Systems

Budeanu[22] (whose work is described in Ref. 22) divided the apparent power, S, into three orthogonal components, that is,

$$S^2 = P^2 + Q_B^2 + D^2 \tag{1.1}$$

where P = active power
Q_B = reactive power as defined by Budeanu
D = distortion power

The definition of reactive power is accepted by the IEC and IEEE,[3] and the complementary power, P_C (which he called fictitious), is given by

$$P_C = \surd(S^2 - P^2) \tag{1.2}$$

Fryze[23] separated electric current into two orthogonal components, i_a (active) and i_b (reactive):

$$i = i_{a+}\, i_b \tag{1.3}$$

and proposed for the reactive power the following definition:

$$Q_F = V\, I_b = \surd(S^2 - P^2) \tag{1.4}$$

Shepherd and Zakikhani[24] proposed the following decomposition for the apparent power:

$$S^2 = S_R^2 + S_x^2 + S_D^2 \tag{1.5}$$

where

$$S_R^2 = \sum_1^n V_n^2 \sum_1^n I_n^2 \cos^2 \varphi_n \tag{1.6}$$

$$S_x^2 = \sum_1^n V_n^2 \sum_1^n I_n^2 \sin^2 \varphi_n \tag{1.7}$$

$$S_D^2 = \sum_1^n V_n^2 \sum_1^p I_p^2 + \sum_1^m V_m^2 \left(\sum_1^n I_n^2 + \sum_1^p I_p^2 \right) \tag{1.8}$$

where the suffix n refers to the harmonics present in both voltage and current, m refers to the harmonics present in voltage only, and p refers only to the harmonics present in current only. In these expressions, S_R^2 is said to be the active apparent power, S_x^2 the reactive apparent power, and S_D^2 the distortion apparent power.

In the literature there are other definitions for reactive power, Q, apparent power, S, and complementary power. The reader is advised to see references by Sharon,[25] Emanuel,[26,27] Klusters and Moore,[28] Czarnecki,[29] and Slonin and Van Wyk,[30] as well as the book by Arrillaga et al.,[3] for more details on this subject.

Most commercial computer programs dealing with harmonic analysis adopt the simple Budeanu approach in either single- or three-phase systems and define the power factor as P/S, that is, total active power in the system/total apparent power. Also, distortion power or distortion volt-amperes are calculated as

$$D = \sqrt{(S^2 - P^2 - Q_B^2)} \tag{1.9}$$

Reference 3 gives two examples to show the variations in P, Q, and S, as well as the power factor with different assumptions:

Example 1. This is a comparative study with four single-phase networks with identical rms values of voltage (113.5 V) and current (16.25 A). With different source voltage waveforms, P and Q values calculated with different definitions of Q and S are tabulated to show how much differences can be obtained in these values.

Example 2. This is a study with pure sinusoidal source voltages and four cases of unbalanced loads (with different values of resistors), three cases without a neutral wire, and one with a neutral wire. Apparent powers and power factors are tabulated to illustrate differences in values with different definitions of Q. Also, the authors have calculated S and the power factor with different sources (with and without harmonics) and loads (linear and nonlinear).

1.20 RELIABILITY

Sustained interruptions are called outages by electric utilities.[1,11] Current trends in power quality standards being established are toward describing any interruption of power longer than 1 min as a sustained interruption. Some of the reliability indices for utility distribution are defined as follows:

SAIFI = system average interruption frequency index = (number of customers interrupted).(number of interruptions)/total number of customers
SAIDI = system average interruption duration index = $\sum$[(No. of customers interrupted).(duration of outage)]/total number of customers
CAIFI = customer average interruption frequency index = total number of customer interruptions/number of customers affected
CAIDI = customer average interruption duration index = $\sum$(customer interruption durations)/total number of customer interruptions
ASAI = average system availability index = customer hours service availability/customer hours service demand

where customer service demand = 8760 for an entire year

Typical target values for these indices are shown in Table 1.5.[1] These are simply design targets, and actual values can, of course, vary significantly from these values, depending on the stage of power system development, utility policy, and required reliability in the system. Elektrotek Concepts Inc.[1] have reported their experience with utilities in which SAIFI is usually around 0.5 and SAIDI is between 2.0 and 3.0 h.

TABLE 1.5
Reliability Indices Table

Index	Target
SAIFI	1.0
SAIDI	1.0–1.5 h
CAIDI	1.0–1.5 h
ASAI	0.99983

Source: From Dugan, R.C. et al., (1996). *Electrical Power Systems Quality*, McGraw-Hill, New York.

1.21 POWER QUALITY DATA COLLECTION

To develop compatibility levels and to deepen the understanding of system performance, data on power quality are being collected by various agencies from around the world.[31] Some of these organizations are Electrical Power Research Institute (EPRI), Canadian Electrical Association, National Power Laboratories (Necedah, Wisconsin), Norwegian Electric Power Research Institute, International Union of Producers and Distributors of Electrical Energy (UNIPEDE), Electricite de France,

Northeast Utilities Service (Berlin, Connecticut), East Midlands Electricity (EME, Nottingham, England), and Consolidated Edison Company of New York, Inc. For details of these power quality surveys see Reference 31.

Because of reasons of cost, power quality monitoring is still in its infancy. However, as a result of deregulation in the utility industry in many countries, they will be forced to compete for customers. One of the ways in which they can achieve this is by monitoring power quality and showing the customers that their power supply reliability is better than that of their competitors. Hence, utilities on their own are likely to undertake more monitoring of the feeders supplying their major customers.

The data required to monitor power quality are usually voluminous. Hence, software must be used to automatically characterize measured events and store the results in a well-defined database. It will be economical to integrate the data collected from power quality and in-plant monitoring with electric power instrumentation, site descriptions, and event information. For details of a power quality database management and analysis system called PQView, developed by Electrotek Concepts, Inc. (Knoxville, Tennessee) see Reference 32.

Over the last several years, the EPRI and one of its contractors (Electrotek Concepts, Inc.) have been developing a vendor-independent interchange format for power quality-related information. For the details of this Power Quality Data Interchange Format (PQDIF), PQView, and PQWeb systems, please see their Web sites.[33–36]

In 1996, EPRI and Electrotek placed PQDIF in the public domain to facilitate the interchange of power quality data between interested parties. EPRI and Electrotek have also offered the format, sample source code, and documentation to the IEEE 1159.3 task force as a possible initial format to meet that group's requirements.[37,38]

In 1991, the Power System Relaying Committee of the IEEE Power Engineering Society developed the standard C37.111.[39] The main purpose of this standard was to define a common format for the data files and exchange medium needed for the interchange of various types of fault, test, or simulation data. Among others, the standard defines as sources of data the following: digital fault recorders, analog tape recorders, digital protective relays, transient simulation programs, and analog simulators.

For the details of indices for assessing the harmonic distortion of power quality, please see Reference 40, D. Daniel Sabin et al. For general comments on the pitfalls of electric power quality indices, see Reference 41, G. Heydt et al.

A power quality database can provide a basis for developing equipment compatibility specifications and guidelines for future equipment enhancements. In addition, a database of the causes for recorded disturbances can be used to make system improvements. By ensuring equipment compatibility, safety hazards resulting from equipment misoperation or failure can be avoided.

Performance indices that measure system reliability in terms of voltage outages are defined in a document (Reference 31) prepared by the EPRI, Palo Alto, California.

1.22 SUMMARY

In this chapter we have discussed the importance of power quality, the common disturbances that occur in power systems, typical characteristics of electromagnetic phenomena, solutions to the power quality problems adopted by the utilities and customers, grounding, reactive power definitions in nonsinusoidal systems, and some common reliability indices used in the utilities. We have also discussed the CBEMA and ITI curves, and how they can be used with field measurements to select suitable power-conditioning equipment. The need to monitor power quality, the software for analysis and storage of data, and the use of PQ databases are discussed.

APPENDIX

ITI (CBEMA) CURVE APPLICATION NOTE

The ITI (CBEMA) curve, included within this application note, is published by Technical Committee 3 (TC3) of the Information Technology Industry Council (ITI, formerly known as the Computer & Business Equipment Manufacturers Association). It is available at http://.www.itic.org/technical/iticurv.pdf.

1 SCOPE

The ITI (CBEMA) curve and this application note describe an ac input voltage envelope that typically can be tolerated (no interruption in function) by most information technology equipment (ITE). The curve and this application note comprise a single document and are not to be considered separate from each other. They are not intended to serve as a design specification for products or ac distribution systems. The curve and this application note describe both steady-state and transitory conditions.

2 APPLICABILITY

The curve and this application note are applicable to 120 V nominal voltages obtained from 120-V, 208-V/120-V, and 120-V/240-V 60-Hz systems. Other nominal voltages and frequencies are not specifically considered, and it is the responsibility of the user to determine the applicability of these documents to such conditions.

3 DISCUSSION

This section provides a brief description of the individual conditions that are considered in the curve. For all conditions, the term "nominal voltage" implies an ideal condition of 120 V rms, 60 Hz.

Seven types of events are described in this composite envelope. Each event is briefly described in the following sections, with two similar line voltage sags being described under a single heading. Two regions outside the envelope are also noted. All conditions are assumed to be mutually exclusive at any point in time and, with the exception of steady-state tolerances, are assumed to commence from the nominal voltage. The timing between transients is assumed to be such that the ITE returns to equilibrium (electrical, mechanical, and thermal) prior to commencement of the next transient.

3.1 Steady-State Tolerances

The steady-state range describes an rms voltage that is either very slowly varying or is constant. The subject range is ±10% from the nominal voltage. Any voltages in this range may be present for an indefinite period and are a function of normal loadings and losses in the distribution system.

3.2 Line Voltage Swell

This region describes a voltage swell having an rms amplitude of up to 120% of the rms nominal voltage, with a duration of up to 0.5 s. This transient may occur when large loads are removed from the system or when voltage is supplied from sources other than the electric utility.

3.3 Low-Frequency Decaying Ring Wave

This region describes a decaying ring-wave transient that typically results from the connection of power factor correction capacitors to an ac distribution system. The frequency of this transient may range from 200 Hz to 5 KHz, depending upon the resonant frequency of the ac distribution system. The magnitude of the transient is expressed as a percentage of the peak 60-Hz nominal voltage (not the rms value). The transient is assumed to be completely decayed by the end of the half-cycle in which it occurs. It is assumed to occur near the peak of the nominal voltage waveform. The amplitude of the transient varies from 140% for 200-Hz ring waves to 200% for 5-KHz ring waves, with a linear increase in amplitude with increasing frequency waveform.

3.4 High-Frequency Impulse and Ring Wave

This region describes the transients that typically occur as a result of lightning strikes. The wave shape applicable to this transient and general test conditions are described in ANSI/IEEE C62.41-1991. This region of the curve deals with both amplitude and duration (energy), rather than rms amplitude. The intent is to provide an 80-J minimum transient immunity.

3.5 Voltage Sags

Two different rms voltage sags are described. Generally, these transients result from the application of heavy loads, as well as fault conditions, at various points in the ac distribution system. Sags of 80% of nominal (maximum deviation of 20%) are assumed to have a typical duration of up to 10 s, and sags of 70% of nominal (maximum deviation of 30%) are assumed to have a duration of up to 0.5 s.

3.6 Dropout

A voltage dropout includes both severe rms voltage sags and complete interruptions of the applied voltage, followed by immediate reapplication of the nominal voltage. The interruption may last up to 20 ms. This transient typically results from the occurrence and subsequent clearing of faults in the ac distribution system.

3.7 No-Damage Region

Events in this region include sags and dropouts that are more severe than those specified in the preceding paragraphs, and continuously applied voltages that are less than the lower limit of the steady-state tolerance range. The normal functional state of the ITE is not typically expected during these conditions, but no damage to the ITE should result.

3.8 Prohibited Region

This region includes any surge or swell that exceeds the upper limit of the envelope. If the ITE is subjected to such conditions, damage to it may result.

REFERENCES

1. Dugan, R.C., McGranaghan, M.F., Beaty, H.W. (1996). *Electrical Power Systems Quality*, McGraw-Hill, New York.
2. Dugan, R.C., McGranaghan, M.F., Santoso, S., Beaty, H.W. (2002). *Electrical Power Systems Quality*, McGraw-Hill, New York.
3. Arrillaga, J., Watson, N.R., Chen, S. (2000). *Power System Quality Assessment,* John Wiley, Chichester, U.K.
4. Ghosh, A., Ledwich, G. (2002). *Power Quality Enhancement Using Custom Power Devices*, Kluwer Academic, New York.
5. Bollen, M. H. J. (2000). *Understanding Power Quality Problems: Voltage Sags and Interruptions*, IEEE Press, New York.
6. Sankaran, C. (2001). *Power Quality*, CRC Press, Boca Raton, FL.
7. Schlabbach, J., Blume, D., Stephanblome, T. (2001). *Voltage Quality in Electrical Power Systems,* IEE, London.
8. Heydt, G.T. (1991). *Electric Power Quality,* Stars in a Circle Publications, west Lafayette, In.
9. Sabin, D., Sundaram, A. (February 1996). Quality enhances reliability, *IEEE Spectrum*, 33(2), 34–41.
10. Douglas, J. (December 1993). Solving problems of power quality, *EPRI Journal*, 8–15.
11. IEEE Standard 1159 (1995), Recommended Practice on Monitoring Power Quality.
12. IEEE Standard 519 (1992), Recommended Practice and Requirements for Harmonic Control in Electric Power Systems.
13. Fluke Corporate Publication: "Harmonics and Power Quality", Part 3—Conducting a Site Analysis.
14. IEEE Standard 446 (1987), IEEE Recommended Practice for Emergency and Standby Power Systems for Industrial and Commercial Applications (IEEE Orange Book), 1996.
15. IEEE Standard 142 (1991), Recommended Practice for Grounding of Industrial and Commercial Power Systems (Green Book).
16. Federal Information Processing Standard (1994), Guidelines on Electrical Power for ADP Installations.
17. IEEE Standard 518, IEEE Guide for the Installation of Electrical Equipment to Minimize Electrical Noise Inputs to Controllers from External Sources.
18. IEEE Standard 1100 (1992), Recommended Practice for Powering and Grounding Sensitive Electronic Equipment (Emerald Book).

19. IEEE (1993). *The New IEEE Standard Dictionary of Electrical and Electronic Terms*, 5th edition, IEEE, New York.
20. EPRI Wiring and Grounding for Power Quality (Publication CU.2026.3.90).
21. National Electrical Code (1993), National Fire Protection Association (NFPA 70).
22. Antoniu, S. (1984). Le regime energetique deformant, Une question de priorite, *RGE*, **6/84,** 357–362.
23. Fryze, S. (1932). Wirk, Blind und Scheinleistung in elektrischen Stromkreisen mit nichtsinusformigen Verlauf von Strom und Spannung, *Elektrotechniche Zeitschrift*, **25,** 596–599, **26,** 625–627, **29**, 700–702.
24. Shepherd, W., Zakikhani, P. Suggested definition of reactive power for nonsinusoidal systems, *Proc. IEE*, **119,** 1361–1362.
25. Sharon, D. (1973). Reactive power definition and power factor improvement in nonlinear systems, *Proc. IEE*, **120,** 704–706.
26. Emanuel, A. E. (1977). Energetical factors in power systems with nonlinear loads, *Archiv fur Elektrotechnik*, **59,** 183–189.
27. Emanuel, A. E. (1990). Power in nonsinusoidal situations. A review of definitions and physical meaning, *IEEE Transactions on Power Delivery*, **PWRD-5**, 1377–1383.
28. Klusters, N. L., Moore, W. J. M. (1980). On definition of reactive power under nonsinusoidal conditions, *IEEE Transactions on Power Apparatus and Systems*, **PAS-99**, 1845–1850.
29. Czarnecki, L. S. (1987). What is wrong with the Budeanu concept of reactive and distortion power and why it should be abandoned, *IEEE Transactions on Instrumentation and Measurement*, **IM-36** (3), 834–837.
30. Slonin, M. A., Van Wyk, J. D. (1988). Power components in a system with sinusoidal and nonsinusoidal voltages and/currents, *Proc. IEE*, **135,** 76–84.
31. Electrical Power Research Institute (EPRI). Report EL-2081, **2**, Project 1356-1, 3-3 to 3-4.
32. Probing Power Quality Data, *IEEE Computer Applications in Power* (1994), 8–14.
33. Web site http://grouper.ieee.org/groups/1159/3/pqdif.html
34. Web site http://www.pqnet.electrotek.com/pqnet/main/measure/pqdif/pqdif.htm
35. Web site http://www.ucaforum.org
36. Web site http://www.remotepq.com
37. IEEE1159.3 (2003) Recommended for the Transfer of Power Quality Data.
38. Moreno, M. (Editor) (2007). *Power Quality: Mitigation Technologies in a Distributed Environment,* Springer-Verlag, New York.
39. IEEE C37.111 (1991). Standard Common Format for Transient Data Exchange (COMTRADE) for Power Systems.
40. Sabin, D.D., Brooks, D.L., Sundaram, A. (1999). Indices for assessing harmonic distortion from power quality measurements: Definitions and Bench Mark Data, *IEEE Transactions on Power Delivery*, **PWRD-14**(2), April, 1377–1383.
41. Heydt, G., Jewell, W.T. (1998). Pitfalls of Electric Power Quality Indices, *IEEE Transactions on Power Delivery*, **PWRD-13**(2), April, 570–578.

2 Static Var Compensators

2.1 INTRODUCTION

The rapid development of the thyristor has led to the development of static var compensator (SVC) systems to provide reactive power, load balancing, power factor improvement, and to reduce voltage variations and associated light flicker due to arc furnace loads. Some of the applications of the SVC systems are listed here.

In transmission systems:

a. To increase active power transfer capacity and transient stability margin[1]
b. To damp power oscillations
c. To achieve effective voltage control

2.1.1 Increase in Transient Stability Margin

Let us try to understand how SVCs contribute to achieve these foregoing objectives.

In a simple two-machine power system separated by a reactance of X ohms, it is assumed that the terminal voltages of both the machines are V volts. If δ is the angle difference between the two sources, then the transmitted power P in watts is given by

$$P = (V^2/X) \sin \delta$$

Let us assume that an SVC is connected at the midpoint of this two-machine system. Then, the reactance between each source and SVC will be $X/2$ ohms. Now, the power transmitted between the two sources is

$$[V^2/(X/2)] \sin \delta/2 = 2(V^2/X) \sin \delta/2$$

Thus, the amount of maximum power transmitted between the two machines is doubled to $2(V^2/X)$ from V^2/X because the maximum value of a sine function is unity. As can be seen from Figure 2.1a,b, the transient stability limit is also increased. This can be seen by applying the equal-area criterion to the two power angle curves with and without SVC at midpoint.

2.1.2 Damping of Power Oscillations

The dynamic behavior of the previous simple two-machine power system will be described by the swing equation

$$M\,(d^2\delta/dt^2) = P_m - P_e \tag{2.1}$$

where M = angular momentum (also sometimes known as inertia constant),
$P_m - P_e$ = accelerating power

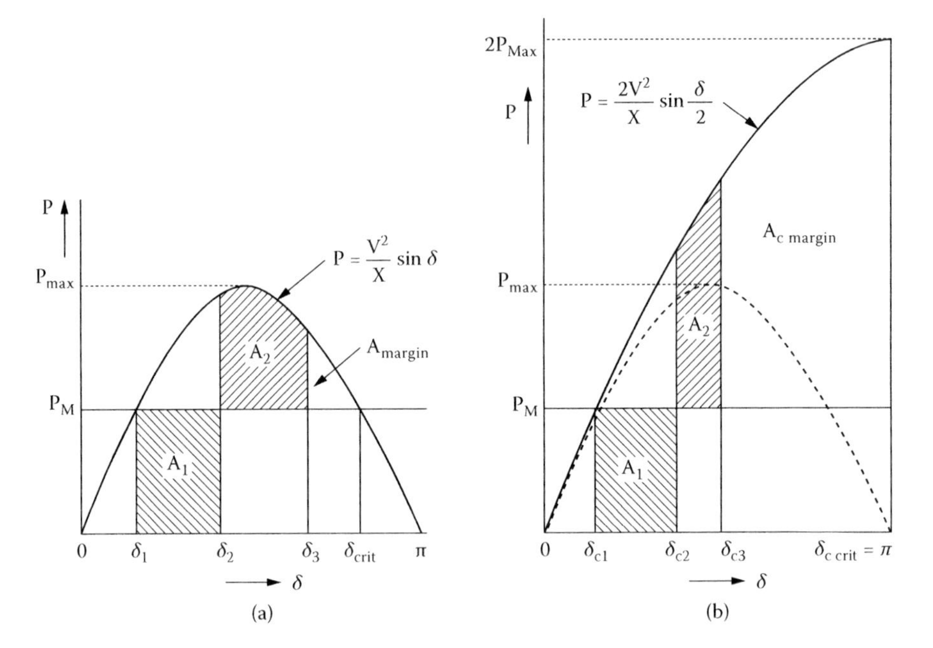

FIGURE 2.1 Equal area illustrating transient stability margin for a two-machine power system: (a) without compensation, and (b) with an ideal midpoint compensator. (From Gyugi, L. (1988). Power Electronics in Electric Utilities: Static Var Compensators, *Proceedings of the IEEE*, 76(4), 483–494. With permission from IEEE.)

Linearizing the previous equation around an operating point, we obtain the following equation:

$$M\frac{d^2(\Delta\delta)}{dt^2}+\frac{\partial P_E}{\partial V_m}\Delta V_m+\frac{\partial P_E}{\partial \delta}\Delta\delta=0. \tag{2.2}$$

In the preceding equation, ΔV_m will be zero if the midpoint voltage was kept constant by connecting an SVC at that point. Under those conditions, the angle $\Delta\delta$ would oscillate undamped with a frequency of

$$\omega_0=\sqrt{\frac{1}{M}\frac{\partial P_E}{\partial \delta}\bigg|_0}. \tag{2.3}$$

Hence, to provide damping, ΔV_m must be varied as a function of $d^{(\Delta\delta)}/dt$, that is,

$$\Delta V_m = K\frac{d(\Delta\delta)}{dt} \tag{2.4}$$

This means that the midpoint voltage is to be increased (by providing capacitive vars) when $d^{(\Delta\delta)}/dt$ is positive (to increase the transmitted electric power and thereby to oppose the acceleration of the generator). When $d^{(\Delta\delta)}/dt$ is negative, the midpoint

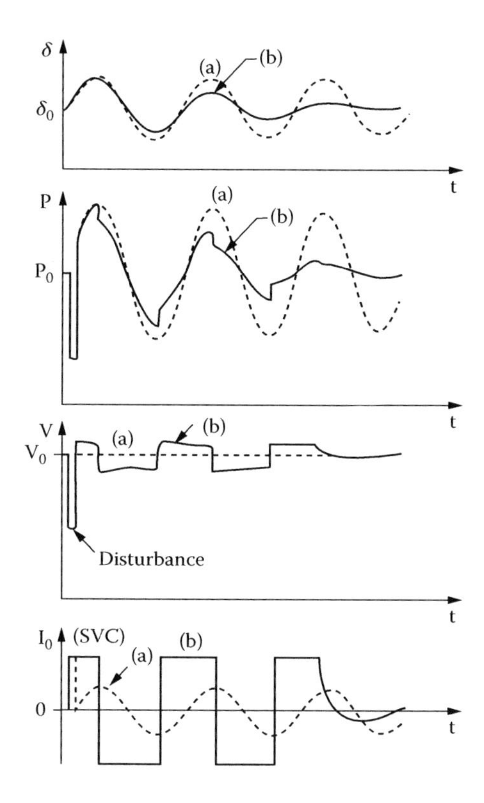

FIGURE 2.2 Power oscillation damping by static var compensator: (a) voltage regulation only, and (b) voltage control as a function of $\frac{d\delta}{dt}$. (From Hammad, A. (1986). Analysis of power system stability enhancement by static var compensators, *IEEE Trans. Power Systems*, **PWRS-1 (4)**, 222–227. Figure 7 on page 226 of the paper. With permission from IEEE.)

voltage is to be decreased (by absorbing inductive vars). This in turn reduces the transmitted electric power, thereby opposing the deceleration of the generator.

Figure 2.2a,b show the results of a computer simulation from Reference 3 and the improvement in damping due to an SVC. Similar results are published in Reference 2.

2.1.3 Voltage Support

Figure 2.3 shows the receiving end voltage characteristics of a typical lossless (radial) 275 kV line for different power factors.[1,2] For different radial feeders of other voltage ratings, the receiving end voltage characteristics also will be similar. It may be seen from this figure that in weak systems, the voltage collapse occurs if the transmitted electrical power exceeds beyond a certain value. This phenomenon is usually referred to as "voltage collapse or instability." Suppose a large load area is supplied from two different sources through different transmission lines. If one of these lines trip, all the load will be transferred to the other line, thus exceeding the limit sometimes for the transmitted electrical power on that line. In those situations, an SVC, which is a rapidly variable source of appropriate rating connected to the receiving

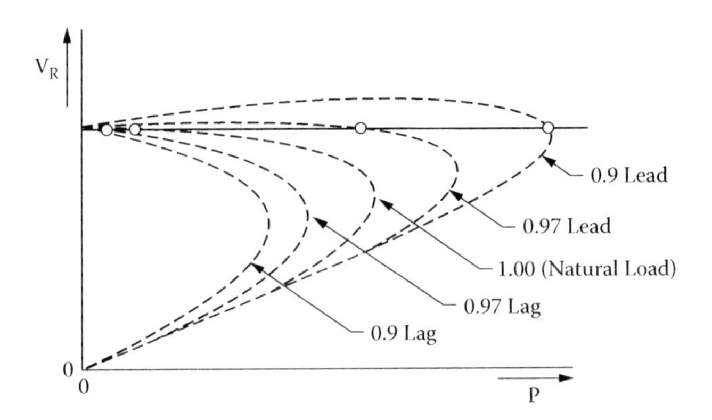

FIGURE 2.3 Amplitude variation of the receiving-end voltage V_R as a function of load P and load power factor (dashed lines). Possible voltage regulation with a variable var source (continuous line). (From Gyugi, L. (1988). Power Electronics in Electric Utilities: Static Var Compensators, *Proceedings of the IEEE*, 76 (4), 483–494. With permission from IEEE.)

end, can prevent the voltage collapse by keeping the terminal voltage constant at the receiving end.

In addition, SVCs are also used

i. In transmission systems
   a. To reduce temporary overvoltages
   b. To damp subsynchronous resonances
   c. To damp power oscillations in interconnected power systems
ii. In traction systems
   a. To balance loads
   b. To improve power factor
   c. To improve voltage regulation
iii. In HVDC systems
   a. To provide reactive power to ac–dc converters
iv. In arc furnaces
   a. To reduce voltage variations and associated light flicker

2.2 STATIC VAR COMPENSATOR SYSTEMS VERSUS SYNCHRONOUS CONDENSERS, CAPACITORS, AND REACTORS

Prior to the development of SVC systems, synchronous condensers, capacitors, and reactors were the only devices available for reactive power control. But today, all these devices are used depending on the specific requirements of a particular application. Reactors and capacitors are used to reduce or increase voltage at a particular bus under light or peak load sinusoidal steady-state operating conditions.

However, these are not suitable for the load balancing of single-phase loads or for smooth control of voltage at a particular bus.

Synchronous condensers (or compensators) can either absorb or supply reactive power to the power system, providing a smooth control of the voltage at a particular bus, have overload capability, and generate negligible harmonics. In the case of HVDC applications, synchronous condensers can provide extra short-circuit capacity in weak ac systems, which SVC systems cannot provide. The disadvantages of the synchronous condensers are high capital costs, maintenance costs, slow control response, and inability to balance single-phase loads.

2.3 SHUNT AND SERIES COMPENSATION

In this chapter, we will be discussing only shunt reactive compensation. For the sake of completeness, we should mention that it is possible to provide a series compensation. In particular, transmission capability can be significantly increased by installing series capacitors in the middle of a line. As is well known, the maximum power transfer between two buses is inversely proportional to the reactance between the two buses. By installing a series capacitor in the middle of the line, we will decrease the reactance and increase the power transfer capability between the two buses. A careful design of series capacitor installation is necessary to avoid the following problems.

a. If there is a power station at one end of the line with series capacitors, shaft damage can occur due to subharmonic resonances in the system.
b. Provision should be made to short circuit the series capacitor if any fault develops in the series capacitors.
c. The series capacitor bank should be properly designed to withstand overvoltages under different transient conditions.

Considerable progress has been made in the design of series capacitor installations, and many of them are operating satisfactorily. Zinc oxide arresters are used in some of these installations to reduce the overvoltages under transient conditions after fault clearing.

2.4 FUNDAMENTALS OF LOAD COMPENSATION

One of the applications of SVCs is to balance single-phase loads such as traction loads, so that the negative-sequence voltage at the point of common coupling (PCC) is within the limits set by the respective national standards. Gyugi et al.[5] have discussed Steinmetz's earlier work, which shows that, in a supply system with phase sequence ABC, a resistive load of P watts connected between phases A and B can be made to appear in the three-phase ac supply system as a balanced resistive load by connecting a capacitive source of $+jP/\sqrt{3}$ vars between phases B and C and an inductive source of $-jP/\sqrt{3}$ vars between phases C and A. We will try to establish this result in the following text.

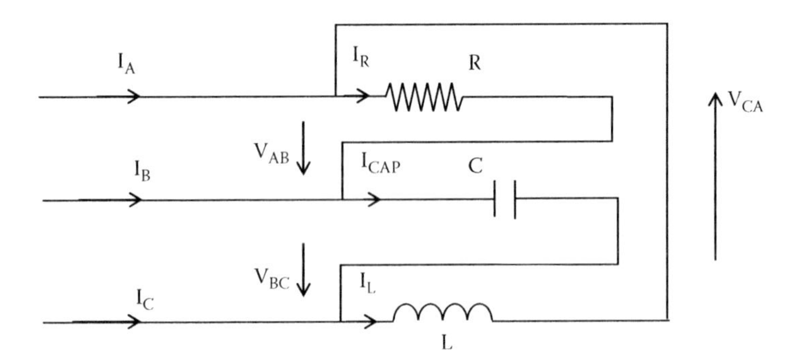

FIGURE 2.4 Steinmetz circuit.

Consider the R, C, and L elements connected across phases AB, BC, and CA, respectively, in a delta configuration, as shown in Figure 2.4. From the phasor diagram in Figure 2.5 and considering V_{AN} as a reference phasor, the following equations can be written:

$$V_{AB} = V_L \angle 30°$$

$$V_{BC} = V_L \angle -90° = -jV_L$$

$$V_{CA} = V_L \angle 150° = (-\sqrt{3}/2)\ V_L + j(V_L)/2$$

$$I_R = V_{AB}/R = V_L/R\ [\sqrt{3}/2 + j/2]$$

$$I_{CAP} = jV_{BC}\omega C = j\ (-jV_L)\ \omega C = V_L \omega C$$

$$I_L = V_{CA}/j\omega L = [V_L/(j\omega L)]\ [1/2 + j\ \sqrt{3}/2]$$

$$I_A = I_R - I_L = [\{\sqrt{3}/(2R) - 1/(2\omega L)\} + j\ \{1/(2R) - \sqrt{3}/(2\omega L)\}]\ V_L \qquad (2.5)$$

$$I_B = I_{CAP} - I_R = V_L\ [\{\omega C - \sqrt{3}/(2R)\} - j1/(2R)] \qquad (2.6)$$

$$I_C = I_L - I_{CAP} = V_L\ [\{1/(2\omega L) - \omega C\} + j\sqrt{3}/(2\omega L)] \qquad (2.7)$$

If I_A, I_B, and I_C are to be balanced three-phase currents drawn by resistive loads (i.e., unity power factor loads), then their magnitudes must be equal with 120° phase angle difference between them. Further, the phase angle of I_A must be zero because it is in phase with V_{AN}, and hence, the imaginary component of the current I_A must be zero in Equation 2.5.

Therefore,

$$1/(2R) = \sqrt{3}/(2\omega L)$$

$$\omega L = \sqrt{3}\ R \qquad (2.8)$$

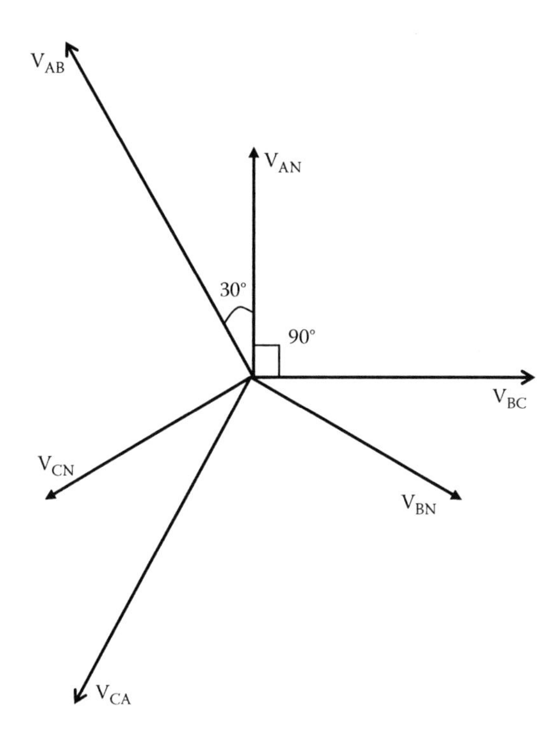

FIGURE 2.5 Phasor diagram showing the line voltages of the circuit in Figure 2.4.

Imposing the condition that the phase angle of I_B must be −120°, we can show that

$$1/\omega C = \sqrt{3}R \tag{2.9}$$

$$I_B = V_L/(\sqrt{3}R) \angle -120^\circ$$

By substituting the values of L and C from Equations 2.8 and 2.9 in Equation 2.7, we can show that

$$I_C = V_L/(\sqrt{3}R) \angle 120^\circ$$

Hence, we have established that a single-phase resistive load connected across phases A and B can be balanced by connecting pure reactive elements C and L across phases BC and CA, respectively, with the values given by the Equations 2.8 and 2.9. If the power consumed by the resistor R is P watts, the reactive powers generated by C and L are $+jP/\sqrt{3}$ and $-jP\sqrt{3}$ vars, respectively.

If there are two single-phase lagging loads $P_1 + j\,Q_1$ and $P_2 + j\,Q_2$ connected across phases AB and BC, respectively, they can be made to appear in the supply system as a pure resistive load by connecting purely reactive loads of the values shown in Table 2.1.[6]

In general, it may be seen from Table 2.1 that the amount of compensation required across different phases, AB, BC, and CA, is not the same for single-phase

TABLE 2.1
Capacities of Reactive Elements

	Capacities of Reactive Elements (Positive Sign Indicates Capacitive Source)		
	Q_{AB}	Q_{BC}	Q_{CA}
To compensate a lagging load $P_1 + j\,Q_1$ between phases A and B	$+jQ_1$	$+jP_1/\sqrt{3}$	$-jP_1/\sqrt{3}$
To compensate a lagging load $P_2 + j\,Q_2$ between phases B and C	$-jP_2/\sqrt{3}$	$+jQ_2$	$+jP_2/\sqrt{3}$
	$jQ_1 - jP_2/\sqrt{3}$	$jP_1/\sqrt{3} + jQ_2$	$jP_2/\sqrt{3} - jP_1/\sqrt{3}$

loads, and hence, some of the SVCs used in electric traction systems may not have symmetrical ranges in all the three phases.

2.5 REACTIVE POWER RELATIONSHIPS BETWEEN WYE- AND DELTA-CONNECTED SYSTEMS

It is common to use wye–delta transformers with SVCs because the delta windings provide a path to circulate zero-sequence components of the fundamental and other harmonic currents. (Note: If all the three-phase currents are balanced even though they are distorted with harmonics, then all the triplen harmonics are of zero-sequence nature. However, if all the three-phase currents do not contain balanced harmonic currents, then the triplen harmonics not only contain zero-sequence components but will also contain both positive- and negative-sequence triplen harmonics. Positive- and negative-sequence components of the triplen harmonics can still flow into the system from nonlinear harmonic-producing loads even with wye–delta transformers.) Hence, it is necessary to be able to express secondary reactive powers in terms of primary reactive powers, and vice versa. Consider the wye–delta transformer with capacitive loads Q_{AB}, Q_{BC}, and Q_{CA} on the wye side in Figure 2.6. As can be seen from the phasor diagram in Figure 2.7, the currents in the capacitive loads will be leading 90° the respective line voltage. As the line current due to the capacitive load Q_{BC} will be in phase with the phase voltage V_{AN}, it will not contribute to any measured reactive power between A phase and grounded neutral.

Let us define $Im\ I_A$ to be the imaginary part of the line current I_A, and V_{AN} be the reference phasor.

$$Q_A = V_{AN}\ Im\ I_A$$

$$I_A = I_{AB} - I_{CA} = (Q_{AB}/V_L) \angle 120° - (Q_{CA}/V_L) \angle -120°$$

$$Im\ I_A = [j\sqrt{3}/(2V_L)]\ (Q_{AB} + Q_{CA}) = j(Q_{AB} + Q_{CA})/(2\ V_{AN})$$

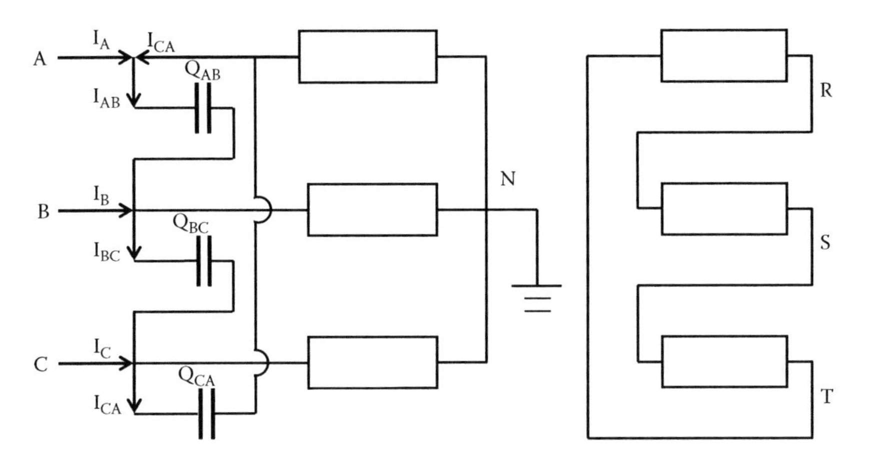

FIGURE 2.6 Wye–delta transformer.

because $V_L/\sqrt{3} = V_{AN}$

$$Q_A = V_{AN}\,|\,Im\,I_A| = 0.5\,(Q_{AB} + Q_{CA}) \tag{2.10}$$

By Symmetry, we can write

$$Q_B = 0.5\,(Q_{BC} + Q_{AB}) \tag{2.11}$$

$$Q_C = 0.5\,(Q_{CA} + Q_{BC}) \tag{2.12}$$

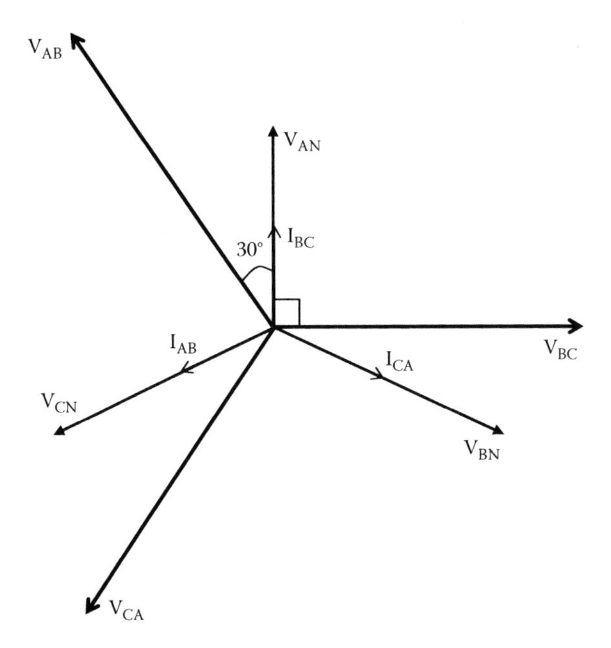

FIGURE 2.7 Phasor diagram of the circuit in Figure 2.6.

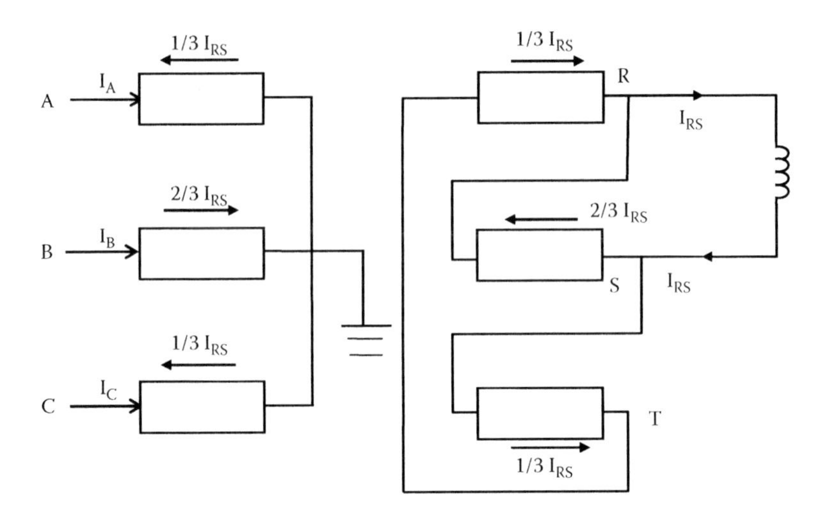

FIGURE 2.8 Wye–delta transformer with single-phase load on delta side.

Adding Equations 2.10, 2.11, and 2.12 yields

$$Q_A + Q_B + Q_C = Q_{AB} + Q_{BC} + Q_{CA} \tag{2.13}$$

From Equation 2.12,

$$2Q_C = (Q_{CA} + Q_{BC}) \tag{2.14}$$

Subtracting Equation 2.14 from Equation 2.13 yields

$$Q_{AB} = Q_A + Q_B - Q_C \tag{2.15}$$

By Symmetry, we can write

$$Q_{BC} = Q_B + Q_C - Q_A \tag{2.16}$$

$$Q_{CA} = Q_C + Q_A - Q_B \tag{2.17}$$

Consider a *y*/*d*11-connected transformer with a reactive load connected across secondary phases *R* and *S* (see Figure 2.8). For reasons of simplicity, assume that the transformer ratio is unity between wye and delta phases. (This implies that the currents in wye and delta windings are numerically equal and V_{line} on wye side/V_{line} on delta side = √3). This assumption will not restrict the application of these results to transformers with a different transformation ratio. From the phasor diagram in Figure 2.9, the following equations can be written.

$$V_{AN} = V_{TR} = V_{AN} \angle 0°$$

$$V_{BN} = V_{RS} = V_{AN} \angle -120°$$

$$V_{CN} = V_{ST} = V_{AN} \angle 120°$$

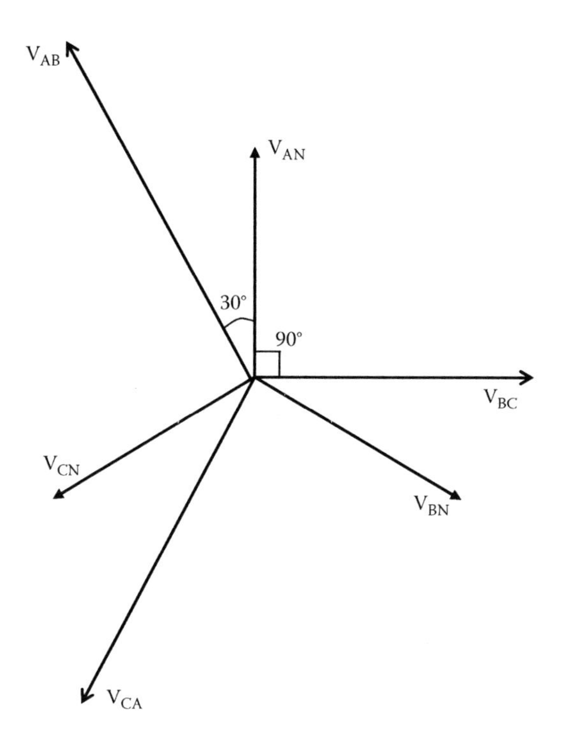

FIGURE 2.9 Phasor diagram of the circuit in Figure 2.8.

$$V_{AB} = \sqrt{3}\ V_{RS}$$

$$I_{RS} = |V_{RS}|/(\omega\, L) \angle (-90°)$$

$$= V_{AN}/(\omega L) \angle (-120° - 90°) = V_{AN}/(\omega L) \angle -210°$$

$$Q_{RS} = |V_{RS}|\ |I_{RS}|$$

$$Q_A = V_{AN}\ I_{AN}$$

Note that in Figure 2.8, I_{RS} is the load current flowing from the delta winding of the transformer from node R to node S. Let the delta-winding impedance per phase be Z. Then, within the delta winding, there are two parallel paths with impedance Z between RS, and impedance $2Z$ in the other path RTS. Thus, the current distribution in these two paths would be $2\ I_{RS}/3$ and $I_{RS}/3$, respectively. If we assume a turns ratio of 1:1, the current I_{AN} would be $I_{RS}/3$.

$$Q_A = -\ |V_{AN}|\ |I_{RS}/3|\ \sin \varphi_A$$

$$\varphi_A = 0° - (-210°) = 210°$$

$$Q_A = |V_{RS}|\ |I_{RS}|/6 = Q_{RS}/6$$

$$Q_B = V_{BN} I_{BN}$$

$$Q_B - |V_{BN}|\ |2I_{RS}/3|\ \sin \varphi_B$$

$$\varphi_B = -120° - (-210°) = 90°$$

$$Q_B = 2|V_{RS}|\ |I_{RS}|/3 = 4Q_{RS}/6$$

$$Q_C = V_{CN} I_{CN}$$

$$Q_C = -\ |V_{CN}|\ |I_{RS}/3|\ \sin \varphi_C$$

$$\varphi_C = 120° - (-210°) = 330°$$

$$Q_C = |V_{RS}|\ |I_{RS}|/6 = Q_{RS}/6$$

By connecting reactive loads across *ST* and *TR* phases, and using superposition, it can be shown that

$$Q_A = (Q_{RS} + Q_{ST} + 4Q_{TR})/6 \tag{2.18}$$

$$Q_B = (Q_{ST} + Q_{TR} + 4Q_{RS})/6 \tag{2.19}$$

$$Q_C = (Q_{TR} + Q_{RS} + 4Q_{ST})/6 \tag{2.20}$$

Solving the preceding equations for Q_{RS}, Q_{ST}, and Q_{TR} in terms of Q_A, Q_B, and Q_C, the following relationships can be derived:

$$Q_{RS} = (5Q_B - Q_A - Q_C)/3 \tag{2.21}$$

$$Q_{ST} = (5Q_C - Q_B - Q_A)/3 \tag{2.22}$$

$$Q_{TR} = (5Q_A - Q_C - Q_B)/3 \tag{2.23}$$

Substituting the preceding equations for Q_A, Q_B, and Q_C from the Equations 2.10, 2.11, and 2.12, respectively, the following equations can be derived:

$$Q_{RS} = (2Q_{AB} + 2Q_{BC} - Q_{CA})/3 \tag{2.24}$$

$$Q_{ST} = (2Q_{BC} + 2Q_{CA} - Q_{AB})/3 \tag{2.25}$$

$$Q_{TR} = (2Q_{CA} + 2Q_{AB} - Q_{BC})/3 \tag{2.26}$$

Solving the preceding equations for Q_{AB}, Q_{BC}, and Q_{CA} in terms of Q_{RS}, Q_{ST}, and Q_{TR}, the following equations can be written:

$$Q_{AB} = (2Q_{RS} + 2Q_{TR} - Q_{ST})/3 \tag{2.27}$$

$$Q_{BC} = (2Q_{ST} + 2Q_{RS} - Q_{TR})/3 \tag{2.28}$$

$$Q_{CA} = (2Q_{TR} + 2Q_{ST} - Q_{RS})/3 \tag{2.29}$$

Summary of Formulas

$$Q_A = 0.5\ (Q_{AB} + Q_{CA}) \tag{2.10}$$

$$Q_B = 0.5\ (Q_{BC} + Q_{AB}) \tag{2.11}$$

$$Q_C = 0.5\,(Q_{CA} + Q_{BC}) \quad (2.12)$$

$$Q_{AB} = Q_A + Q_B - Q_C \quad (2.15)$$

$$Q_{BC} = Q_B + Q_C - Q_A \quad (2.16)$$

$$Q_{CA} = Q_C + Q_A - Q_B \quad (2.17)$$

$$Q_A = (Q_{RS} + Q_{ST} + 4Q_{TR})/6 \quad (2.18)$$

$$Q_B = (Q_{ST} + Q_{TR} + 4Q_{RS})/6 \quad (2.19)$$

$$Q_C = (Q_{TR} + Q_{RS} + 4Q_{ST})/6 \quad (2.20)$$

$$Q_{RS} = (5Q_B - Q_A - Q_C)/3 \quad (2.21)$$

$$Q_{ST} = (5Q_C - Q_B - Q_A)/3 \quad (2.22)$$

$$Q_{TR} = (5Q_A - Q_C - Q_B)/3 \quad (2.23)$$

$$Q_{RS} = (2Q_{AB} + 2Q_{BC} - Q_{CA})/3 \quad (2.24)$$

$$Q_{ST} = (2Q_{BC} + 2Q_{CA} - Q_{AB})/3 \quad (2.25)$$

$$Q_{TR} = (2Q_{CA} + 2Q_{AB} - Q_{BC})/3 \quad (2.26)$$

$$Q_{AB} = (2Q_{RS} + 2Q_{TR} - Q_{ST})/3 \quad (2.27)$$

$$Q_{BC} = (2Q_{ST} + 2Q_{RS} - Q_{TR})/3 \quad (2.28)$$

$$Q_{CA} = (2Q_{TR} + 2Q_{ST} - Q_{RS})/3 \quad (2.29)$$

In the preceding equations, Q_{RS}, Q_{ST} and Q_{TR} are reactive powers measured across phases on the delta side of the transformer. Q_{AB}, Q_{BC}, and Q_{CA} are reactive powers measured across phases on the wye side of the transformer. Q_A, Q_B, and Q_C are reactive powers measured between each phase and neutral on the wye side of the transformer.

2.6 STATIC VAR COMPENSATORS FOR TRANSMISSION SYSTEMS

There are several variations of SVCs discussed in the literature (References 7–11, and 18) making use of thyristor-controlled reactors (TCR). For reasons of space, we will consider only the following types of SVCs in this chapter:[7,18]

- SVC using a TCR and a fixed capacitor (FC)
- SVC using a TCR and thyristor-switched capacitors (TSC)
- SVC using forced commutation inverters
- SVC using a saturated reactor (SR)

For details of other types of SVCs such as high-impedance thyristor-controlled transformer, mechanically switched capacitors (MSC), and TCR see Reference 7.

We will briefly consider the relative merits of the previously listed four SVCs.

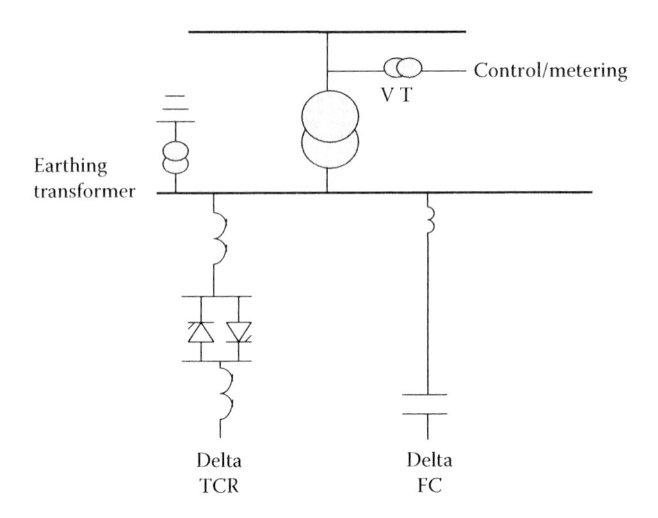

FIGURE 2.10 SVC of the FC/TCR type.

2.6.1 SVC Using a TCR and an FC

In this arrangement, two or more FC (fixed capacitor) banks are connected to a TCR (thyristor controlled reactor) through a step-down transformer (see Figure 2.10). The rating of the reactor is chosen larger than the rating of the capacitor by an amount to provide the maximum lagging vars that have to be absorbed from the system. By changing the firing angle of the thyristor controlling the reactor from 90° to 180°, the reactive power can be varied over the entire range from maximum lagging vars to leading vars that can be absorbed from the system by this compensator.

The main disadvantage of this configuration is the significant harmonics that will be generated because of the partial conduction of the large reactor under normal sinusoidal steady-state operating condition when the SVC is absorbing zero MVAr. These harmonics are filtered in the following manner. Triplex harmonics are canceled by arranging the TCR and the secondary windings of the step-down transformer in delta connection. The capacitor banks with the help of series reactors are tuned to filter fifth, seventh, and other higher-order harmonics as a high-pass filter. Further losses are high due to the circulating current between the reactor and capacitor banks. The losses in these types of SVCs are shown in Figure 2.11.

These SVCs do not have a short-time overload capability because the reactors are usually of the air-core type. In applications requiring overload capability, TCR must be designed for short-time overloading, or separate thyristor-switched overload reactors must be employed.

By segmenting the fixed reactor and also by using 12-pulse operation by employing two coupling transformers or one transformer with two secondary windings (one wye connected, the other delta), other configurations of the SVC can be obtained. Both these modifications reduce harmonics but increase costs due to the requirement of more thyristor switches. In the latter case, further a complex transformer and additional complexity in the thyristor firing angle control, as a result of the 30° angle difference in the wye and delta secondary windings, are required.

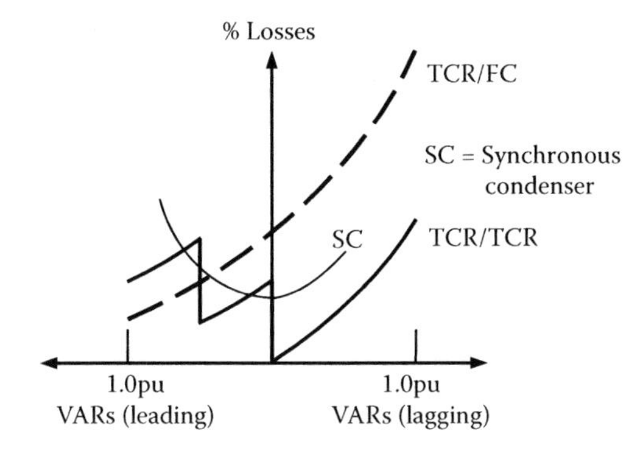

FIGURE 2.11 Comparison of the loss characteristics of TSC–TCR, TCR–FC compensators and synchronous condenser. (Adapted from Byerly, R.T., Poznaniak, D.T., and Taylor, E.R. Jr. (1982). Static Reactive Compensation for Power Transmission Systems, *IEEE Transactions on Vol-PAS101* (10), 3997–4005. With permission from IEEE.)

Steady-State Characteristics of an SVC Using a TCR and an FC

The sinusoidal steady-state characteristics such as voltage–current (or voltage and MVArs supplied by the SVC) relationship of an SVC is shown in Figure 2.12. It consists of three parts. In the regulated region, the voltage and current are linearly related. Outside the regulated interval, output current (var) versus voltage characteristic of the compensator is the same as that of the capacitor (low voltage) or an inductor (high voltage).

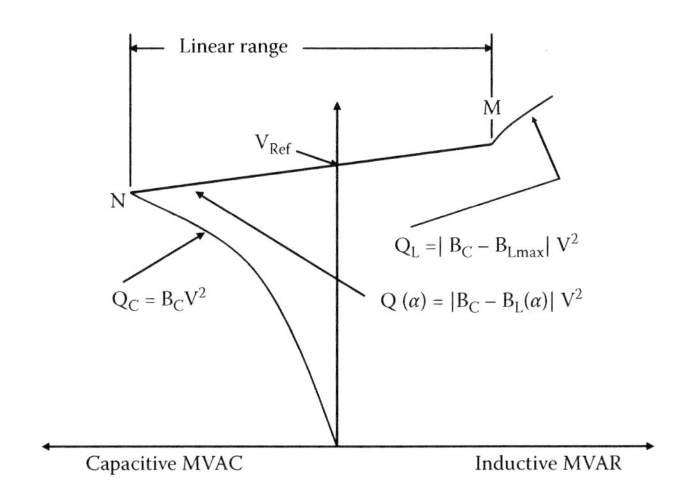

FIGURE 2.12 Steady-state control characteristics of SVC responding to bus voltage changes. (Adopted from Gyugi, L., and Taylor, E.R., Jr. Characteristics of Static, Thyristor-controlled shunt compensators for Power Transmission System Applications, *IEEE Transactions on Vol–PAS99* (5), 1795–1804. With permission from IEEE.)

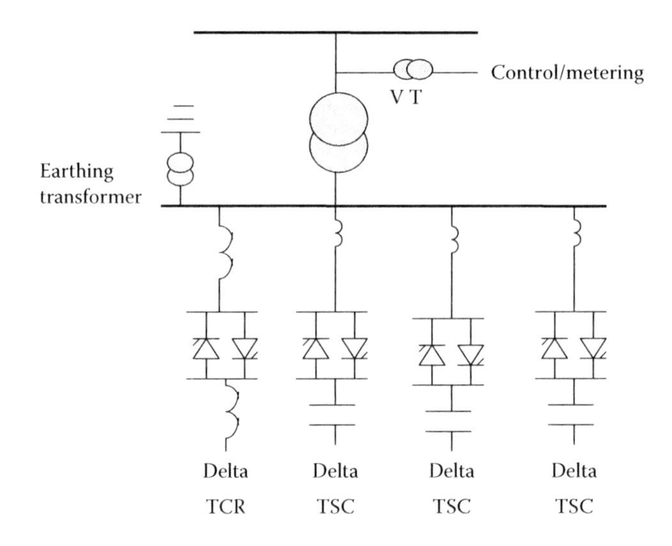

FIGURE 2.13 SVC of combined TSC and TCR type.

2.6.2 SVC Using a TCR and TSC

This compensator overcomes two major shortcomings of the earlier compensators by reducing losses under operating conditions and better performance under large system disturbances. Figure 2.13 shows the arrangement of this SVC with a TCR in parallel with several TSC banks (say, *n*). In view of the smaller rating of each capacitor bank, the rating of the reactor bank will be 1/n times the maximum output of the SVC, thus reducing the harmonics generated by the reactor. In those situations where harmonics have to be reduced further, a small amount of FCs tuned as filters may be connected in parallel with the TCR.

When large disturbances occur in a power system due to load rejection, there is a possibility for large voltage transients because of oscillatory interaction between system and the SVC capacitor bank or the parallel. The L C circuit of the SVC in the FC compensator. This type of compensator has been discussed in an earlier section (see Figure 2.10). In the TSC–TCR scheme, due to the flexibility of rapid switching of capacitor banks without appreciable disturbance to the power system, oscillations can be avoided, and hence the transients in the system can also be avoided. The capital cost of this SVC is higher than that of the earlier one due to the increased number of capacitor switches and increased control complexity.

2.6.3 STATCOM (SVC Using Self-Commutated Inverters)

This SVC consists of an inverter (dc-voltage-sourced converter, i.e., VSC) using gate turn-off (GTO) thyristors. These devices can be turned off as well as on.[10,11] For these inverters, the dc source can be a battery or a capacitor whose terminal voltage can be raised or lowered by controlling the connected inverter. This inverter

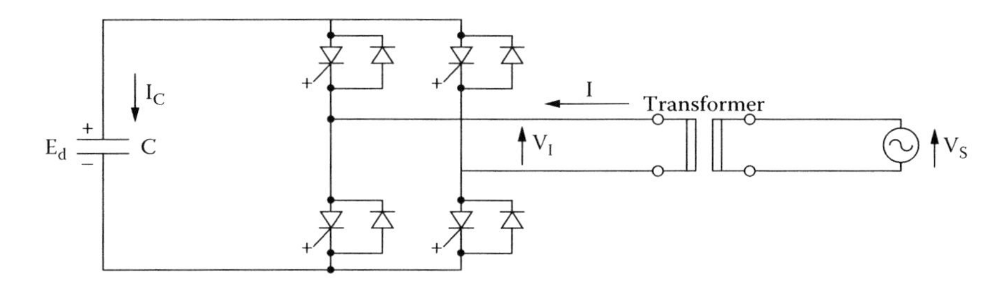

FIGURE 2.14 Basic circuit of SVC. (From Mori, S. et al., "Development of a large static var generator using self commutated inverters for improving power system stability," IEEE Transactions on Power Systems Vol. 8, No. 1, February 1993, pp. 371–377. With permission from IEEE.)

is connected to the supply system through a commutating reactance and an output transformer. When the inverter voltage V_I is equal to the system voltage, the SVC is floating. When V_I is greater than the system voltage, the SVC acts similar to a capacitor, and if V_I is less than the system voltage, the SVC acts similar to an inductor. By using several inverters with phase-angle difference between them, a higher-order pulse operation can be achieved.

Figure 2.14 shows the basic circuit diagram of this SVC, and Figure 2.15 the voltage and current waveforms of the SVC. Figure 2.16 shows the system configuration of an 80-MVA SVC installed in the Inuyama switching station of the Kansai Electric Co. in Japan in 1991. The specification details of this 48-pulse SVC are also given in Ref. 10.

Originally, for start-up purposes, a separate start-up converter was used to provide the required dc supply, but this method was found to be slow. In this new system, a gapped-core design is used for the eight phase-displacement transformers to reduce the effect of dc magnetization, decrease magnetic impedance, and improve the uniformity of voltage sharing between windings.

A ±100-MVAr prototype has been commissioned in the United States.[12] Relocatable GTO-based SVCs have been commissioned in National Grid Company (NGC) Substations.[13] The range of these SVCs are +225/−52 MVAr, including ±75 MVAr STATCOM.

These types of SVCs are compact, and losses are low, typically about 2% of the output under maximum or minimum operating conditions. Under unbalanced or undervoltage conditions, commutation failure is likely in these types of SVCs, and such operating conditions are not recommended.

2.6.4 SVC Using a Saturated Reactor (SR)

Historically, saturated reactor compensators were developed first before the advent of TCR SVCs. A saturated reactor consists of a six- or nine-limb core saturated reactor of conventional transformer-type construction.[7,9,11] In the "Treble–Tripler" saturated reactor with nine-limb core underbalanced system voltage conditions, the harmonics are of the order (18n ± 1), giving the lowest order of harmonic as 17th or 19th for

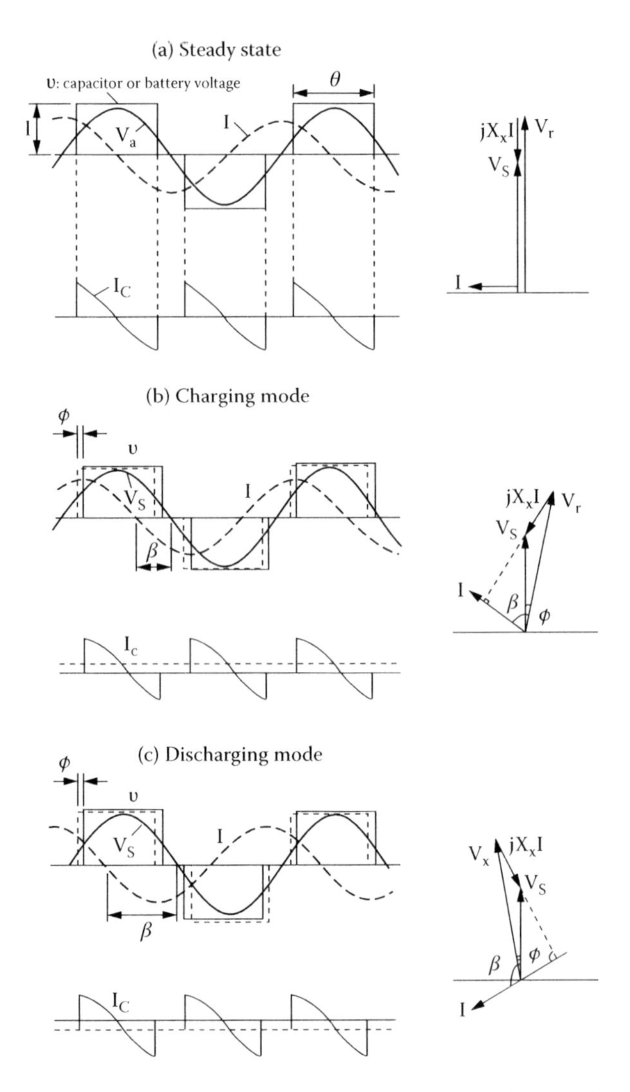

FIGURE 2.15 Voltage and current waveform of SVC. (From Mori, S. et al., "Development of a large static var generator using self commutated inverters for improving power system stability," IEEE Transactions on Power Systems Vol. 8, No. 1, February 1993, pp. 371–377. With permission from IEEE.)

n = 1,2,3, etc. By connecting the delta windings loaded with inductances, further reduction of the harmonics can be achieved (Figure 2.17).

Further, slope-correcting capacitors are connected in series with the saturated reactor to reduce the internal reactance, which sometimes leads to harmonic instability. Damped by-pass filters are always applied across these filters to damp oscillations at subsynchronous frequencies. These compensators, similar to transformers, have a considerable temporary overload capability. They are useful primarily for transmission applications but not for load balancing such as traction loads.

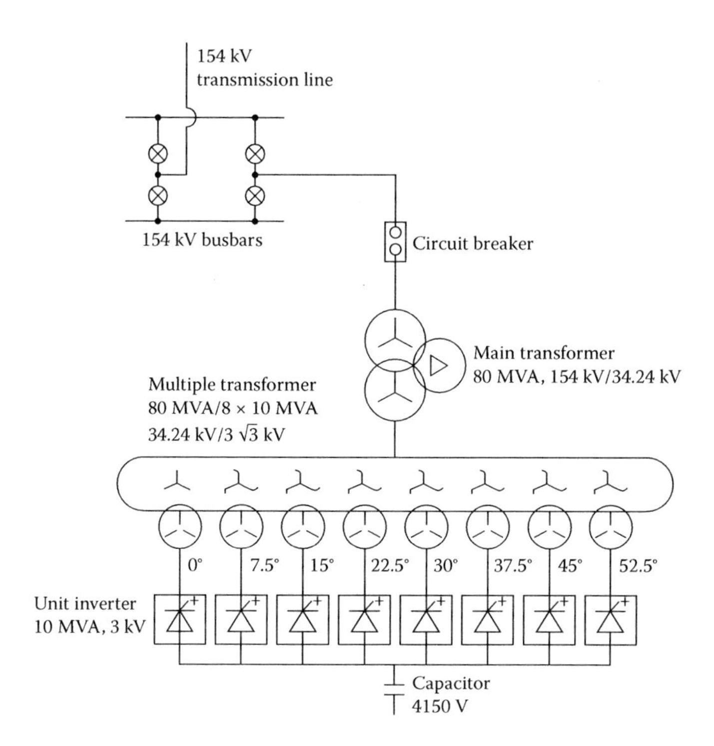

FIGURE 2.16 System configuration of 80-MVA SVC. (From Mori, S. et al., "Development of a large static var generator using self commutated inverters for improving power system stability," IEEE Transactions on Power Systems Vol. 8, No. 1, February 1993, pp. 371–377. With permission from IEEE.)

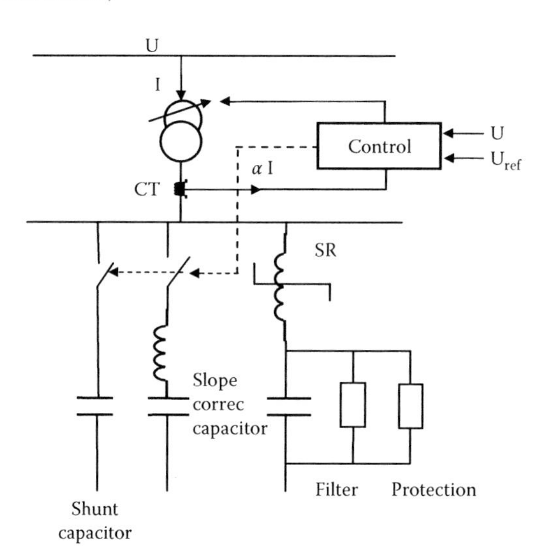

FIGURE 2.17 Basic scheme of saturated reactor compensator. (From CIGRE Report on Static Var Compensators, (1986) prepared by Working Group 38-01, Task Force No. 2, Edited by I.A. Erinmez. With permission from CIGRE.)

TABLE 2.2
Comparison of Different Types of Static Var Compensators

	TCR-Type SVC	TCR–TSC–FC-Type SVC	Self-Commutated Inverter SVC	Saturated-Reactor–Type SVC
Voltage control adjustability (set point)	Yes	Yes	Yes	Limited
Individual phase balancing	Yes	Yes	Yes, if unbalance is small	No
Speed of response	Fast, but system dependent	Fast, but system dependent	Very fast	Fast, but system dependent and slowed by slope-correction capacitors
Generation of harmonics	Filters may be necessary depending on system conditions	Very low, filters may be necessary depending on system conditions	Low	Very low
Limitation of overvoltages	Moderate	Very limited	Poor	Good, within limitation of slope-correction capacitors
Losses	Medium, increases with lagging current	Small–medium, depending on layout	Moderate	Moderate

2.6.5 Comparison of Static Var Systems

Table 2.2 compares different compensators showing their relative merits regarding some of their important properties. The expensive components of an SVC system are thyristors, nonconventional transformer, and reactor designs, as well as HV and EHV filters. However, the total cost of losses of an SVC system during its lifetime is much more than its initial capital cost. Hence, the losses of an SVC system at its normal operating point, in most cases at zero MVAr, play a pivotal role in the economic comparison of alternative SVC systems.

2.6.6 Specification of SVCs

Before specifying the compensators, compensator and system details, and operation and maintenance requirements must be specified. Some of the important items are listed in the following text even though this list is not comprehensive.

a. System details
   1. System frequency variation under normal operating conditions, fault conditions, and generator outage conditions

2. Voltage regulation that is required and its precision
3. Maximum harmonic distortion with the compensator in service
4. Coordination of system protection with compensator protection and reactive power limits
5. Compensator energization details, including any necessary precautions

b. Compensator details
1. Maximum continuous reactive power requirements: capacitive and inductive
2. Overload rating and duration
3. Normal-rated voltage and limits of voltage between which the reactive power ratings must not be exceeded
4. Response times of the compensator for different system disturbances
5. Control requirements
6. Reliability and redundancy of components

c. Operation, maintenance, and installation requirements
1. Spare parts, provision for future expansion
2. Performance with unbalanced voltages or with unbalanced load
3. Cabling details, access, enclosure, and grounding

2.7 A CASE STUDY (AUSTRALIA): CENTRAL QUEENSLAND RAILWAY PROJECT

In the specification of the SVCs, one of the important items that the customer has to provide the supplier is the maximum continuous reactive power requirements, capacitive, and inductive.[6,14] To provide this information, the customer has to perform several system studies such as load flows, dynamic simulation studies of different system operating conditions, etc. The customer also has to provide system harmonic impedances to enable the manufacturer of the SVC to design suitable harmonic filters. The software for these studies is commercially available in the case of the SVCs for transmission systems. Further, the SVCs for the transmission systems are invariably symmetrical, that is, the ranges for the var compensation in all the phases are the same. This need not be so in the case of the SVCs to compensate single-phase railway loads. Hence, in some cases, SVCs have unsymmetrical ranges, that is, the capacitive and inductive ranges across *AB*, *BC*, and *CA* need not be the same.

Most commercially available programs for load flow studies use the positive-sequence network and assume all the system loads to have a constant active power P and a reactive power Q. When we have single-phase railway loads, we need a three-phase load flow program (sometimes referred to as independent-phase load flow) to determine the negative-sequence voltages at the PCC. To illustrate the special problems in specifying SVCs in projects with single-phase loads, we will briefly discuss the Central Queensland Railway Project, which was already completed in the 1980s in two stages and is in operation now.

Figure 2.18 shows the location of Queensland railway substations in the Central Queensland system in 1987. Figure 2.19 shows the typical traction system arrangement, and Figure 2.20 the typical traction substation arrangement. As can be seen from

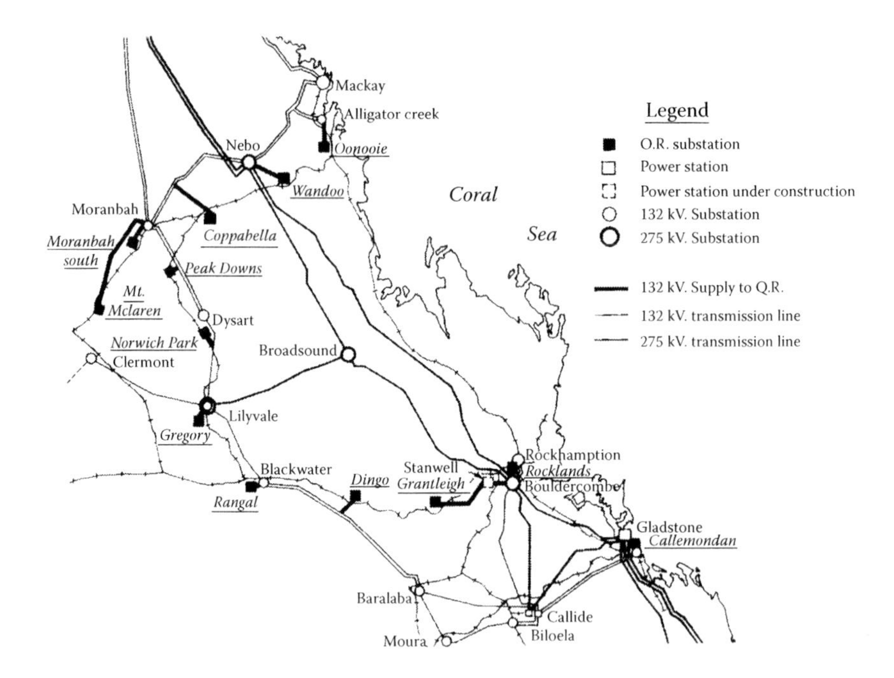

FIGURE 2.18 Location of Queensland Railway substations in the Central Queensland system in 1987. (From Sastry, V.R., Hill, D.V., and Lee, C.J. "Static Var Compensators for balancing the single-phase traction loads in Central Queensland," *Journal of Electrical and Electronic Engineering, Australia*, Vol. 6, No. 3, September 1986, pp. 184–190. With permission from IE (Australia).)

Figures 2.19 and 2.20, the supply is from a 132-kV system through two 132/50-kV, 30-MVA transformers. From this 50-kV supply to the feeders, a catenary voltage of 25 kV is obtained through ±25-kV autotransformers with the central point being earthed.

The traction loads at some substations can be as high as 40 MW at 0.9 power factor lagging and the short-circuit levels as low as 350 MVA. This indicated that some compensation was needed at these substations to maintain an acceptable quality of power supply.

2.7.1 Limits for Voltage Unbalance

Before conducting the three-phase loads to determine the ranges for SVCs, one had to decide the permitted values for the negative-sequence voltages at the PCC.

At present there is no internationally accepted standard for voltage unbalance for time-varying loads. Based on the considerations in P.O. Wright's paper,[14] Australian and NEMA (National Electrical Manufacturer's Association) standards for rotating machines, the following target levels for negative-sequence voltages at the PCC due to traction loads were set at the following:

- 0.7% for half-hour maximum demands
- 1.0% for 5-min peak loads
- 2.0% for 1-min peak loads

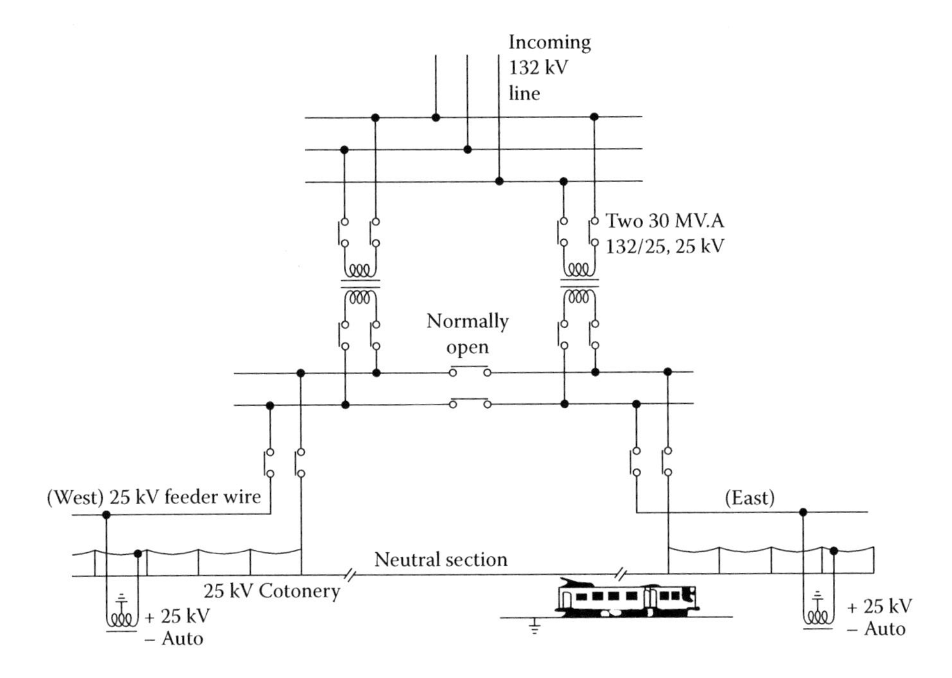

FIGURE 2.19 Typical traction system arrangement. (From Wright, P.O., "Planning of Electricity Supply to A.C. Electrified Railways from a Weak Network in Central Queensland," *Journal of Electrical and Electronic Engineering, Australia*, Vol. 6, No. 2, June 1986, pp. 148–155. With permission from IE (Australia).)

2.7.2 Three-Phase Power Flow Studies

The dynamic range of the SVCs were chosen to satisfy the voltage imbalance limits specified in the earlier section. A three-phase power flow program developed locally was used to conduct the necessary studies.[6] This program uses generator and transformer models as described in Laughton's papers (1968, 1969, 1980).[15–17] The important differences of this three-phase power flow program and other power flow programs that use only positive-sequence representation are

- In the models for generators and transformers at a voltage-controlled bus, average terminal voltages of all the three phases are represented instead of just the positive-sequence voltage because the voltages in phases *A*, *B*, and *C* are unbalanced, and there is some negative-sequence voltage at a generator bus.
- Transformer taps are set to control only one of the line voltages. The other line voltages will be slightly different based on the amount of negative-sequence voltage due to unbalance in the system at that bus.
- System loads were modeled as balanced constant impedances, and traction loads as constant *P*, *Q* loads. If all loads are represented as constant *P*, *Q* loads, then there will be convergence problems in three phase power flow programs due to unbalanced voltages in phases *A*, *B*, and *C*.

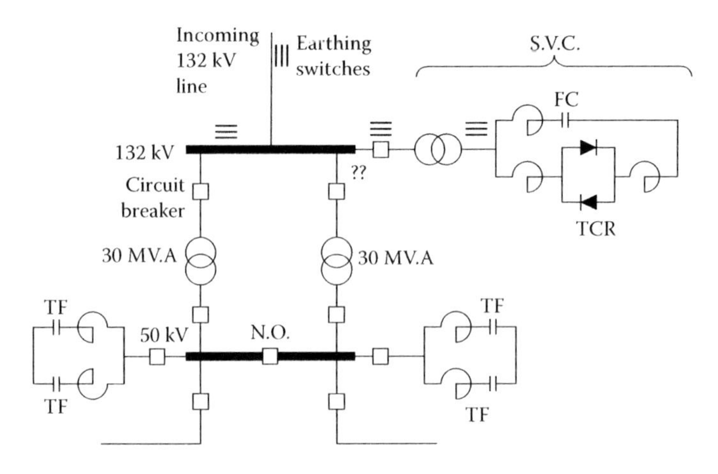

FIGURE 2.20 Typical traction substation arrangement. (From Wright, P.O., "Planning of Electricity Supply to A.C. Electrified Railways from a Weak Network in Central Queensland." *Journal of Electrical and Electronic Engineering, Australia*, Vol. 6, No. 2, June 1986, pp. 148–155. With permission from IE (Australia).)

- When 1-min, 5-min, or half-hour maximum demands occur at one substation, the other substations were assumed to have half-hour diversified loads.
- As the half-hour maximum demands at the railway substations do not coincide, a diversity factor of 0.4–0.6 was assumed. The diversity factor was based on the overall train density.

As can be appreciated from the previous discussion, certain assumptions are quite specific to each traction load compensation project. Further, it is uneconomical to provide full compensation for all peak loading conditions. In view of the different peak loads on different phases such as A, B, and C, we will obtain unsymmetrical ranges for SVCs. To optimize the number of thyristors used in SVCs, nonstandard secondary transformer voltages such as 5.7 kV, 10.4 kV, and so on, are used instead of the standard 11 kV. As wye–delta transformers with 132/10.4 kV or similar ratios are used to supply power to SVCs, and the railway P, Q loads are specified on the 50-kV feeders across phases A, B, and C. One has to calculate the required ranges of compensation at the delta side of the 10.4-kV bus from the values on wye side of the 132-kV bus using the formulae discussed in Section 2.5. Thus, considerable planning studies are necessary before one can specify the ranges for SVCs to compensate traction loads.

In addition harmonic impedance values at the 13 PCC had to be calculated for different possible operating conditions by considering transformer, line outages, and so on, to enable suitable harmonic filter design.

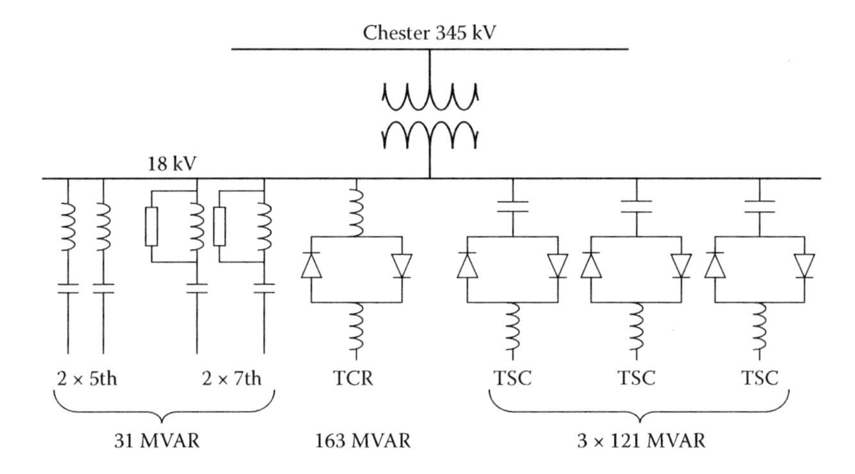

FIGURE 2.21 Single-line diagram of Chester 345-kV SVC. (From Dickmander, D., Thorvaldsson, B., Stromberg, G., and Osborn, D. (1992). Control System Design and Performance Verification for The Chester, Maine Static Var Compensator, *IEEE Transactions on Power Delivery*, 7 (3), 1492–1503. With permission from IEEE.)

2.8 CHESTER–MAINE SVC PROJECT

In the last section, we discussed a project in which SVCs are used to balance the load currents in a traction system. In this section we describe an SVC used to control voltages in a transmission system.

The Chester SVC has been installed at the Chester 345-kV station in the Maine Electric Power Company operating area.[19] A single-line diagram of the SVC is shown in Figure 2.21. The SVC bus is connected to the Chester 345-kV bus via three single-phase power transformers. The equipment connected to the SVC bus includes three TSCs rated 121 MVAr each (at 18.0 kV), one TCR rated 163 MVAr, and 31 MVAr of fifth and seventh harmonic filters. With the transformer impedance taken into account, the overall rating of the SVC is from 125 MVAr inductive to 425 MVAr capacitive, referred to 1.0 pu primary voltage. By phase-angle control of the TCR and switching of the TSC steps, a continuously varying output through the entire MVAr range is obtained.

Because of its importance for voltage control during large disturbances in the New England area and in the Maritime provinces New Brunswick and Nova Scotia, the Chester SVC has been designed to satisfy stringent performance requirements and to use some unique control strategies not previously applied to SVCs. These are discussed in detail in Reference 19.

We will be discussing the general control concepts of the SVCs in the next chapter.

2.9 CONCLUSIONS

In this chapter we have discussed how SVCs improve steady-state and transient stability margins, improve damping of power oscillations, and also assist in preventing voltage collapse.

We have derived the necessary formulae to convert reactive powers from the delta side to the wye side of a transformer and vice versa, when delta–wye transformers are used to supply electrical power to the SVCs.

Further, we have discussed the configuration and relative merits of the following SVCs:

- SVC using a thyristor-controlled reactor (TCR) and a fixed capacitor (FC)
- SVC using a thyristor-controlled reactor (TCR) and thyristor-switched capacitors (TSCs)
- SVC using forced commutation inverters
- SVC using a saturable reactor (SR)

We have also compared the planning aspects of the Central Queensland Railway Project to illustrate the differences in the planning of general transmission SVCs, and SVCs to compensate traction loads from a weak ac power system. We have also discussed briefly the Chester–Maine SVC.

REFERENCES

1. Gyugi, L. (1988). Power Electronics in Electric Utilities: Static Var Compensators, *Proceedings of the IEEE*, 76 (4), 483–494.
2. O'Brien, M. and Ledwich, G. (1987). Static reactive-power compensator controls for improved system stability, *Proceedings of the IEEE*, 134 (1), Pt C, 38–42.
3. Hammad, A. (1986). Analysis of power system stability enhancement by static var compensators, *IEEE Trans. Power Systems*, **PWRS-1 (4)**, 222–227.
4. Weedy, B.M. (1968). Voltage Stability of Radial Power Links, *Proceedings of the IEEE,* 115 (4), 528–536.
5. Gyugi, L., and Taylor E.R. Jr. (1980), Characteristics of Static Thyrister-controlled shunt compensators for Power Transmission System Applications, IEEE Transactions on Vol-PAS 99(5), 1795.
6. Sastry, V.R., Hill, D.V., and Lee, C.J. (1986). Static Var Compensators for balancing the single phase traction loads in Central Queensland, *Journal of Electrical and Electronic Engineering Australia*, 6 (3), 184–190.
7. Byerly, R.T., Poznaniak, D.T., and Taylor Jr., E.R. (1982). Static Reactive Compensation for Power Transmission Systems, *IEEE Transactions on Vol - PAS101* (10), 3997–4005.
8. Miller, T.J.E. (1982). *Reactive Power Control in Electric Systems,* John Wiley, New York.
9. CIGRE Report on Static Var Compensators, Prepared by Working Group 38–01, Task Force No 2, Edited by I.A. Erinmez.
10. Mori, S. et. al. (1993). Development of a large static var generator using self commutated inverters for improving power system stability, *IEEE Transactions on Power Systems*, 8 (1), 371–377.
11. Thanawala, H.L., Young, D.J., and Baker, M.H. (1999). Shunt Compensation: SVC and STATCOM (Ch.4), *Flexible AC Transmission Systems*, IEE, London, 146–197.
12. Schauder, C. et al. (1995). Development of a ± 100 Mvar static condenser for voltage control on transmission systems, *IEEE Transactions on Power Delivery*, 10 (3), 1986–1993.

13. Knight, R.C., Young D.J., and Trainer, D.R. (1998). Relocatable GTO-based static var compensators for NGC substations, *CIGRE Paper.* 14.06.
14. Wright, P.O. (1986). Planning of Electricity Supply to A.C. Electrified Railways from a Weak Network in Central Queensland, *Journal of Electrical and Electronic Engineering Australia*, 6 (2), 148–155.
15. Laughton, M.A. (1968). Analysis of Unbalanced Polyphase Networks by the Method of Phase Coordinates, Part 1. System Representation in Phase Frame of Reference, *IEEE* Proc. 115 (8), 1163–1172.
16. Laughton, M.A. (1969). Analysis of Unbalanced Polyphase Networks by the Method of Phase Coordinates, Part 2. Fault Analysis, *IEEE* Proc, 116 (5), 857–865.
17. Laughton, M.A. and Saleh, A.O.M. (1980). Unified Phase-Coordinate Load Flow and Fault Analysis of Polyphase Networks, *Electrical Power and Energy Systems*, 181–192.
18. Petersson, T. (1985). Thyristor-Controlled Static Var Compensators for Transmission Systems: An overview for prospective users, *ASEA Report*, Vasteras, Sweden.
19. Dickmander, D., Thorvaldsson, B., Stromberg, G., and Osborn, D. (1992). Control System Design and Performance Verification for The Chester, Main Static Var Compensator, *IEEE Transactions on Power Delivery*, 7 (3), 1492–1503.

3 Control of Static Var Compensators

3.1 INTRODUCTION

The applications of the static var compensator (SVC) systems listed in Chapter 2 are reproduced here for easy reference.

I. In transmission systems
   - To achieve effective voltage control
   - To increase active power-transfer capacity
   - To increase transient stability margin
   - To damp subsynchronous resonances
   - To damp power oscillations in interconnected power systems

II. In traction systems
   - To balance loads and to reduce negative-sequence voltages at the point of common coupling (PCC)
   - To improve power factor
   - To improve voltage regulation

III. In high-voltage direct current (HVDC) systems
   - To provide reactive power to ac–dc converters

IV. In arc furnaces
   - To reduce voltage variations and associated flicker

It may be seen from these applications that the control system for each SVC has to be designed to suit that application.

3.2 CONTROL SYSTEMS FOR SVCS IN TRANSMISSION SYSTEM APPLICATIONS

In transmission applications, the control system must perform the following functions.

3.2.1 Voltage Regulation

The control system must keep the bus voltage at which it is connected within the limits specified by the customer. This is essentially positive-sequence voltage regulation of the customer bus. To perform this function, it will have a voltage regulation loop with voltage and current feedback loops providing the necessary slope from the no-load to full-load condition. If X_e is the equivalent system reactance characterizing the network, which continually varies depending on system operating conditions such as generation, load,

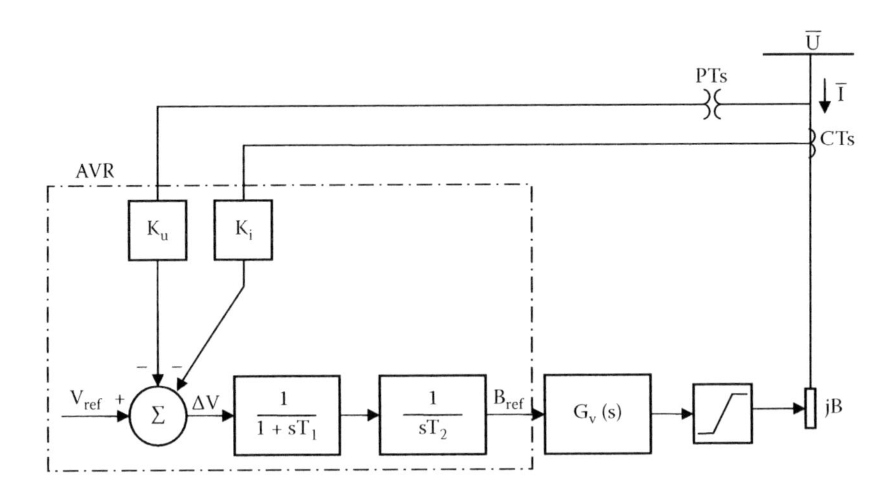

FIGURE 3.1 Block diagram for an SVS automatic voltage regulator. (From Engberg, K. and Ivner, S. (1979), "Static VAR system for voltage during steady-state and transient conditions," Paper presented at EPRI/IREQ international symposium on controlled reactive compensation, 252–265. With permission from ABB.)

and transmission line configuration, we can write a relationship (system characteristic) between the actual bus voltage V and the bus voltage E without SVC in operation as

$$V = E - X_e I_{svc} \tag{3.1}$$

where I_{svc} is the current drawn by the SVC.

The relationship between E and V can also be expressed in terms of the SVC rating Q_{svc} and the three-phase short-circuit level S_c at the bus at which the SVC is connected.

$$V = E\,(1 - (Q_{svc}/S_c)) \tag{3.2}$$

Figure 3.1 shows this voltage regulator control circuit.[1]

3.2.2 Gain Supervision[2,3]

Let us denote the gain ratio K_N by Q_{SVC}/S_c, where Q_{SVC} = the SVC rating in megavolt ampere (MVA) and S_C = the three-phase short-circuit rating in MVA at the point of connection of the SVC, S_{Cmin} = minimum value of S_C, and K_I = slope.

It is common to adopt SVC voltage control system settings when the system short-circuit conditions correspond to S_{Cmin}. The variations in the response time of such a control system as a function of S_C and Q_{SVC} are shown graphically in References 2 and 3. The following conclusions can be drawn regarding the effect of network short-circuit power on the response times of the SVC control system.

I. For large slope values, an increase in network short-circuit power has a smaller effect on the response time.
II. For weaker network states, a small short-circuit power shows sharper increases in response time.

To avoid SVC control system instability and consequent system voltage control problems, an automatic gain supervision control can be included within the SVC voltage regulator. However, an automatic gain supervision control cannot handle special circumstaces, such as those that arise when rebuilding the network after a blackout or when lines are lost. Hence provision must be made in the control system to identify such cases and then the gain must be manually adjusted to a lower value when rebuilding the network. When the short-circuit power reaches more than the minimum value, the system can function at normal settings.

Figures 11 and 12 in Reference 3 show the different control blocks used in the voltage control and volts amperes reactive (var) control circuits of the thyristor-switched capacitor (TSC) and thyristor-controlled reactor (TCR) SVCs.

3.2.3 Reactive Power Control and Coordination[2,3]

A continuously variable reactive power source is a relatively costly system in comparison with shunt compensators. Hence, it is necessary for the SVC to stabilize the network by means of rapid changes in reactive power; other sources can cover the need for vars resulting from changes in power flow, which are relatively slow changes. If this coordination is not carried out, the SVCs would tend to operate very often at their extreme capability ranges due to the speed of their response, whereas the other reactive power sources remain relatively quiescent. The SVCs would therefore be providing cover for normal network operational changes, leaving little for dynamic reserve duty under both large and small disturbances.

The reactive power control loop is a slower-acting loop than the voltage control loop. In its simplest form, it brings the SVC back to a predetermined operating point within its control range following each system change. This is illustrated in Figure 3.2. If the SVC is operating at point K with Q_{ref}, after a system change it will move to point

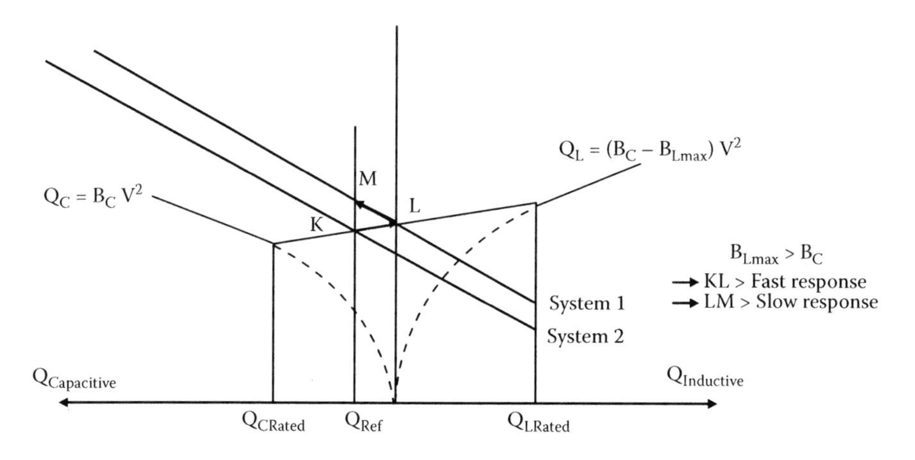

FIGURE 3.2 Behavior of SVC with slow var regulation. (From Gyugi, L. and Taylor Jr., E.R. (1980), Characteristics of Static Thyristor-Controlled Shunt Compensators for Power Transmission Applications, *IEEE Transactions on Vol. PAS99* (5), 1795–1804. With permission from IEEE.)

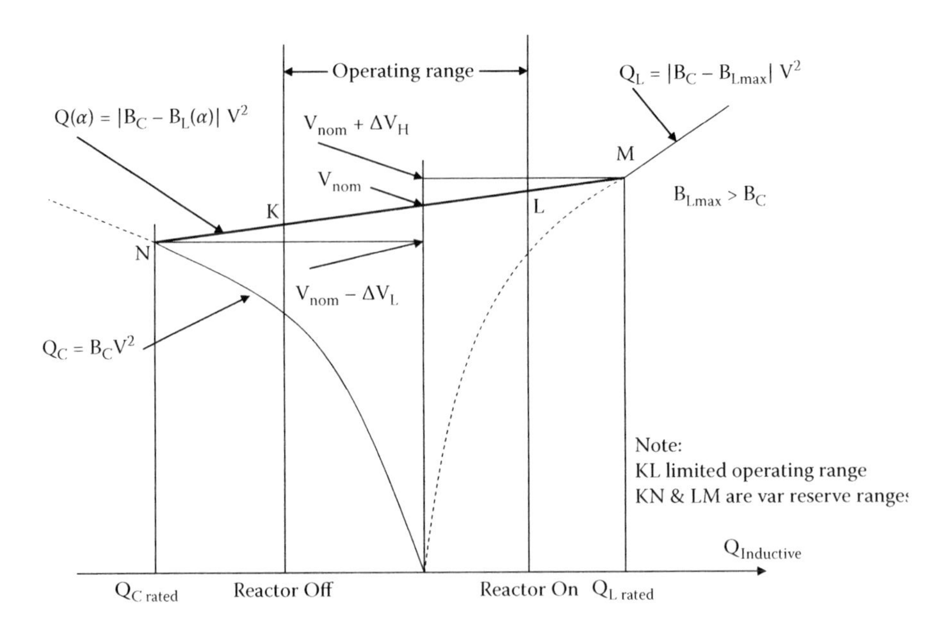

FIGURE 3.3 The steady-state reactive power versus terminal voltage; characteristics of a static var compensator. (From Gyugi, L. and Taylor Jr., E.R. (1980), Characteristics of Static Thyristor-Controlled Shunt Compensators for Power Transmission Applications, *IEEE Transactions on Vol. PAS99* (5), 1795–1804. With permission from IEEE.)

L along the line *KL*; this is because of the fast voltage regulator action. At point *L*, because Q is different from Q_{ref}, it will move further along line *LM*, and finally settle down at point *M*; this is because of the slow reactive power control action.

The switching of a reactor and the movement of the operating points within the limited operating range of the SVC are shown in Figure 3.3.

Examples of the SVC reactive power control strategies are given below:

I. Keeping a TCR in a continuous state of waiting to ensure the availability of maximum reactive power absorption when a cable section to which the TCR is connected is energized.[2]

II. Initiation of continuous conduction of TCRs on detecting system faults in the vicinity of the installation so that voltage rises can be limited through fault clearance.[2]

III. Blocking the capacitor bank in a TSC system and inductive part of the SVC through voltage regulator action immediately after a fault, when large undervoltages occur. This has the effect of reducing overvoltages in the recovery phase after fault clearance.[3]

3.2.4 Control Signals for System Transient Stability, Power Oscillation Damping, and Subsynchronous Resonance Damping Enhancement

The control signals used in these applications are shown in Table 3.1.

TABLE 3.1
Control Signals Used for Additional Feedback Loops

Application	Control Signals Used
Transient stability, power oscillation damping	Additional feedback loops based on system acceleration $\frac{\Delta P}{\Delta t}$ or frequency Δf are used
Subsynchronous resonance damping	Additional feedback loop based on the generator rotor speed deviation signal $\Delta\omega$ can be used

3.3 CONTROL SYSTEMS FOR SVCS IN TRACTION APPLICATIONS

Individual control of each SVC phase susceptance is required for load-balancing applications (e.g., single-phase railway and arc furnace loads). The development of the controller described here is reported in Reference 4 and was based on a railway electrification project in which all railway substations were compensated. The specifications, in the order of priority, required

a. Load balancing and power factor correction of single-phase loads within two cycles
b. Regulation of system positive-sequence voltages to a selectable range of reference voltages and voltage droop characteristics within four cycles
c. Control of the negative-sequence voltage magnitude to a selectable maximum limit within four cycles of the power frequency

It was further stated by the customer that the system frequency experienced frequent and large excursions, and that if all the three foregoing requirements could not be met, then the last two (*b* and *c*) should be satisfied. The following discussion will show that the specifications under (*a*) and (*c*) are rather conflicting. Hence, a compromise in priorities in the control system is required to achieve the requirements of (*a*) and (*c*). It has been shown in Section 2.1 (fundamentals of load compensation) that it is possible to balance the load and adjust the power factor to unity so that only in-phase balanced currents flow in the system. However, if we want to minimize the negative-sequence voltages in $V^{a,b,c}$, then unbalanced currents $I^{a,b,c}$ will have to flow in the system impedance ($Z^{a,b,c}$) because the source voltages ($E^{a,b,c}$) and system impedances ($Z^{a,b,c}$) are balanced, and the load is unbalanced. (Note that $V^{a,b,c} = E^{a,b,c} - Z^{a,b,c}\, I^{a,b,c}$.) In other words, to satisfy requirement (*a*), in-phase balanced currents will have to flow in the system, and to satisfy requirement (*c*), unbalanced currents $I^{a,b,c}$ will have to flow in the system impedance ($Z^{a,b,c}$).

In a Queensland Railway project in Australia, in view of the limited capacity of the SVCs, negative-sequence voltage control was given priority over positive-sequence voltage regulation.[5]

3.3.1 Load Compensation

In Figure 3.4, let B_{AB}, B_{BC}, and B_{CA} be the variable susceptances in the delta-connected load.[4] They may be either capacitive (+) or inductive (–). Let V_{1A} be the

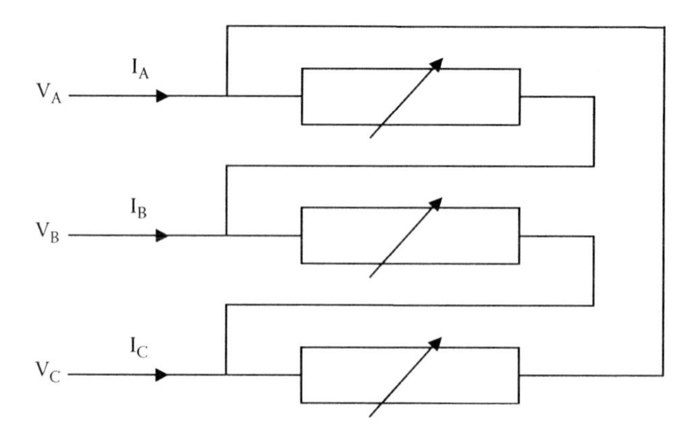

FIGURE 3.4 Basic compensation circuit.

positive-sequence phase-A voltage used as the reference phasor. Assume that the system voltages are balanced. Then, we can write the following expressions for the currents in the delta-connected load:

$$\begin{aligned} I_{AB} &= j\sqrt{3}\, V_{1A} \angle\, 30^\circ\, B_{AB} \\ I_{BC} &= j\sqrt{3}\, V_{1A} \angle -90^\circ\, B_{BC} \\ I_{CA} &= j\sqrt{3}\, V_{1A} \angle\, 150^\circ\, B_{CA} \end{aligned} \tag{3.3}$$

The line current $I_A = I_{AB} - I_{CA}$ and similar equations can be written for I_B and I_C.

In matrix notation,

$$\begin{bmatrix} I_A \\ I_B \\ I_C \end{bmatrix} = j\sqrt{3}\, V_{1A} \begin{bmatrix} 1\angle 30^\circ & 0 & 1\angle -30^\circ \\ 1\angle -150^\circ & 1\angle -90^\circ & 0 \\ 0 & 1\angle 90^\circ & 1\angle 150^\circ \end{bmatrix} \begin{bmatrix} B_{AB} \\ B_{BC} \\ B_{CA} \end{bmatrix} \tag{3.4}$$

The following matrix equation shows the relationship between the symmetrical components I_1, I_2, and I_0 of the A-phase current and the three-phase currents I_A, I_B, and I_C:

$$\begin{bmatrix} I_1 \\ I_2 \\ I_0 \end{bmatrix} = (1/3) \begin{bmatrix} 1 & a & a^2 \\ 1 & a^2 & a \\ 1 & 1 & 1 \end{bmatrix} \begin{bmatrix} I_A \\ I_B \\ I_C \end{bmatrix} \tag{3.5}$$

where $a = 1 \angle 120^\circ$. Substitution of Eq. (3.4) for the line currents I_A, I_B, and I_C yields

$$\begin{bmatrix} I_1 \\ I_2 \\ I_0 \end{bmatrix} = V_{1A} \begin{bmatrix} 1\angle 90^\circ & 1\angle 90^\circ & 1\angle 90^\circ \\ 1\angle 150^\circ & 1\angle -90^\circ & 1\angle 30^\circ \\ 0 & 0 & 0 \end{bmatrix} \begin{bmatrix} B_{AB} \\ B_{BC} \\ B_{CA} \end{bmatrix} \tag{3.6}$$

Resolving I_2 into its real and imaginary components $Re\,(I_2)$ and $\mathrm{Im}\,(I_2)$, we can write

$$\begin{bmatrix} \mathrm{Im}(I_1) \\ \mathrm{Re}(I_2) \\ \mathrm{Im}(I_2) \end{bmatrix} = V_{1A} \begin{bmatrix} 1 & 1 & 1 \\ -\sqrt{3/2} & 0 & \sqrt{3/2} \\ 1/2 & -1 & 1/2 \end{bmatrix} \begin{bmatrix} B_{AB} \\ B_{BC} \\ B_{CA} \end{bmatrix} \tag{3.7}$$

Inversion of the matrix in Equation 3.7 yields

$$\begin{bmatrix} B_{AB} \\ B_{BC} \\ B_{CA} \end{bmatrix} = \frac{1}{3} \begin{bmatrix} 1 & -\sqrt{3} & 1 \\ 1 & 0 & -2 \\ 1 & \sqrt{3} & 1 \end{bmatrix} \begin{bmatrix} Im\,(I_1)/V_{1A} \\ Re\,(I_2)/V_{1A} \\ Im\,(I_2)/V_{1A} \end{bmatrix} \tag{3.8}$$

Fast and accurate calculation of these current components—$Im(I_1)$, $Re(I_2)$, and $Im(I_2)$—along with accurate control of the firing instants of the TCR provide a fast feed-forward control for load compensation without the need for feedback.

3.3.2 Voltage Regulation and Balancing

The regulation of the system positive-sequence voltage is usually based on three-phase rms line voltage measurements,[6] whereas voltage balancing is based on individual phase-voltage measurements.[7] As in the earlier section, we will use symmetrical components to calculate the values of susceptances for positive-sequence voltage regulation and negative-sequence voltage reduction (i.e., voltage balancing).

From Figure 3.5, we can write

$$V_A = E_A - I_L Z_A - I_C Z_A \tag{3.9}$$

where E_A and V_A are the source and load voltages, and I_L and I_C are the load and compensator currents, respectively, and Z_A is the source impedance, assumed to be balanced and purely inductive. Then, $Z_A = j\,X_N$, where X_N is the balanced network reactance. If I_C is also reactive, then $I_C Z_A$ is either in phase with V_A or out of phase by 180°, and this voltage can be controlled by varying the susceptances (and hence, currents) of the compensator.

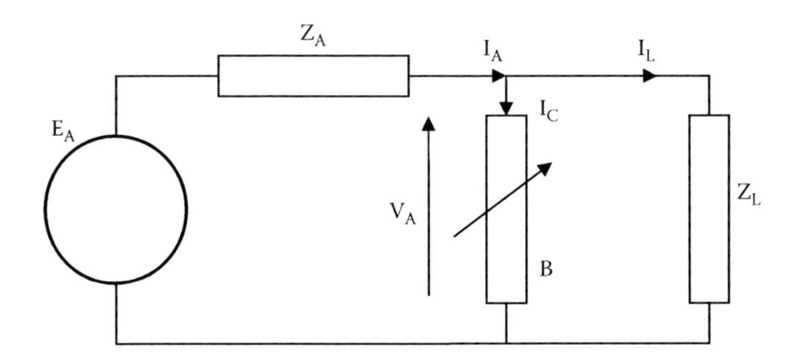

FIGURE 3.5 Single-phase system for voltage control.

Now, let $\Delta V_A = I_C Z_A$. With the help of the compensator line currents in Equation 3.2, we can write down the equations relating the controllable voltages[4] ΔV_A, ΔV_B, and ΔV_C to the susceptances B_{AB}, B_{BC}, and B_{CA}:

$$\begin{bmatrix} \Delta V_A \\ \Delta V_B \\ \Delta V_C \end{bmatrix} = \sqrt{3}\, X_N\, V_{1A} \begin{bmatrix} 1\angle 30^\circ & 0 & 1\angle -30^\circ \\ 1\angle -150^\circ & 1\angle -90^\circ & 0 \\ 0 & 1\angle 90^\circ & 1\angle 150^\circ \end{bmatrix} \begin{bmatrix} B_{AB} \\ B_{BC} \\ B_{CA} \end{bmatrix} \quad (3.10)$$

Symmetrical component transformation of ΔV_A, ΔV_B, and ΔV_C yields

$$\begin{bmatrix} \Delta V_1 \\ \Delta V_2 \\ \Delta V_0 \end{bmatrix} = X_N\, V_{1A} \begin{bmatrix} 1\angle 0^\circ & 1\angle 0^\circ & 1\angle 0^\circ \\ 1\angle 60^\circ & 1\angle 180^\circ & 1\angle -60^\circ \\ 0 & 0 & 0 \end{bmatrix} \begin{bmatrix} B_{AB} \\ B_{BC} \\ B_{CA} \end{bmatrix} \quad (3.11)$$

This equation shows that a change can be produced in the positive-sequence voltage by changing all susceptances equally.

Resolving ΔV_2 into real and imaginary components, and inverting the relevant portion of the matrix obtained from Equation 3.11, yields

$$\begin{bmatrix} B_{AB} \\ B_{BC} \\ B_{CA} \end{bmatrix} = 1/(3X_N) \begin{bmatrix} 1 & 1 & \sqrt{3} \\ 1 & -2 & 0 \\ 1 & 1 & -\sqrt{3} \end{bmatrix} \begin{bmatrix} \mathrm{Re}(\Delta V_1)/V_{1A} \\ \mathrm{Re}(\Delta V_2)/V_{1A} \\ \mathrm{Im}(\Delta V_2)/V_{1A} \end{bmatrix} \quad (3.12)$$

In this case, because X_N is unknown and variable, a feedback controller must be used. If the required corrections in the voltages $\mathrm{Re}(\Delta V_1)$, $\mathrm{Re}(\Delta V_2)$, and $\mathrm{Im}(\Delta V_2)$ are calculated from system individual phase-voltage measurements, then the necessary corrections in the susceptances B_{AB}, B_{BC}, and B_{CA} can be calculated from Equation 3.12.

3.3.3 Measurement of Sequence Components

We have seen in earlier sections that sequence components of voltages and currents are necessary to calculate the necessary corrections in compensator susceptances for achieving the objectives listed in Section 3.3, namely,

a. Load balancing and power-factor correction
b. System positive-sequence voltage regulation
c. Reduction of negative-sequence voltage at the bus at which the SVC is connected

We will now discuss how these sequence components are measured.[4] If we assume that all our unbalance is due to the delta-connected or ungrounded wye-connected loads, then we can assume that no zero sequence currents exist in the system because

$I_A + I_B + I_C = 0$. Under these conditions, Clarke's transformation yields the following instantaneous α, β, 0 components:

$$\begin{bmatrix} i_\alpha \\ i_\beta \\ i_0 \end{bmatrix} = (2/3) \begin{bmatrix} 1 & -1/2 & -1/2 \\ 0 & \sqrt{3}/2 & -\sqrt{3}/2 \\ 1/2 & 1/2 & 1/2 \end{bmatrix} \begin{bmatrix} i_A \\ i_B \\ i_C \end{bmatrix} \tag{3.13}$$

which can be simplified as $i_a = i_A$; $i_\beta = (1/\sqrt{3})\,(i_B - i_C)$; and $i_0 = 0$

Next, phasor currents I_α and I_β can be transformed into phasor-positive and phasor-negative sequence currents I_1 and I_2, respectively, by the following transformation:

$$\begin{bmatrix} I_1 \\ I_2 \end{bmatrix} = \frac{1}{2} \begin{bmatrix} 1 & j \\ 1 & -j \end{bmatrix} \begin{bmatrix} I_\alpha \\ I_\beta \end{bmatrix} \tag{3.14}$$

Instantaneous values of I_1 and I_2 can be obtained by multiplying the matrix in Equation 3.14 by the reference phasor defined by $\sin \omega t + j \cos \omega t$. Then, we can express the positive-sequence current i_1 in terms of i_α and i_β:

$$i_1 = i_\alpha\,(\sin \omega t + j \cos \omega t) + j\, i_\beta (\sin \omega t + j \cos \omega t) \tag{3.15}$$

or

$$Re\,(i_1) = i_\alpha \sin \omega t - i_\beta \cos \omega t \tag{3.16a}$$

$$Im\,(i_1) = i_\alpha \cos \omega t + i_\beta \sin \omega t \tag{3.16b}$$

The following standard transformation between the α and β components, and d and q axes, is usually available as a standard computer programming package:

$$d = i_\alpha \cos \varepsilon + i_\beta \sin \varepsilon \tag{3.17a}$$

$$q = -i_\alpha \sin \varepsilon + i_\beta \cos \varepsilon \tag{3.17b}$$

The outputs d and q give the real and imaginary parts of the positive-sequence currents or voltages when the input signals $\cos \varepsilon = \sin \omega t$ and $\sin \varepsilon = -\cos \omega t$.

To calculate the negative-sequence values, either interchange input phases B and C or reverse the output value of phase β. Under balanced conditions, the $d - q$ transformation yields *dc* values, but with any unbalance in the input, a second harmonic term also appears in the output. This term must be filtered, as only *dc* values are required.

The conventions used in the following equations are listed below. Alpha (α) axis leads beta (β) axis by 90°. Quadrature (q) axis leads direct (d) axis by 90°. A,B,C phase sequence is assumed.

In the following equations, $\underline{V}_\alpha$ and $\underline{V}_{AB}$, $\underline{V}_{BC}$ and $\underline{V}_{CA}$ are phasors $\underline{V}_\alpha$ and $\underline{V}_{AB}$ are in phase.

The negative sequence voltages can be measured alternatively as follows.

Because there cannot exist any zero sequence component in the line to line voltages V_{AB}, V_{BC}, and V_{CA} the line to line voltage system can be represented, by the following transformation by a two phase α, β system. Let

$$\underline{V}_{\alpha} = \underline{V}_{AB}/\sqrt{3} \tag{3.18a}$$

$$\underline{V}_{\beta} = (\underline{V}_{BC} - \underline{V}_{CA})/3 \tag{3.18b}$$

The three-phase system is converted into a two-phase system.

The positive and negative sequence voltages V_{pos} and V_{neg} of the equivalent two phase α,β system are

$$V_{pos} = 0.5\ (\underline{V}_{\alpha} + j\ \underline{V}_{\beta}) \tag{3.19a}$$

$$V_{neg} = 0.5\ (\underline{V}_{\alpha} - j\ \underline{V}_{\beta}) \tag{3.19b}$$

Figure 3.6(a) shows a block diagram for the calculation of V_{α} and V_{β}. The amplifiers A1 and A2 are used to perform the transformation from three to two phases.

Figure 3.6(b) shows a block diagram of a negative sequence detector based on these principles. The signals cos ωt and sin ωt are generated by a phase-locked oscillator.

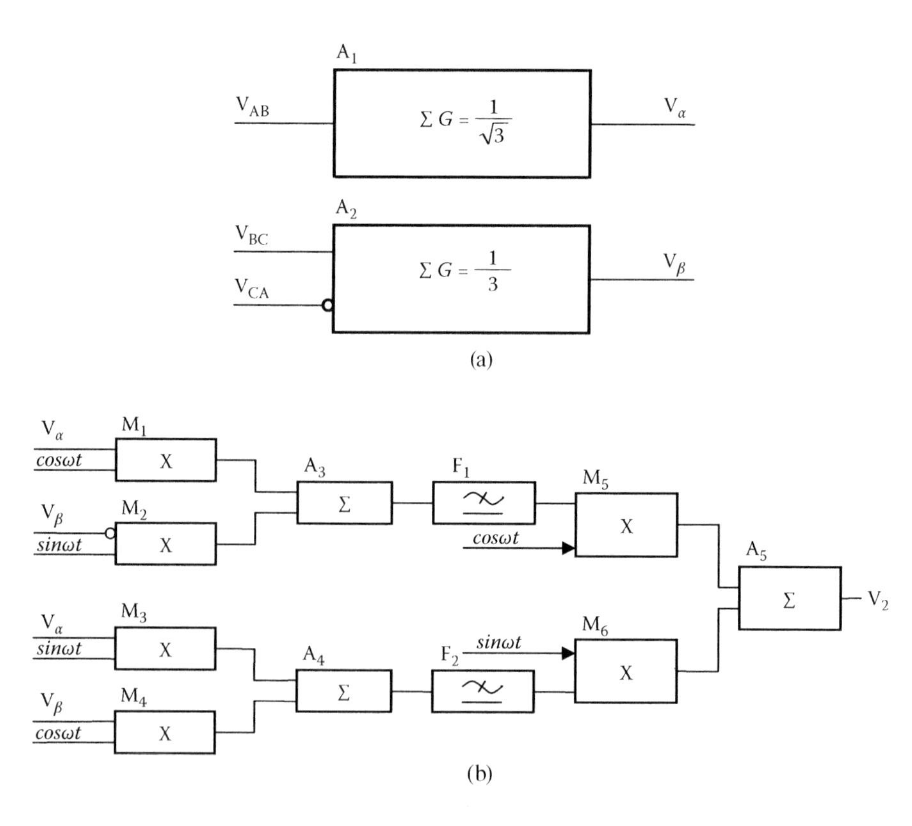

FIGURE 3.6 (a) Negative-sequence detector (Part 1); (b) negative-sequence detector (Part 2).

By the multipliers $M1$ to $M4$, which could be realized by digital to analogue converters, a multiplication of the voltages V_α and V_β with cos ωt and sin ωt is performed.

The voltages V_α and V_β can be written as

$$V_\alpha(t) = V_1 \cos\omega t + V_2 \sin\omega t \tag{3.20a}$$

$$V_\beta(t) = V_3 \sin\omega t - V_4 \cos\omega t \tag{3.20b}$$

(Note that if the three-phase system is completely balanced, then the magnitudes of V_α and V_β are the same. Further, if V_d and V_q are direct and quadrature-axis voltages, then $V_d = V_1 = V_3$, and $V_q = V_2 = V_4$.)

The output signal from amplifier A3, when there is a negative-sequence voltage present in the system, is given by

$$V_{A3}(t) = V_\alpha \cos \omega t - V_\beta \sin \omega t \tag{3.21a}$$

$$V_{A3}(t) = V_1 \cos^2\omega t + V_2 \sin \omega t \cos \omega t - V_3 \sin^2\omega t + V_4 \sin\omega t \cos\omega t \tag{3.21b}$$

$$V_{A3}(t) = 0.5(V_1 - V_3) + 0.5(V_1 + V_3)\cos 2\omega t + 0.5(V_2 + V_4) \sin 2\omega t \tag{3.21c}$$

The notation $V_{A3}(t)$ comprises a dc component 0.5 $(V_1 - V_3)$ and components that are two times as in Fig. 3.6 the system fundamental frequency.

In a similar way, it is found that the dc component of the output from amplifier A4 is 0.5 $(V_2 - V_4)$. The dc components are filtered out by low-pass filters F1 and F2. The voltage formed by the multipliers M5 and M6, and amplifier A5 is

$$V_{A5} = 0.5\,(V_1 - V_3) \cos \omega t + 0.5\,(V_2 - V_4) \sin \omega t \tag{3.22}$$

This is the negative-sequence voltage component of the phase voltages. The rms value of this voltage can be measured by a voltmeter, which will directly show the magnitude of the negative-sequence voltage.

3.4 PHASE-LOCKED OSCILLATOR CONTROL SYSTEM

Another phase-locked oscillatory control system for TCR with individual phase control and common oscillatory control has been described by J.D. Ainsworth. See References 8–11 for details.

3.5 IMPLEMENTATION DETAILS OF A PROGRAMMABLE HIGH-SPEED CONTROLLER

The programmable high-speed controller (PHSC) described in Reference 4 is based on bit-slice microprocessor technology that provides very high control speeds and a user-friendly function block programming language.

In these types of controllers, the phase-locked-loop (PLL) synchronization circuit is used to synchronize the controller to the positive-sequence voltage and control the sequencing of the programs.

The input stage transforms the three-phase system voltages and currents into positive- and negative-sequence components. This conversion program runs 24 times per cycle of the system voltage (or every 15°), with 50-Hz system frequency, which corresponds to 833 μs. The reference pulse from the PLL sets the sine–cosine generator to initial conditions of each power cycle to eliminate any accumulated error in the angular increments.

For load compensation, B_{AB}, B_{BC}, and B_{CA} are directly calculated according to Equation 3.8. Allowing for transformer impedance and transformer delta–wye connection, and so on, the per-unit susceptance of a TCR required for a particular load condition is calculated. Suppose this is B_L per unit in the AB phase of the delta connection of the compensator. Then, the per-unit susceptance of a TCR with respect to the firing angle is given as

$$B_L = (1/\pi)\,(2\pi - 2\alpha + \sin 2\alpha) \text{ for } \pi/2 < \alpha < \pi \tag{3.23}$$

As no accurate polynomial expression could be found for the value of α in terms of B_L, a linear interpolation of the function was used with nine break points. These break points were chosen to eliminate the integral squared error between the piecewise linear and actual curves. Each firing angle will be calculated only once per power cycle, a few degrees before the voltage of the TCR reaches the positive maximum.

3.5.1 Priority Logic

Priority logic checks whether any of the TCRs are close to their operating limits so that voltage control requirements can take priority.

Priority logic consists of three parts: (1) for load compensation, (2) for the reduction of the negative-sequence voltage, and (3) for positive-sequence voltage regulation with droop. Each part calculates the value of the susceptance B in each phase to perform its function, and all the three values of B are summed up. The following multipliers and adders ramp up the value of B from a preset value to the individual values following start-up or a system fault.

If the value of the negative-sequence voltage is within the specified limit, then the output of the regulator is zero, and the values of B from this section are also zero.

For voltage regulation, the droop characteristic requires that the positive-sequence current be calculated using Equation 3.7 rather than being measured.

3.5.2 Detection of Faults

Positive- and negative-sequence voltage values of V_1 and V_2 are used to detect faults. (See Table 3.2.)

The values of V_1, V_2, and V_2/V_1 in Table 3.2 are used to detect the faults and send a signal to the PLL to keep its frequency in the firing unit constant. Then, the input to the integrators in the controller are set to zero, and the values of all susceptances are reduced to B_0, currently preset at −100% (fully inductive to prevent overvoltages on clearing the fault).

After the fault is cleared, the values of B are ramped from B_0 toward their pre-fault values in six cycles. The frequency of the PLL is released after one cycle, and the integrators are released after six cycles.

TABLE 3.2
Identification of Faults from Expected Voltage Ratios

Type of Fault	Faulted Phase Voltage to Ground	Expected V1	Expected V2/V1
Three-phase fault	50%	50%	0%
Single-line-to-ground fault	50%	83%	20%

If the value of B in any phase approaches the maximum or minimum limits, an integrator begins to ramp down the gain of the load compensation section in the priority logic, as priority must be given to voltage control. The gain is ramped back to 100% when the values of B fall below the limits plus a small (5%) hysteresis band.

3.5.3 Program Sequence

The program "input stage" transforms the three-phase systems for voltage and current into positive and negative sequence components. "Priority logic" described in section 3.2.1 runs 24 times per cycle while the alpha algorithm described in section 3.2.3 runs only once for each phase per cycle. The monitoring program starts to run each time after the "priority logic" but is interrupted by the higher priority alpha algorithm whenever they run.

3.5.4 Special Features of the Programmable High-Speed Controller

I. Use of the phase-locked loop
  a. Provides an accurate reference for generating the real and imaginary components of the sequence voltages and currents
  b. Eliminates any error in averaging filters during frequency excursions
  c. Provides hardware interrupts to start specific programs at definite times in a power cycle

II. Voltage regulation and balancing are achieved using an algorithm that uses the sequence components of the system voltages and currents.

III. The accurate calculation of the firing angle, α, allowing for unbalanced capacitance, inductance, and/ transformer leakage impedance, eliminates the need for feedback of the compensator currents.

The setting of the firing angle of each TCR only once per cycle eliminates all even harmonics and hence reduces the cost of filtering.

3.6 CONCLUSIONS

In this chapter, at the beginning, the applications of the static var compensators for transmission systems, single phase traction systems, HVDC and arc furnace applications are listed. Each of the above types have their own specific control requirements.

In transmission systems one of the important objectives is to damp power oscillations in interconnected power systems; in traction systems the primary objective is to reduce negative sequence voltages at the point of common coupline; in arc furnace loads, it is to reduce the voltage flicker.

As explained in section 2.1.2 in chapter 2, in a transmission static var compensator to provide damping the ΔV_m must be varied as a function of

$$d(\Delta\delta)/dt \text{ i.e } \quad \Delta V_m = K\, d(\Delta\delta)/dt$$

This means that midpoint voltage is to be increased, by providing capacitive vars, when $d(\Delta\delta)/dt$ is positive; this increases the transmitted electric power and opposes the acceleration of the generator. When $d(\Delta\delta)/dt$ is negative, the midpoint voltage is to be decreased by absorbing inductive vars; this in turn reduces the transmitted electric power and thereby opposes the deceleration of the generator. In the control system of this static var compensator, we need a signal proportional to $d(\Delta\delta)/dt$.

Alternatively such signal is not necessary for static var compensators used in either traction systems or for arc furnace loads. Signals proportional to negative phase sequence voltages are required in traction applications. In static var compensators to control flicker due to arc furnace loads signals proportional to but voltages are required. Hence based on the primary objective of a static var compensator, suitable control signals have to be obtained and the control system designed.

We have discussed an SVC control system with a programmable high speed controller with facilities for individual phase control, load compensation, voltage balancing and regulation. Measurement of negative sequence voltages is also described.

REFERENCES

1. Engberg, K. and Ivner, S. (1979). *Static VAR system for voltage during steady-state and transient conditions*, Paper presented at EPRI/IREQ international symposium on controlled reactive compensation, 252–265.
2. CIGRE Report on Static Var Compensators, (1986) Prepared by Working Group 38-01, Task Force No. 2, Edited by I.A. Erinmez.
3. Romegialli, G. and Beeler, H. (1981). Problems and concepts of static compensator control, *Proceedings of the IEE*, 128 (6), Pt. C. No. 6., pp. 382–388.
4. Gueth, G., Ensted, G.P., Ray, A., and Menzies, R.W. (1987). Individual phase control of a static compensator for load compensation and voltage balancing and regulation, *IEEE Transactions on Power Systems*, 2(4), 898–905.
5. Sastry, V.R., Hill, D.V., and Lee, C.J., (1986) Static Var Compensators for balancing the single phase traction loads in Central Queensland. *Journal of Electrical and Electronic Engineering, Australia,* 6(3), September 184–190.
6. Schweickardt, H. and Romegialli, G. (1978). The static VAR source in EHV transmission systems and its control. *Brown Boveri Review*, 65 (9), 585–589.
7. Goosen, P.V. et al. (1984). FC/TCR Type Static Compensators in ESCOM's 132 kV Network. *CIGRE Paper*, 38 (9).
8. Ainsworth, J.D. (1988) Phase-locked oscillator control system for thyristor-controlled reactors. *IEE Proceedings*, 135, Pt C, (2) 146–156.

9. Ainsworth, J.D., (1968) The phase-locked oscillator—a new control system for controlled static convertors. *IEEE Transactions*, PAS-87, 859–865.
10. Ainsworth, J.D., (1985) Developments in the phase-locked oscillator control system for HVDC and other large converters. *IEE Conference Publication 255 on AC and DC Power Transmission*, 98–103.
11. British Patent Specification 1 551 089 (1977).

4 Harmonics

In this chapter we will discuss the harmonic sources and undesirable effects of the harmonics in the power system, harmonics in a thyristor-controlled reactor, and the K-factor.

Different categories of harmonic-producing loads are supplied by the electric utilities, such as

1. Domestic loads such as fluorescent lamps, light dimmers, etc.
2. Ripple control systems for regulating hot-water loads
3. Medium-sized industrial loads such as several adjustable speed drives in a cement mill, paper mill, etc.
4. Large loads such as high-voltage direct current (HVDC) converters, aluminum smelters, static var compensators (SVCs), heavy single-phase ac traction loads for hauling coal trains, etc.

The undesirable effects of the harmonics produced by the aforementioned loads are listed as follows[1–5]:

1. Capacitors may draw excessive current and prematurely fail from increased dielectric loss and heating.
2. Harmonics can interfere with telecommunication systems, especially noise on telephone lines.
3. Transformers, motors, and switchgear may experience increased losses.
4. Induction motors may refuse to start (cogging) or may run at subsynchronous speeds.
5. Circuit breakers may fail to interrupt currents due to improper operation of blowout coils.
6. The time–current characteristics of fuses can be altered, and protective relays may experience erratic behavior. In particular, maloperation of the relays associated with ripple control systems can occur.
7. Errors happen in induction kilowatt-hour meters.
8. Excitation problems cause generator failure.
9. Interference occurs with large motor controllers.
10. Overvoltages and excessive currents in the system happen due to resonances of harmonics in the network and consequent dielectric instability of insulated cables.

We will briefly consider the following harmonic sources:

1. Converters (6-pulse and 12-pulse)
2. Single-phase power supplies, including switch-mode types

3. dc drives
4. ac drives
5. Pulse-width modulation (PWM)
6. Telecontrol signals
7. Cycloconverters
8. Transformers
9. Harmonics in rotating machines
10. Arc furnace loads

4.1 CONVERTER HARMONICS

In converters the pulse number is defined as the number of pulsations (cycles of ripple) of the direct voltage per cycle of the alternating voltage.[1] This will be determined by the nonsimultaneous commutations per cycle of the alternating voltage in the bridge circuit. Usually, 6- or 12-pulse configurations are used in HVDC transmission.

Usually, the following assumptions are made to calculate the expressions for the harmonics in the converter circuits:

The alternating voltage has no harmonics except the first. However, there can be higher harmonic currents on the ac side.

The direct current has no harmonics, but harmonic voltages can be present on the dc side.

A converter of pulse number p generates characteristic harmonics of order

$$n = pq \tag{4.1}$$

on the dc side and

$$n = pq \pm 1 \tag{4.2}$$

on the ac side, q being an integer.

In large power converters there is a large inductance on the dc side. In view of this, direct current is constant, and during converter operation the dc side acts like a harmonic voltage source. In contrast, in view of the balanced ac voltage sources, the ac side acts like a harmonic current source and supplies the harmonic currents demanded by the converter operation. Further, if we make the assumption that the positive and negative sides of the current waveforms are symmetrical, then no even harmonics can exist.

If $f(\theta)$ is a periodic function of θ, the general trigonometric form of the Fourier series is

$$f(\theta) = a_0 + \sum_{n=1}^{\infty} a_n \cos n\theta + b_n \sin n\theta \tag{4.3}$$

The constant a_0 can be evaluated by integrating both sides of the function over a period, that is, between the limits of 0 and 2π radians of the value θ. Because over a period the definite integral of the cosine and sine terms yield zero,

$$a_0 2\pi = \int_0^{2\pi} f(\theta) d\theta$$

$$a_0 = (1/2\pi) \int_0^{2\pi} f(\theta) d\theta \tag{4.4}$$

Multiplying Equation 4.3 by $\cos(n\theta)$, where n takes different values from 1 to ∞, and using the orthogonal property of sine and cosine functions, we can write

$$\int_0^{2\pi} f(\theta)\cos(n\theta)d\theta = \int_0^{2\pi} a_n \cos^2(n\theta)d\theta$$

$$= \int_0^{2\pi} a_n[(1+\cos 2n\theta)/2]d\theta = a_n.(2\pi/2) = a_n\pi$$

$$\text{Hence } a_n = (1/\pi)\int_0^{2\pi} f(\theta)\cos(n\theta)d\theta \quad (4.5)$$

Similarly, multiplying Equation 4.3 by $\sin(n\theta)$, and integrating over a period, we can show that

$$b_n = (1/\pi)\int_0^{2\pi} f(\theta)\sin(n\theta)d\theta \quad (4.6)$$

In Equations 4.4, 4.5, and 4.6, the limits on the definite integrals can be changed from 0 to 2π to $-\pi$ to $+\pi$ because both these limits cover one complete period.

Harmonic analysis of the waveform of the alternating current shows that no characteristic harmonics of order $3q$ (triplen harmonics) can exist.

Let us consider the analysis of a train of rectangular pulses of unit height and arbitrary width of w radians of duration w/ω second (see Figure 4.1). These pulses might represent current through one valve. The relevant Fourier coefficients with reference to a one per unit direct current are

$$a_0 = 1/(2\pi)\int_{-w/2}^{w/2} d\theta = w/2\pi \quad (4.7)$$

$$a_n = 1/(\pi)\int_{-w/2}^{w/2} \cos(n\theta)d\theta = \frac{2}{\pi\, n}\sin\frac{nw}{2} \quad (4.8)$$

The corresponding Fourier series for the positive current pulses is

$$F_1\theta = (2/\pi)\,[(w/4 + \sin(w/2)\cos\theta + (1/2)\sin(2w/2)\cos 2\theta + (1/3)\sin(3w/2)\cos 3\theta$$
$$+ (1/4)\sin(4w/2)\cos 4\theta + \cdots)] \quad (4.9)$$

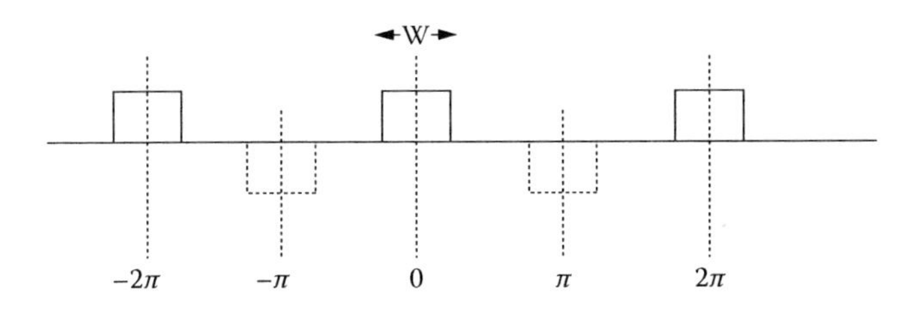

FIGURE 4.1 Trains of positive and negative pulses. (From Kimbark, E.W., *Direct Current Transmission* (1971), Vol. 1, Harmonics and Filters chapter, John Wiley, New York.)

Similarly, the corresponding Fourier series for the negative current pulses is

$$F_2\theta = (2/\pi)\ [(-w/4 + \sin(w/2)\cos\theta - (1/2)\sin(2w/2)\cos 2\theta + (1/3)\sin(3w/2)\cos 3\theta - (1/4)\sin(4w/2)\cos 4\theta + \cdots)] \quad (4.10)$$

The Fourier series for the train of alternately positive and negative rectangular current pulses is obtained by adding the two foregoing equations:

$$F_3 = (4/\pi)\ [(\sin(w/2)\cos\theta + (1/3)\sin(3w/2)\cos 3\theta + (1/5)\sin(5w/2)\cos 5\theta + \cdots)] \quad (4.11)$$

With six-pulse rectification (and inversion), the Fourier series for the phase currents can be obtained by substituting $w = (2\pi/3)$ in the Equation 4.11 for F_3 and inserting the actual value I_d for the direct current in the place of one per unit value.

The ac current in phase A is

$$i_a = [2\sqrt{3}/\pi]\ I_d\ [\cos\theta - (1/5)\cos 5\theta + (1/7)\cos 7\theta - (1/11)\cos 11\theta + (1/13)\cos 13\theta - (1/17)\cos 17\theta + (1/19)\cos 19\theta \cdots] \quad (4.12)$$

A study of Equation 4.12 reveals the following facts regarding the waveform of i_a:

1. There are no triplen harmonics.
2. The harmonics (characteristic harmonics) present are of the order $6k \pm 1$ for integer values of k.
3. In balanced systems (i.e., when i_a, i_b, i_c magnitudes are the same and they are displaced by 120°), those harmonics of the order $6k + 1$ are of positive sequence.
4. In balanced systems (i.e., when i_a, i_b, i_c magnitudes are the same and they are displaced by 120°), those harmonics of the order $6k - 1$ are of negative sequence.
5. The root-mean-square (rms) magnitude of the fundamental frequency is

$$I_1 = (1/\sqrt{2})\ (2\sqrt{3}/\pi)\ I_d = (\sqrt{6}/\pi)\ I_d \quad (4.13)$$

6. The rms magnitude of the n-th harmonic is

$$I_n = I_1/n$$

4.1.1 Effect of Transformer Connections

If both the windings of the transformer have a wye–wye or delta–delta connection and the ratios 1:1, the line currents on the network side and valve side have the same wave shape. Under these conditions, the ac current in phase A is

$$i_a = [2\sqrt{3}/\pi]\ I_d\ [\cos\theta - (1/5)\cos 5\theta + (1/7)\cos 7\theta - (1/11)\cos 11\theta + (1/13)\cos 13\theta - (1/17)\cos 17\theta + (1/19)\cos 19\theta \cdots] \quad (4.14)$$

Let the transformers be connected in wye on the valve side and delta on the network side. Further, let us assume that the bank ratio is made 1:1 instead of the individual transformer. Then the ac current in phase A is

$$i_a = [2\sqrt{3}/\pi]\ I_d\ [\cos\theta + (1/5)\cos 5\theta - (1/7)\cos 7\theta - (1/11)\cos 11\theta + (1/13)\cos 13\theta - (1/17)\cos 17\theta - (1/19)\cos 19\theta] \quad (4.15)$$

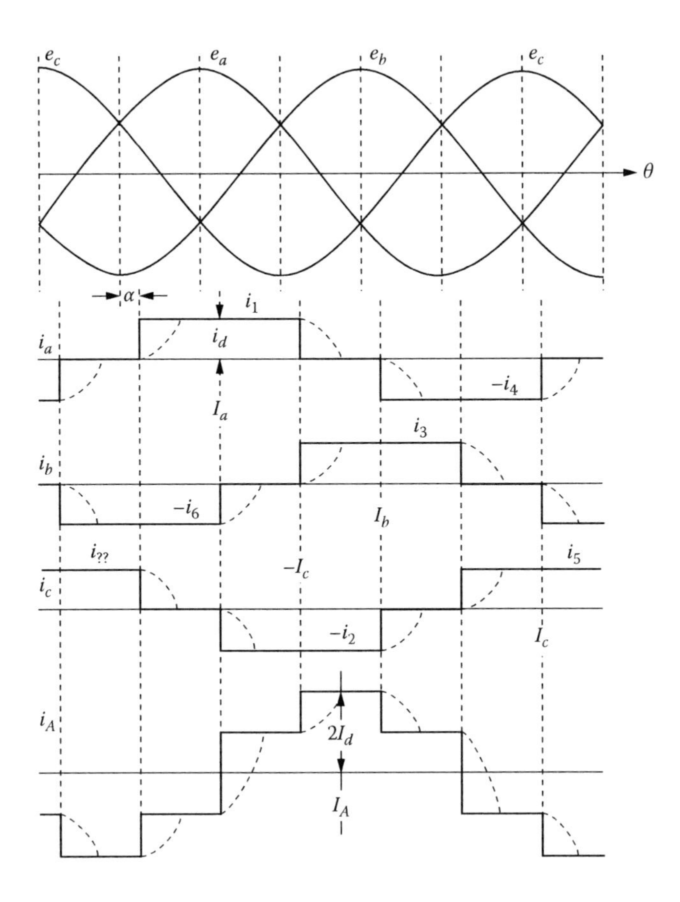

FIGURE 4.2 Waveforms in a 6-pulse bridge circuit. (From Kimbark, E.W., *Direct Current Transmission* (1971), Vol. 1, Harmonics and Filters chapter, John Wiley, New York.)

The waveforms in the two foregoing equations are rather different because of difference in the signs of certain harmonics. Adding the two equations, the waveform of the *A* phase in a 12-pulse converter can be obtained.

Waveforms of voltage and currents in a 6-pulse bridge are shown in Figure 4.2.[1] Waveforms of the current and electrical circuit of a 12-pulse bridge circuit are shown in Reference 2.

$$i_a = 2\,[2\sqrt{3}/\pi]\,I_d[\cos\theta$$
$$-(1/11)\cos 11\,\theta + (1/13)\cos 13\theta - (1/23)\cos 23\,\theta + (1/25)\cos 25\,\theta \cdots] \quad (4.16)$$

The series contains only harmonic of order 12k ± 1.

4.1.2 Harmonics When There Is Overlap in the Commutation Process

In practical HVDC systems, the reactance in the commutation circuit causes conduction overlap of the incoming and outgoing phases.[1,4,6] Using as a reference the corresponding commutating voltage (i.e., the zero-voltage crossing) and assuming a purely inductive commutation circuit, the following expression defines the commutating current[5,6]:

$$i_c = (E/\sqrt{2}\, X_c)\ [\cos\alpha - \cos\theta] \tag{4.17}$$

where X_c is the reactance (per phase) of the commutation circuit, which is largely determined by the transformer leakage reactance.

At the end of the commutation, $i_c = I_d$ and $\theta = u$, and Equation 4.17 is expressed as

$$I_d = (E/\sqrt{2}\, X_c)\ [\cos\alpha - \cos(\alpha + u)] \tag{4.18}$$

Dividing Equation 4.17 by Equation 4.18 yields

$$i_c = I_d\{[\cos\alpha - \cos\theta]/[\cos\alpha - \cos(\alpha + u]\} \quad \text{for} \quad \alpha < \theta < \alpha + u \tag{4.19}$$

Also,

$$i = I_d \quad \text{for} \quad \alpha + u < \theta < \alpha + 2\pi/3 \tag{4.20}$$

and

$$i = I_d - I_d[\cos(\alpha + 2\pi/3) - \cos\theta]/[\cos(\alpha + 2\pi/3) - \cos(\alpha + 2\pi/3 + u)]$$

for

$$(\alpha + 2\pi/3) < \theta < (\alpha + 2\pi/3 + u) \tag{4.21}$$

The negative current pulse still possesses half-wave symmetry, and therefore, only odd harmonics are present. As can be seen from the foregoing expressions, these harmonics are dependent upon firing angle α and overlap angle u, and their magnitudes can be expressed as a percentage of the fundamental component of the ac phase current I_1.

Kimbark in Reference 1 gives the necessary formulae to calculate all these odd harmonics, both their magnitude and phase angle, and also graphically the variation of these harmonics with firing angle α and overlap angle u, as a percentage of the fundamental component of the ac phase current I_1.

Figure 4.3 shows the fifth-harmonic current on the ac side of a six-pulse converter as a function of converter angles.

4.1.3 Direct-Voltage Harmonics

For the three-phase six-pulse configuration, the orders of the harmonic voltages $h = 6k$. The repetition interval of the waveform for a six-pulse interval is $\pi/3$, and it contains the following three different functions:

$$v_d = \sqrt{2}V_c\cos[\theta + \pi/6] \quad \text{for} \quad 0 < \theta < \alpha \tag{4.22}$$

$$v_d = \sqrt{2}V_c\cos[\theta + \pi/6] + (1/2)\sqrt{2}V_c\sin\theta = (\sqrt{6}/2)V_c\cos\theta \quad \text{for} \quad \alpha < \theta < \alpha + u \tag{4.23}$$

$$v_d = \sqrt{2}V_c\cos[\theta - \pi/6] \quad \text{for} \quad \alpha + u < \theta < \pi/3 \tag{4.24}$$

where V_c is the commutating phase-to-phase rms voltage.

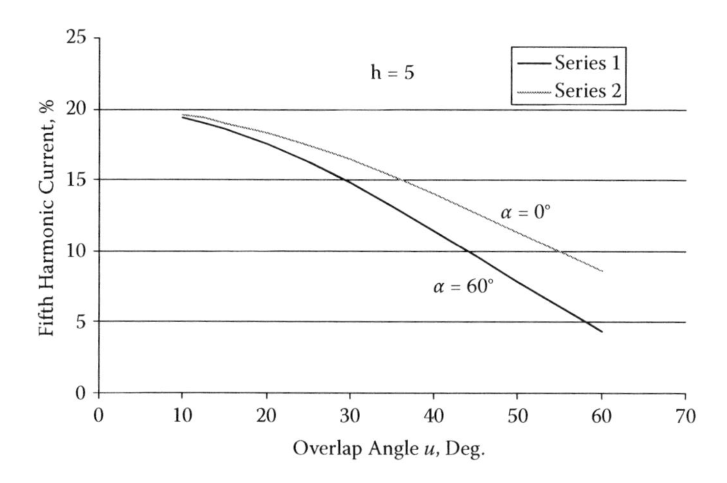

FIGURE 4.3 Fifth-harmonic current as a function of converter angles. (Adapted from Kimbark, E.W., *Direct Current Transmission* (1971), Vol. 1, Harmonics and Filters chapter, John Wiley, New York.)

Figure 4.4 shows the sixth harmonic of direct voltage of a six-pulse converter as a function of converter angles. For the variation of the other harmonics, please see Kimbark, Reference 1.

4.1.4 Imperfect System Conditions

In practical power systems, the ac system is never balanced, and there can be asymmetry in firing angles and other parts of the control system. Further, the dc current may be modulated in the rectifier due to the presence of an inverter. Under these conditions, **noncharacteristic** harmonics may be present. All those harmonics that are not **characteristic** (predicted earlier by Fourier analysis in the earlier equations) are referred to as **noncharacteristic harmonics.** Sometimes these are referred to as **uncharacteristic harmonics.**

The following terms are also used in the literature:

Interharmonics[2]**:** Sinusoidal oscillations whose frequency is not a whole-number multiple of the fundamental frequency

Subharmonics: Sinusoidal oscillations whose frequency is lower than the fundamental frequency, but the fundamental frequency divided by the sub-harmonic frequency is a whole number

From field measurements, noncharacteristic harmonic magnitudes can be determined. Kimbark reports in Reference 1 the measurements at Lydd (at the British end of the Cross Channel link), whereas Arrillaga et al.[4] report the results of the back-to-back tests of the New Zealand HVDC converters. These will give you typical values of noncharacteristic harmonics obtained in practical HVDC systems. Sometimes, these can cause more problems than characteristic harmonics because their

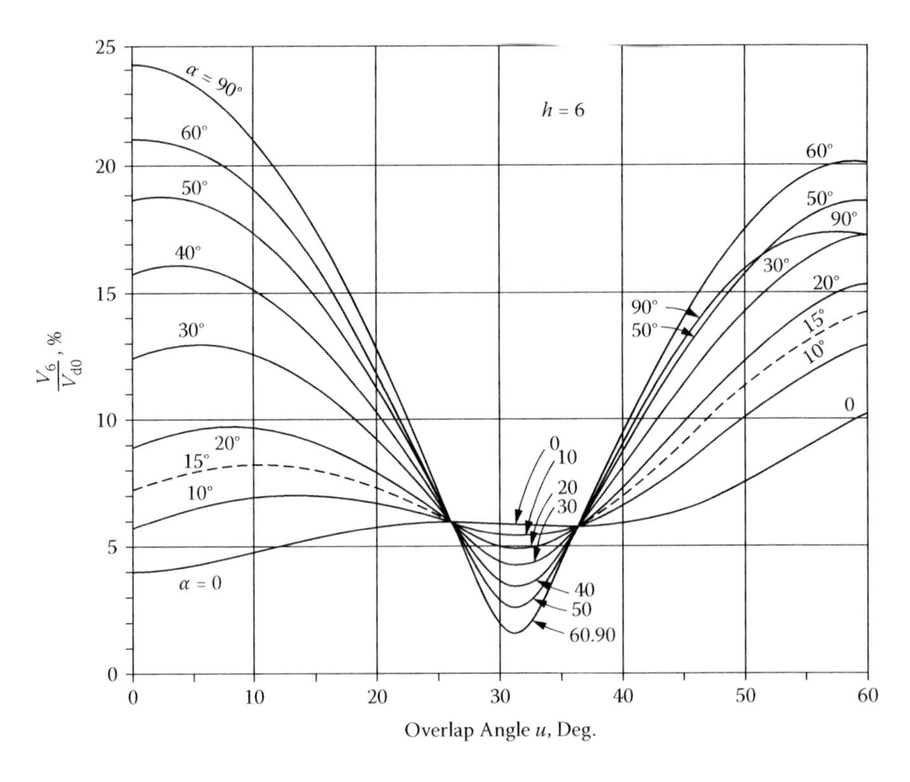

FIGURE 4.4 Sixth harmonic of direct voltage of six-pulse converter as a function of converter angles. (From Kimbark, E.W., *Direct Current Transmission* (1971), Vol. 1, Harmonics and Filters chapter, John Wiley, New York.)

magnitudes cannot be predicted easily by digital simulation of the system as the unbalances present in the system are rather hard to estimate.

Due to firing asymmetry, both even and triplen harmonics can be present in an HVDC system. These have been estimated by Kimbark.[1] If the relative displacement is 2ε between the positive and negative pulses, the ratio of an even harmonic of order h to the fundamental wave at small overlap is

$$I_h/I_1 = 2\sin h\ \varepsilon/2h\ \cos\varepsilon \approx \varepsilon \tag{4.25}$$

For example, for $\varepsilon = 1°$, the second and fourth harmonics are 1.74% of the fundamental current.

Let us assume that the widths of the positive and negative pulses are slightly different. In the Fourier analysis of the converter waveforms, the ratio of the odd harmonic to the fundamental is

$$I_h/I_1 = \sin\,(hw/2)/h\sin\,(w/2) \tag{4.26}$$

Now substituting $w = 2\ \pi/3$ and $h = 3q$ yields

$$\frac{I_h}{I_1} = \sin(q\pi \pm 1.5q\varepsilon)/3q\sin(\pi/3 \pm \varepsilon/2) \tag{4.27}$$

For example, for $\varepsilon = 1° = 0.0174$ rad, I_3 is 1% of the fundamental.

4.2 SINGLE-PHASE POWER SUPPLIES

Single-phase power supplies are used to supply power to the following equipment[7,8]:

- Personal computers
- Copying machines
- Lighting
- Audio equipment
- TVs
- Electronic ballasts for fluorescent lamps

Before the advent of switch-mode power supplies (please see Figure. 4.5) transformers were used to reduce the voltage on the secondary side and then rectified to obtain suitable dc voltage. The inductance of the transformer was smoothing the waveform of the input current, thus reducing the harmonic content.

Some older generation of TV receivers used half-wave rectification, thus producing considerable levels of dc and even-ordered harmonics. Color receivers demand a peak current that is two to three times as large as that taken by a monochrome receiver. The third-harmonic current is 0.32 A in a black-and-white transistor receiver but increases to 0.82 A in a thyristor receiver.[4]

Newer technology has provided switch-mode power supplies to achieve a smooth dc output. The main advantages of these power supplies are

- Lightweight.
- Compactness.
- Efficient operation.
- Lack of the need for a transformer with 110 V or 220 V primary because the input diode bridge is directly connected to the ac line.
- These power supplies use dc/dc conversion techniques. The voltage across capacitor C_1 in Figure 4.6 is rather coarse. This dc is again converted back to ac at a very high frequency, that is, 20 kHz to 100 kHz.

Unlike conventional single-phase power supplies, because there is no large inductance of the transformer, the input current to the power supply comes in very short pulses. A distinctive characteristic of switch-mode power supply is a very high third-harmonic content in the current. In older installations, neutral conductors may not

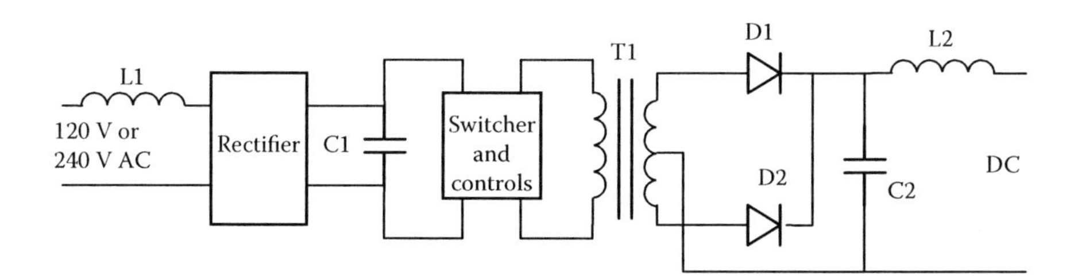

FIGURE 4.5 Switch-mode power supply.

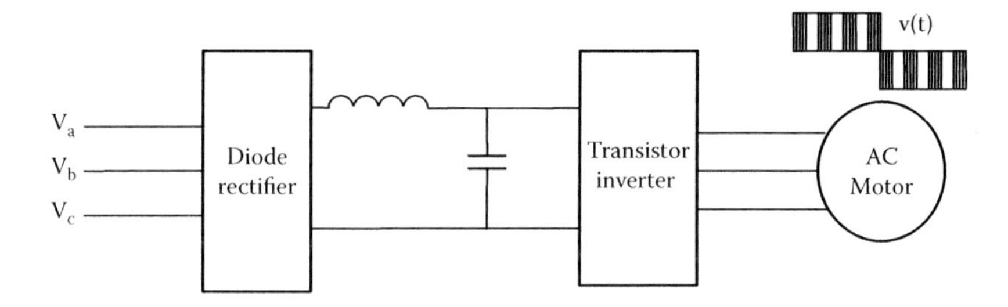

FIGURE 4.6 PWM ASD (pluse width modulated adjustable speed drives).

be adequately rated to handle a large third-harmonic current in the neutral because all third-harmonic currents in the three phases will be added in the neutral. This can cause overheating of the neutral and, in extreme cases, burning.

Switch-mode power supplies have a poor power factor. One of the companies developed a PowerMiser model to improve the power factor to 0.98 from 0.65, also reducing the harmonic content.[8] The effect of the power factor correction is shown in Table 4.1.

This 1000-W model can be operated from a standard 115-V, ac 15-A outlet without exceeding the 12-A limit imposed by Underwriters Laboratories regulations.

There are PowerMiser designs that can supply power up to 4000 W. Such larger units will need a three-phase delta–wye-connected transformer and cannot be directly connected to the standard 110-V, 15-A single-phase outlet. Because third-harmonic currents circulate in the delta winding, they will not appear in the line currents. Hence, the kVA input to these delta–wye transformers appears less than the kVA output. If the transformer is not designed for the output kVA, including the circulating kVA, then it can get heated up and, in some extreme cases, might fail in service.[8]

TABLE 4.1
Effect of Power Factor Capacitor in Switch-Mode Power Supplies

	With No Power Factor Correction	With Power Factor Correction
Load, watts	1000	1000
Volts, rms	116.64	117.96
Amperes, rms	18.35	12.1
Power factor	0.651	0.987
Fundamental, amperes, rms	12.389	11.990
Harmonic Order	**Percentage Harmonic**	**Percentage Content**
2	0.8	0
3	84	7.8
5	59	10
7	31	6.5
9	9.5	0.6
11	7	3.2
13	9	1.5

In fluorescent lamp systems with electronic ballasts, the high-frequency, controlled-output voltage that is possible with transistorized inverters increases fluorescent tube efficiency and permits more sophisticated control, such as dimming. In commercial buildings, the fluorescent lighting load typically accounts for 40 to 60% of a building load. Hence, in some cases without power factor correction and a cleaner power supply waveform, problems can occur with excessive third-harmonic current flowing in the neutral.

4.3 DC DRIVES

Dc motors have a wide speed range and higher starting torques with simple control systems, compared with ac drives,[9,4] but the initial purchase and maintenance costs are high. However, the cost of power electronics used with ac drives is dropping significantly in recent years. Hence, the application of dc drives is limited to those applications where the speed and torque characteristics of the dc motor are essential.

Six-pulse converters are usually used to obtain the necessary dc power supply for these drives. But the earlier assumptions used in HVDC converters, such as constant direct current, are not applicable to these relatively small converters because the dc motor armature inductance is very low, and the variation in the firing angle is rather large. Representing the dc motor as a series R L circuit with the back EMF of the motor, in References 4 and 9 harmonic currents are estimated as a function of the ripple ratio $r = I_r / I_d$, where I_r is the alternating ripple of the dc motor, and I_d is the mean direct current. References 4 and 9 show the harmonic content of the supply current for the six-pulse converter with finite inductive load.

4.4 AC DRIVES

Ac drives require a variable-frequency supply that is obtained from a rectifier–inverter combination.[7,10] These normally use standard squirrel-cage induction motors because of their low cost and maintenance.

Inverters can be broadly classified as voltage source inverters (VSIs) or current source inverters (CSIs). The ac side of both inverters is connected to a load or to an ac system. The dc input to a VSI is a dc voltage source. A large capacitor or an *LC* filter with a small inductor and relatively large shunt capacitor connected to the output of the rectifier bridge provides the constant dc voltage output for VSIs. The dc input to a CSI is a dc current source, which is usually realized by a controlled dc source that is connected to a large inductor in series.

Some drive configurations use a VSI employing PWM techniques to synthesize an ac waveform as a train of variable-width dc pulses (Figure 4.7). The inverter uses either silicon-controlled rectifiers (SCRs), gate turn-off (GTO) thyristors, or power transistors for this purpose. The advantages of VSI PWM drives are

- They offer the best energy efficiency for drives up to at least 500 HP.
- It is not necessary to vary rectifier output voltage to control motor speed.

Hence, rectifier thyristors can be replaced with diodes, allowing the thyristor control circuitry to be eliminated.

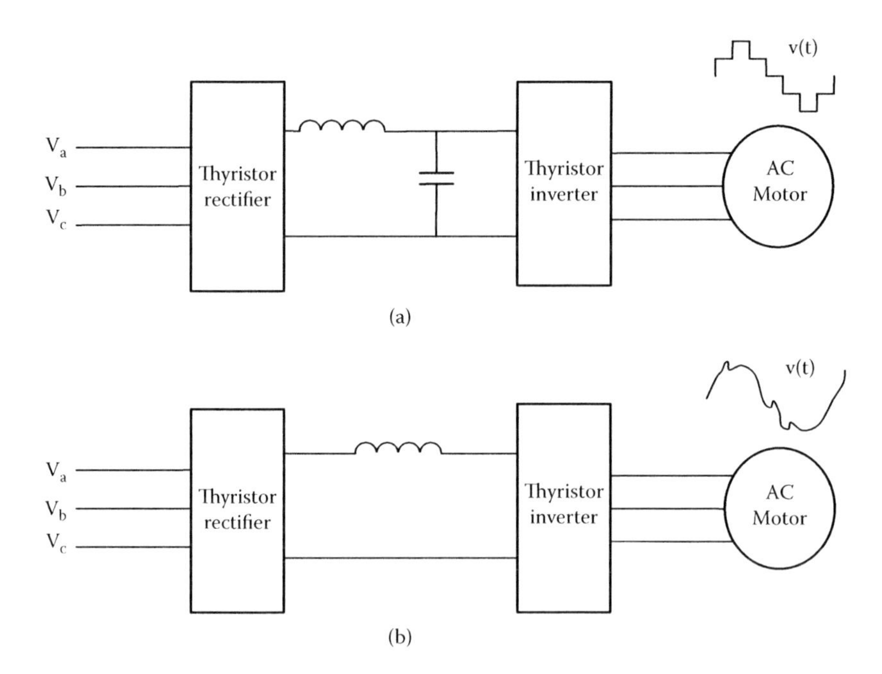

FIGURE 4.7 Large ac adjustable speed drives (ASDs); (a) voltage source inverters (VSIs), (b) current source inverters (CSIs).

Some very-high-power drives employ 12 pulses instead of 6. VSIs are limited to applications that do not require rapid change in speed (Figure 4.7a).

A CSI is more reliable and fault tolerant than a VSI because the large inductor limits the rate of rise of current in the event of a fault. The thyristor-based CSIs are used for high-power electric motor drives because they have good acceleration/deceleration characteristics. However, they have higher losses compared with VSIs, and further require a motor with leading power factor (synchronous or induction with capacitors) or added control circuitry to commutate the inverter thyristors. A CSI drive must be designed for use with a specific motor. Thyristors in CSIs must be protected against inductive voltage spikes, which increases the cost of this drive (Figure 4.7b).

4.5 PULSE-WIDTH MODULATION (PWM)

PWM is used in inverters to supply power to the ac motors more commonly in the industry now.[4,10,11]

In PWM,[11] a sawtooth wave is used to modulate the chops as shown in Figure 4.8. The sawtooth wave has a frequency that is a multiple of three times the sine-wave frequency, allowing symmetrical three-phase voltages to be generated from a three-phase sine-wave set and one sawtooth waveform. The PWM control signal is generated by feeding the sawtooth carrier signal and sine-wave-modulating signal to a

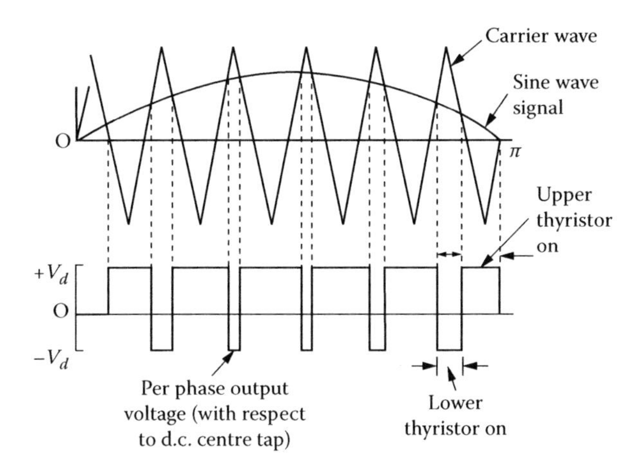

FIGURE 4.8 The principle of pulse-width modulation (PWM). (From Arrillaga, J., Bradley, J.D., and Bodger, P. (1985), *Power System Harmonics,* John Wiley, Chichester, U.K. With permission.)

comparator circuit. In this simple form, this method controls line-to-line voltage from zero to full voltage by increasing the magnitude of the sawtooth, with little regard to the harmonics generated.

More efficient PWM techniques have been developed to control the fundamental and harmonic voltages simultaneously.[12–16] Using suitable switching times with five on/off actions per cycle, one can eliminate both fifth and seventh harmonics together. See Reference 11 for more details.

Generally, at any fundamental switching frequency, each chop per half-cycle of the inverter phase voltage waveform can eliminate one harmonic of the waveform or reduce a group of harmonic amplitudes.[13] Assuming that there are m chops per half-cycle, one chop can be utilized to control the fundamental voltage and the other (m − 1) to reduce the other specified low-order harmonics or to minimize power losses caused by a specified range of harmonics within the motors.

However, it must be noted that, as the total rms harmonic voltage cannot change, the portion of the rms voltage that was provided by the elimination of harmonics will be spread over the remaining harmonic magnitudes. Hence, the motor designer must take this fact into account. Also, the integrating filter characteristic of the motor will reduce some of these higher-order-current harmonics.

Because of the number of switchings, the usable fundamental rating and the converter efficiency are reduced as GTOs require significant energy (resulting in increased thyristor losses and heating effects) for each switching operation. In contrast, devices such as IGBTs (Integrated state Bipolar Transistors) and IGCTs (Integrated State Commutated Thyrestors) require much lower switching energy and are better suited to the use of PWM techniques.[10] However, with the use of chain circuits that require reduced switching frequency, that is, once per two cycles of frequency, high-power inverters can be constructed using GTOs.

4.6 TELECONTROL SIGNALS

For tariff metering, for lighting control, or for alerting personnel in telecontrol systems, control signals are transmitted via the mains to the telecontrol receivers.[2] Older systems operate in the frequency range of 110 Hz–3 kHz, whereas modern systems operate in the range of 110–500 Hz. The operating frequency range below 500 Hz lies mainly between the typical harmonics and, in the range above 500 Hz, at harmonic frequencies that are not generated by three-phase bridge circuits in steady-state condition. Telecontrol signals are transmitted as short-duration impulse telegrams containing the relevant telecontrol frequencies. The total duration of the telegram is one minute.

4.7 CYCLOCONVERTERS

Cycloconverters are used for producing very-low-frequency output.[17,18,4] They consist of a dual-converter configuration controlled through time-varying phase-modulated firing pulses so that it produces an alternating voltage per phase. The cycloconverter waveforms contain frequencies that are not integer multiples of the main output frequency. When the output frequency f_0 is a submultiple of the input frequency f_i, we can clearly have a well-defined output frequency. Otherwise there is no clearly defined output frequency.

The following points may be noted:

- To obtain symmetrical output voltage waveform, the output must be an integral submultiple of the input frequency.
- The highest frequency that can be obtained using a cycloconverter is one-third of the input frequency.
- Output power control is obtained by changing the firing angle of the SCRs.

Space will not permit a detailed discussion of cycloconverters here. The reader is advised to see References 17 and 18 for more information on this subject.

4.8 TRANSFORMERS

4.8.1 HARMONICS IN NO-LOAD EXCITING CURRENT

Because transformer cores are made of ferromagnetic materials, the exciting current of the transformer under no-load conditions is not sinusoidal even though the applied voltage and flux are sinusoidal due to the nonlinear nature of the B-H curve, and similarly the relationship between flux and magnetomotive force (MMF).[19,4] Figure 4.9 shows how the exciting current can be determined once the applied voltage flux, waveforms are drawn together with the flux versus the MMF curve. For any value of flux, there are two values of MMF, depending on whether the flux is increasing or decreasing. From these MMFs, the magnetizing current can be computed and plotted corresponding to these values of flux, depending on whether the flux is rising

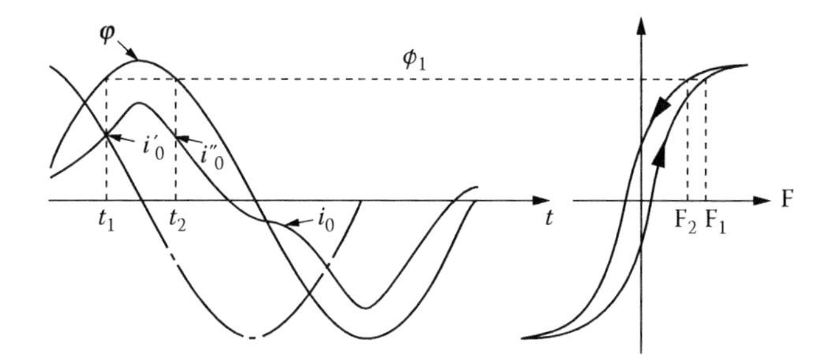

FIGURE 4.9 Waveform of the no-loading exciting current of a transformer. (From Sarma, M.S. (1985), *Steady-State Theory and Dynamic Performance*, Wm. C. Brown, Dubuque, Iowa. With permission from Thomson Engineering.)

or falling. In Figure 4.9, at t_1, the flux is increasing, and at t_2, the flux is decreasing, even though the instantaneous value of flux is the same. Analysis of this waveform yields of a large value for the third harmonics.

With three-limb transformers, as the triplen harmonics return through air (or rather through oil and tank), the higher reluctance of the path reduces the third harmonics to a lower value (about 10% of the value in independent core phases). The fifth and seventh harmonics may also be large enough, around 5–10%.

4.8.2 Harmonics due to Inrush Current

The transient current drawn by a transformer when it is first energized depends upon the instant of switching.[4,19] If the transformer is energized when the applied sinusoidal voltage is at its peak, usually there is no transient. However, as can be seen from Figure 4.10c, if the transformer is energized when the applied voltage is zero and it has a residual flux ϕ_r with unfavorable polarity, then the total flux that would be required to counteract the applied voltage according to Lenz's law would be 2 ϕ_m + ϕ_r. To produce such a flux, the transformer may have to draw many times (say, 100 times) the normal exciting current of the transformer. In actual practice, the inrush current may be 8 to 10 times the full-load current of the transformer, depending upon the instant of switching and polarity of the applied voltage.

Depending upon the transformer, the inrush current phenomenon may persist for a few seconds, thus creating harmonics during this period.

4.8.3 DC Magnetization

Under unbalanced conditions, the transformer excitation current can contain odd and even harmonics.[4,20] The transformers feeding half-wave rectifiers or supplying power to a three-phase converter with unbalanced firing fall into this category.

It has been shown[20] that the magnitude of the harmonic components of the excitation current in the presence of the direct current on the secondary side of the transformer increase almost linearly with the dc content. The linearity is better for the lower-order harmonics. As the harmonics generated by the transformer under dc magnetization are

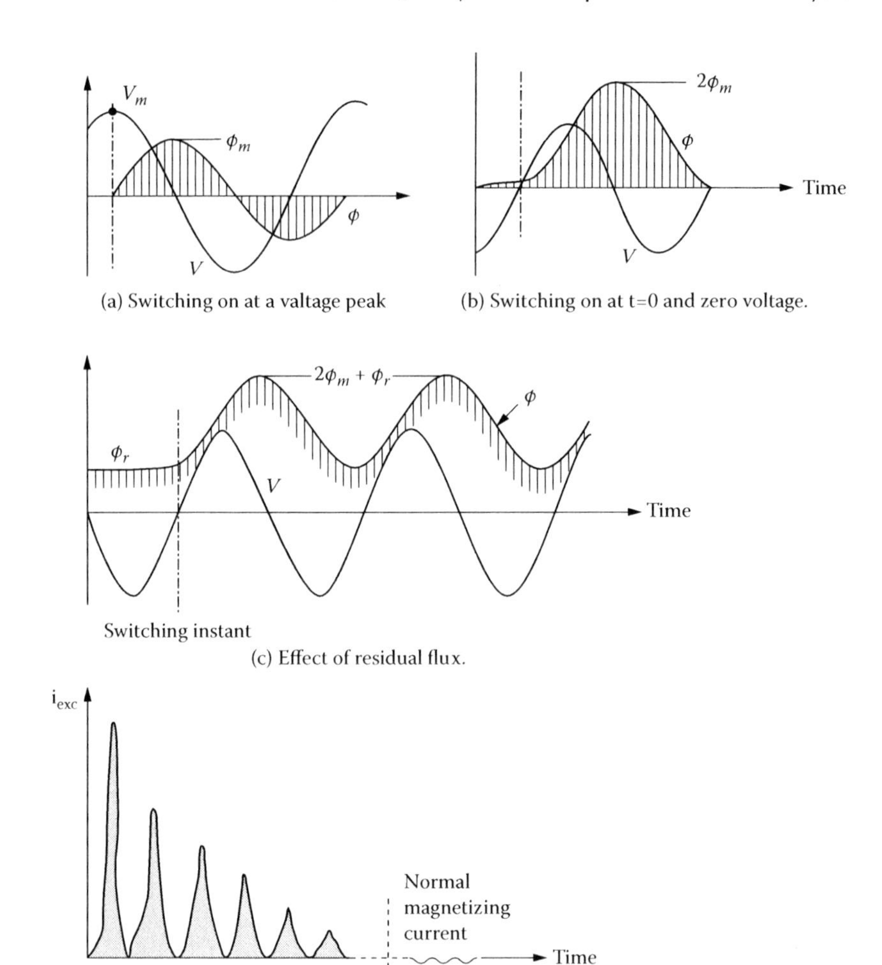

FIGURE 4.10 Inrush current phenomenon. (From Sarma, M.S. (1985), *Steady-State Theory and Dynamic Performance*, Wm. C. Brown, Dubuque, Iowa. With permission from Thomson Engineering.)

largely independent of the ac excitation, there appears to be no advantage in designing a transformer to run "underfluxed" in the presence of direct current.

4.9 HARMONICS IN ROTATING MACHINES

The distribution of the armature windings and the presence of slots in the machines cause spatial harmonics in them.[19,4] These in turn produce time harmonics in the induced voltages, which appear at the terminals.

Most of the power station generators are wye-connected. In such machines, triplen harmonic voltages do not appear in line-to-line voltages. Also, triplen harmonics can be eliminated even in phase-to-neutral voltages by using two-third pitch winding. Usually, the most significant harmonics to be minimized by the use of fractional pitch windings are the fifth and seventh. Higher harmonics than the ninth are so small that they require little attention except in rare cases.

4.10 HARMONICS IN ARC FURNACE LOADS

Voltage flicker is caused by arc furnace loads at the point of common coupling. For steel making, arc furnaces of the rating 100 MW and above are used.[10,20,21] Usually, a melting cycle can be divided into three distinct steps—drilling period, melting period, and reheating period. The full voltage is applied during the melting period, and the time required for melting is rather large compared with other steps. Due to uneven arc length, the bus voltage fluctuates continuously during the melting cycle.

For large arc furnaces, reactive voltage controllers such as SVCs are required to keep the voltage within the specified limits.

We will discuss international flicker meter in the chapter dealing with harmonic measurements.

4.11 HARMONICS IN A THYRISTOR-CONTROLLED REACTOR

Consider a thyristor-controlled reactor shown in Figure 4.11. Assume that the firing angle of the thyristor is α, and the conduction angle is σ (see Figure 4.12).

When the thyristor is in partial conduction, let us calculate the harmonics.[22] The voltage across the inductor v_L is given by

$$v_L = L\, di/dt = \sqrt{2}V \sin \omega t \tag{4.28}$$

Hence, $i(t) = (\sqrt{2}/L) \int_{\alpha}^{wt} V \sin \omega t \; dt$

$$i(t) = (\sqrt{2}V/\omega L)\,(\cos \alpha - \cos \omega t) \tag{4.28a}$$

for $\alpha < \omega t < \alpha + \sigma$

$$i(t) = 0 \tag{4.28b}$$

for $\alpha + \sigma < \omega t < \alpha + \pi$.

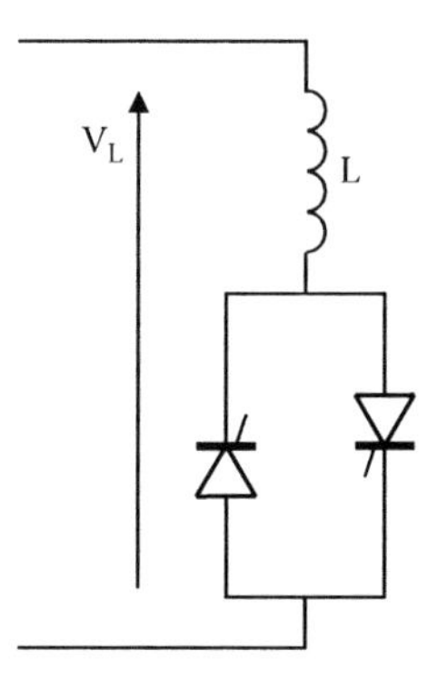

FIGURE 4.11 Thyristor-controlled reactor.

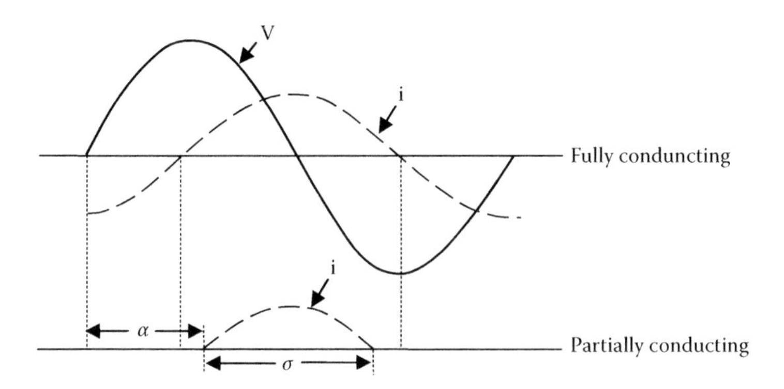

FIGURE 4.12 Waveforms of the thyristor current.

By Fourier analysis of the preceding waveform, we can obtain the following expressions for the fundamental and n-th harmonic component of the current through the reactor:

$$I_1 = \frac{(\sigma - \sin\sigma)}{\pi X_L} \quad \text{Amps, rms} \tag{4.29}$$

$$I_n = \frac{4V}{\pi X_L}\left[\frac{\sin(n+1)\alpha}{2(n+1)} + \frac{\sin(n-1)\alpha}{2(n-1)} - \frac{\cos\alpha \sin n\alpha}{n}\right] \text{Amps, rms} \tag{4.30}$$

where $n = 3, 5, 7, \ldots$

In the literature (CIGRE publication),[23] with some algebraic manipulation, the following alternative expression for I_n is obtained:

$$I_n = \frac{4V[\cos\alpha \sin(n\alpha) - n\sin\alpha \cos(n\alpha)]}{\pi X_L n(n^2 - 1)} \quad \text{Amps, rms} \tag{4.31}$$

where $n = 3, 5, 7, \ldots$

In practical SVCs, the thyristor reactors are connected in delta and, in each phase, the reactor is split into two parts with the thyristors located in the middle to provide better protection for them in the event of a fault in the reactor (see Figure 4.13).

4.12 THE K-FACTOR

The K-factor is a weighting of the harmonic load currents according to their effects on transformer heating, as derived from ANSI/IEEE C57.110. A K-factor of 1.0 indicates a linear load (no harmonics). The higher the K-factor, the greater the harmonic heating effects.[5,24]

If a company with many offices were to install poor-quality electronic ballasts having poor K-factor, a larger transformer would be needed than is apparent from the overall power consumption calculation. The K-factor is a number derived from a numerical calculation based on the summation of harmonic currents generated by

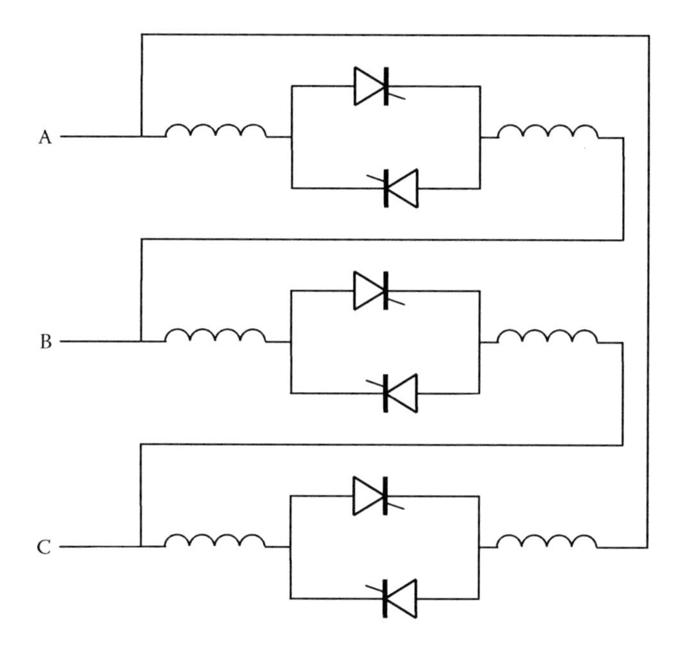

FIGURE 4.13 Three-phase thyristor-controlled reactor.

the nonlinear load. The higher the K-factor, the more significant the harmonic current content.

The algorithm used to compute the K-factor is[24]

$$K = \frac{\sqrt{\sum_h (I_h^2 h^2)}}{\sqrt{\sum_h I_h^2}} \tag{4.32}$$

where h is the harmonic number.

Details of the calculation method can be found in IEEE Standard 1100-1992.

In terms of the foregoing K-factor, the following expression is used to determine the derated (or maximum allowed) current:

$$I_{max} = \sqrt{\frac{1 + P_{EC.R}}{1 + KP_{EC.R}}}(I_R) \tag{4.33}$$

where I_R is the fundamental rms current under rated load conditions, and $P_{EC.R}$ is the ratio of eddy-current loss to rated $I^2 R$ loss (I being the total rms current).

The K-factor rating is an index of the transformer's ability to withstand harmonic content while operating within the temperature limits of its insulating system. Transformers come in basic K-factors such as 4, 9, 13, 20, 30, 40, and 50. The strategy is to calculate the K-factor for your load and then specify a transformer with a K-factor of an equal or higher value. In this way, the transformer can be sized to the load without derating. The advantage of using a K factor transformer is that it is usually more economical than using a derated, oversized transformer.

4.13 CONCLUSIONS

In this chapter we discussed the different sources of the harmonics and derived expressions for characteristic harmonics in a converter of pulse number "p." In particular, if we are dealing with a 6-pulse system, the harmonics generated will be of the order $6k \pm 1$, the lowest harmonic being the fifth. With wye–delta and wye–wye transformers we can build a 12-pulse converter. We have also discussed the effect of overlap on the harmonics and dc harmonics. If there is any asymmetry, for example, due to inaccurate firing angles, in general all the harmonics will be produced, including positive- and negative-sequence systems of all the harmonic orders 2,3,4,5,6 ,7,8,9,10, …, etc., which can pass through transformers. It is not economical to attempt filtering of all these; consequently, high accuracy of firing-angle generation in the control system of the order of $\pm 1°$ of the fundamental frequency waveform is necessary.[25]

Calculation of amplitudes at higher frequencies above the 25th harmonic (1500 Hz) is subject to errors due to asymmetries, but these amplitudes are usually negligible for practical purposes. Also, we derived expressions for the harmonic currents generated by a thyristor-controlled reactor (in a single-phase system) in terms of its firing angle. Further, we have explained the K-factor and the algorithm for its calculation.

REFERENCES

1. Kimbark, E.W., *Direct Current Transmission* (1971). Vol. 1, John Wiley, New York.
2. Schlabbach, J., Blume, D., and Stephanblome, T. (2001). *Voltage Quality in Electrical Power Systems,* IEE, London.
3. Mahmoud, Aly A. (1984). *Power System Harmonics: An Assessment.* IEEE Tutorial Course Text, EHO221-2-PWR, 1–9.
4. Arrillaga, J., Bradley, J.D., and Bodger, P. (1985). *Power System Harmonics,* John Wiley, Chichester, U.K.
5. Arrillaga, J. and Watson, N.R. (2003). *Power System Harmonics*, John Wiley, Chichester, U.K.
6. Adamson, C. and Hingorani, N.G. (1960) *High Voltage Direct Current Power Transmission*, Chapter 3, Garraway, London.
7. Dugan, R.C., McGranaghan, M.G., and Beaty, H.W., (1996). *Electrical Power Systems Quality,* McGraw-Hill, New York.
8. Feature Article: "Reducing the Safety Hazard of Switchers," *Electronics Australia*, February 1990, 124–125.
9. Dobinson, L.G. (1975). Closer accord on Harmonics, *Electronics and Power,* 567–572, May 15, 1975.
10. Ghosh, A. and Ledwich, G. (2002). *Power Quality Enhancement Using Custom Power Devices*, Kluwer Academic, New York.
11. Song, Y.H. and Johns, A.T. (Editors) (1999). *Flexible AC Transmission Systems* IEE, London. See Chapter 4 on "Shunt Compensation: SVC and STATCOM " by H.L. Thanawala, D.J. Young, and M.H. Baker, 146–197.

12. Schonung, A. and Stemmler, H. (1973). Reglage d'unmotor triphase reversible l'aide d'unconvertisseur static de frequence commandesuivant *le procede dela sous-oscillation*, *Revue Brown Boveri*, (51) 557–576.
13. Patel, H.S. and Hoft, R.G. (1973). Generalised technique of harmonic elimination in voltage control in thyristor inverters. Part 1: Harmonic elimination, *IEEE Transactions*, IA, 9(3) 310–317.
14. Patel, H.S. and Hoft, R.G. (1974). Generalised technique of harmonic elimination in voltage control in thyristor inverters. Part 2: Voltage control techniques, *IEEE Transactions*, IA, 10 (5) 666–673.
15. Buja G.S and Indri G.B (1977). Optimal pulse width modulation for feeding ac motors, *IEEE Transactions*, IA, 13(1) 38–44.
16. Byers, D.J. and Harman, R.T.C. Control of ac motors in a new concept of electric town car, Paper presented to *the Institution of Engineers*, Australia.
17. Ramamoorty, M. (1978). *An Introduction to Thyristors and Their Applications*, Macmillan Press, London.
18. Pelly, B.R. (1971). *Thyristor Phase-controlled Converters and Cycloconverters*, John Wiley, Chichester, U.K.
19. Sarma, M.S. (1985). *Steady-State Theory and Dynamic Performance*, Wm.C. Brown, Dubuque, Iowa.
20. Yacamini, R. and De Oliveria, J.C. (1978). Harmonics produced by direct current in converter transformers. *Proceedings IEE*, 125, 873–878.
21. Acha, E. and Madrigal, M. (2001). *Power System Harmonics*, John Wiley, Chichester, U.K.
22. Miller, T.J.E. (1982). *Reactive Power Control in Electric Systems,* John Wiley, New York.
23. CIGRE Report on Static Var Compensators, Edited by I.A. Erinmez, Prepared by Working Group 38-01, Task Force No 2.
24. Application Note AN102, Xitron Technologies at the website support @xitrontech. com.
25. Ainsworth, J.D. (1965). Filters, Damping Circuits, and Reactive Volt-Amps in HVDC Converters, Chapter 7, pp. 137–174, in *High Voltage Direct Converters and Systems*, edited by B.J. Cory, Macdonald, London.

5 Utility Harmonic Regulations and Standards

5.1 INTRODUCTION

In this chapter we will first point out the undesirable effects of the harmonics in the power system. Thereafter we will discuss IEEE Standard 519-1992 that is, IEEE recommended practices and requirements for harmonic control in power systems, and IEC 61000-series standards dealing with electromagnetic compatibility. We will discuss the two different philosophies based on which these standards are developed. We will also point out many practical issues that have to be solved to apply these standards, such as, (1) designing an efficient procedure to evaluate the acceptability of an installation, (2) specifying rational conditions to perform an evaluation, and (3) defining equitable limits to check acceptability. We will conclude with a discussion of the salient features in the B.C. Hydro's utility harmonic regulations based on the IEEE Std 519-1992.

5.2 UNDESIRABLE EFFECTS OF THE HARMONICS

Different categories of harmonic-producing loads are supplied by the electric utilities such as

1. Domestic loads like fluorescent lamps, light dimmers, etc.
2. Ripple control systems for regulating hot-water loads
3. Medium-sized industrial loads like several adjustable speed drives in a cement mill, paper mill, etc.
4. Large loads like high-voltage direct current (HVDC) convertors, aluminum smelters, static var compensators, heavy single-phase ac traction loads for hauling coal trains, etc.

The undesirable effects of the harmonics produced by these loads are listed as follows:[1–3]

1. **Capacitors:** These may draw excessive current and prematurely fail from increased dielectric loss and heating. Also, under resonance conditions, considerably higher voltages and currents can be observed than would be the case without resonance.

   IEEE Std 18-1992[4] gives limits on voltage, current, and reactive power for capacitor banks based on their ratings. These can be used to determine the maximum allowable harmonic levels.
2. **Power Cables:** In systems with resonant conditions, cables may be subjected to voltage stress and corona, which can lead to dielectric (insulation) failure. Further harmonic currents can cause heating.

3. **Telephone Interference:** Harmonics can interfere with telecommunication systems, especially noise on telephone lines. A "standard" human ear in combination with a telephone set has a sensitivity to audio frequencies that peaks at about 1 kHz. Two systems with slightly different weighting systems are used to obtain a reasonable indication of the interference from each harmonic. The two systems are
   - **C-message weighting** by Bell Telephone system (BTS) and Edison Electric Institute used in the United States and Canada[5]
   - **Psophometric weighting** by the International Consultative Commission on Telephone and Telegraph System[6]

   The **C-message weighting** system uses the telephone influence factor (TIF). It is a dimensionless quantity used to describe the interference of a power transmission line on a telephone line. It is expressed as

$$\mathrm{TIF} = (1/V)\sqrt{\sum_{n=1}^{x} (K_f P_f V_f)^2} \quad (5.1)$$

where

V = root-mean-square (rms) voltage of the transmission line
V_f = harmonic voltage of frequency f
$K_f = 5f$, the coupling coefficient
P_f = weight of the harmonic frequency f, the maximum being 1 for $f = 1000$ Hz

I.T and kV.T products

In practice, telephone interference is often expressed as a product of the current and the TIF, that is, the I.T product (where I is the rms current and T is TIF). Alternatively, it is sometimes expressed as a product of the voltage and the TIF weighting, where the voltage is in rms kV, that is, the kV.T product.

See References 7 and 1 for details of the CCITT (International Telegraph and Telephone Consultative Committee) system. Table 5.1 shows the corresponding quantities in BTS-EEI (Bell Telephone System-Edison Electric Institute) and CCITT systems.[7,1]

4. **Rotating Equipment (Motors and Generators):** Harmonic voltages and currents contribute to increased copper and iron losses, leading to the heating of machines and thus reducing their efficiency.

   Converters with the p pulse number produce characteristic harmonics of the order $pq \pm 1$. Such harmonics are of the positive-sequence type, whereas $pq - 1$ harmonics are of the negative-sequence type. Harmonics of the positive-sequence type are of the order pq + 1, whereas harmonics of the negative-sequence type are of the order pq – 1. These harmonic currents flow in the stator; further, they in turn induce even characteristic harmonics in the rotor. The resultant effect of all the positive-and negative-sequence harmonics in the stator and even harmonics in the rotor is to produce forward and backward torques resulting in pulsated or reduced torques in the machines. Rotor harmonics cause additional rotor heating.

TABLE 5.1
Corresponding Quantities in BTS-EEI and CCITT Systems

BTS-EEI	CCITT
C-message weighting	Psophometric weighting
Telephone influence factor (TIF)	Telephone harmonic form factor (TFF)
C-message-weighted voltage	
Longitudinal	Psophometrically weighted voltage
Transverse	Psophometric voltage
IT product	Equivalent disturbing current
KVT product	Equivalent disturbing voltage

Source: Kimbark, E.W. (1971). *Direct Current Transmission*, Vol. 1. John Wiley, New York.

The pulsating torques can cause mechanical oscillations, which in turn can contribute to shaft fatigue and accelerated aging. Further, induction motors may refuse to start (cogging) or may run at subsynchronous speeds.

5. **Transformers:** Harmonic currents increase copper losses and stray load losses, and harmonic voltages cause an increase in iron losses. Higher-frequency harmonics increase losses because they are dependent on frequency, but in general higher harmonics are smaller in magnitude. Further, harmonics are responsible for increased audible noise.

   IEEE C57-1200-1987[8] proposes an upper limit of 5% for the current distortion factor at rated current. Also the recommended maximum rms overvoltages that the transformer should be able to withstand in the steady state is 5% at rated load and 10% at no load. The harmonic currents in the applied voltage must not result in a total rms voltage exceeding these ratings.

6. **Electronic Equipment:** Computers and allied equipment such as programmable controllers frequently require ac sources that have no more than a 5% harmonic voltage distortion factor, with the largest single harmonic below 3% of the fundamental voltage. Higher levels of harmonics result in erratic functioning or malfunctioning of the equipment. Hence, many medical instruments are provided with line conditioners.

   Voltage notches can introduce frequencies in the radio frequency (RF) range, resulting in signal interference in logic or communication circuits.

7. **Metering:** Induction disk devices, such as watthour meters, can give erroneous readings in systems with severe distortion. Even a small error in metering can cost a significant amount of money in settlement bills in cases where a large number of units of energy are involved.

8. **Relaying**[9]: As with other equipment, switchgears also can experience increased losses due to harmonics.

   Maloperation of relays can occur if distortion factors are in the range of 10–20%. Most harmonic standards do not recommend such levels of harmonics in power systems. See References 3 and 9 for a detailed Canadian study that documents the effects of harmonics on relay operation. Maloperation of the relays associated with ripple control systems can also occur.

Circuit breakers may fail to interrupt currents because of improper operation of blowout coils. The time–current characteristics of fuses can be altered.

5.3 SPECIFICATION OF THE HARMONIC LIMITS

When a utility receives an application for electricity supply, it must be able to respond quickly regarding the approval for utility connection, at the same time satisfying itself that there will not be any unacceptable harmonic problems caused by the effects previously listed. In practice, some empirical rules will have to be developed to deal with small and medium loads. When supply to large harmonic-producing loads is involved, some detailed harmonic analysis at the design stage of the project, and field measurements before and after commissioning of the project, are essential to verify the compliance of the limits (harmonic voltage, current, THD, etc.) specified in the supply agreement.

The specifications of the harmonic limits and later field measurements to ensure compliance are rather difficult subjects due to the following reasons:

1. The precise harmonic (voltage or current) levels at which different electrical or electronic equipment maloperates is generally unknown.
2. There is considerable uncertainty of the harmonic impedances of the different components in a power system, such as loads, transformers, rotating equipment, etc., and consequently, the ac harmonic impedance of the system at the point of common coupling[10,11] cannot be determined with high accuracy.
3. The harmonics vary in a random fashion, and if there are several harmonic sources in the system, some criterion must be adopted (such as using phasor sum, or arithmetic sum, or root sum squared [rss]) to obtain the combined effect of these sources.[12,13]
4. If the new harmonic-producing load connected to the power system is the dominant harmonic source compared with other harmonic sources in the system, it is possible to obtain an estimate of its contribution by measuring harmonic currents entering into the system at the point of common coupling. In all practical situations, this condition may not be satisfied. In these situations, in view of the interaction between system and load harmonics, it is impossible to separate the harmonic current just due to the load.
5. The requirements of different power systems in different countries (or even in the same country in different states) can vary. In Australia, with sparse population in rural areas, large mining loads, aluminum-smelter loads, and single-phase traction loads have to be supplied with very low short-circuit levels at different substations. Further, several ripple control systems are in use to control loads consisting of hot-water systems. In contrast, the U.K. has a highly interconnected network with distributed generation serving a large number of load centers often close to the generation. Hence, harmonic penetration problems in view of the long transmission lines and low short-circuit levels in Australia will be rather

different from that of the U.K. Thus, the nature of the different power systems has to be taken into account in the specification of harmonic limits to similar harmonic-producing loads in different countries.

5.4 PHILOSOPHICAL DIFFERENCES BETWEEN IEEE 519-1992 AND IEC 61000-SERIES STANDARDS

The appendix shows the list of IEC 61000-series standards. In defining harmonic standards, the term point of common coupling (PCC) is widely used. The point of common coupling is defined as the point in the utility service to a particular customer where another customer can be connected.

In IEEE 519-1992, a series of harmonic voltage limits are specified at different buses at different voltage levels where multiple nonlinear loads are connected. Also limits on harmonic currents are specified based on the ratio I_{SC}/I_L at different voltage levels,

where

I_{SC} = Available short-circuit current

I_L = 15- or 30-min (average) maximum demand current

Another term that is used in the harmonic standard is total demand distortion (TDD). It is identical to THD (defined in Chapter 1 earlier), except I_L is used instead of the fundamental current component.

This standard divides the responsibility for limiting harmonics between customers and utility. Customers will be responsible for limiting the harmonic current injections, whereas the utility will be responsible for limiting voltage distortion in the supply. (Please see the discussion in Section 5.7.1 regarding the interaction between a power system and a nonlinear load.)

The IEC 61000 series deals with the broad area of low-frequency electromagnetic phenomena including conduction and radiation aspects. It considers harmonics and interharmonics as a part of this whole field. IEC 61000 uses the concepts of compatibility levels, emission limits, and planning levels using probability in the estimation of harmonic levels. It specifies harmonic levels for the greatest 95% probability daily value and 99.9% weekly value.

Although all countries still have their own special provisions to suit their local conditions, depending upon the development of their electricity supply systems and their operating practices, the need for international standards has arisen to ensure the growth of international trade and commerce. With the development of the European Union, equipment manufacturers want to sell their equipment in other countries in Europe and to work satisfactorily in that environment. Hence, the 61000 series of IEC Standards have been developed. Australia, New Zealand, and South Africa also have adopted some parts of IEC 61000-series standards.

5.5 IEEE 519-1992

We will briefly discuss the contents of this standard in this section. Initially, some references, definitions, and symbols are given. Later, sources of harmonic generation, namely, converters, arc furnaces, static var compensators (SVCs), inverters for dispersed generation, electronic phase control, cycloconverters, switch-mode power supplies, and pulse width-modulated drives are discussed. Then, system response characteristics dealing with resonant conditions and the effect of system loading are considered. The undesirable effects of harmonics on different components of the power system are described (see Section 5.2). Reactive power compensation and harmonic control are the subject of Section 7 in the IEEE 519 standard. In analysis methods, details of harmonic current calculations, system frequency response calculations, modeling guidelines for harmonic analysis, telephone interference, line-notching calculations for low-voltage systems, THD, and displacement power-factor improvement calculations are provided. The measurements section deals with requirements for instruments, representation of harmonic data, and transducers for harmonic measurements. Then two sections consider the limits for harmonic currents and voltages. When new disturbing loads are proposed to be connected to the system, initial harmonic measurements and preliminary and detailed harmonic analysis are discussed. Some examples and bibliography are provided at the end.

IEEE Std 519-1992-recommended limits for harmonic current distortion are listed in Tables 5.2, 5.3, and 5.4.

TABLE 5.2
Current Distortion Limits for General Distribution Systems (120 V through 69,000 V) (IEEE Standard 519-1992)

	Maximum Harmonic Current Distortion in Percent of I_L					
	Individual Harmonic Order (Odd Harmonics)					
I_{sc}/I_L	<11	$11 \le h < 17$	$17 \le h < 23$	$23 \le h < 35$	$35 \le h$	TDD
<20*	4.0	2.0	1.5	0.6	0.3	5.0
20 < 50	7.0	3.5	2.5	1.0	0.5	8.0
50 < 100	10.0	4.5	4.0	1.5	0.7	12.0
100 < 1000	12.0	5.5	5.0	2.0	1.0	15.0
>1000	15.0	7.0	6.0	2.5	1.4	20.0

Notes:
Even harmonics are limited to 25% of the odd harmonic limits.

Current distortions that result in a dc offset, for example, half-wave converters, are not allowed.

* All power generation equipment is limited to these values of current distortion, regardless of actual I_{sc}/I_L, where
I_{sc} = maximum short-circuit current at PCC,
I_L = maximum demand load current (fundamental frequency component at PCC).

TOD: Total Demand and Distortion

Source: With permission from IEEE Std. 519-1992, IEEE Recommended Practices and Requirements for Harmonic Control in Electric Power Systems.

TABLE 5.3
Current Distortion Limits for General Distribution Systems (69,001 V through 161,000 V) IEEE Standard 519-1992

	Maximum Harmonic Current Distortion in Percent of I_L					
	Individual Harmonic Order (Odd Harmonics)					
I_{sc}/I_L	<11	$11 \le h < 17$	$17 \le h < 23$	$23 \le h < 35$	$35 \le h$	TDD
<20*	2.0	1.0	0.75	0.3	0.15	2.5
20 < 50	3.5	1.75	1.25	0.5	0.25	4.0
50 < 100	5.0	2.25	2.0	0.75	0.35	6.0
100 < 1000	6.0	2.75	2.5	1.0	0.5	7.5
>1000	7.5	3.5	3.0	1.25	0.7	10.0

Notes:
Even harmonics are limited to 25% of the odd harmonic limits.

Current distortions that result in a dc offset, for example, half-wave converters, are not allowed.

* All power generation equipment is limited to these values of current distortion, regardless of actual I_{sc}/I_L, where
I_{sc} = maximum short-circuit current at PCC,
I_L = maximum demand load current (fundamental frequency component at PCC).

Further notice that the values in this table are half of the corresponding values in Table 5.2.

TOD: Total Demand Distortion

Source: With permission from IEEE Std. 519-1992, IEEE Recommended Practices and Requirements for Harmonic Control in Electric Power Systems.

Table 5.5 shows the harmonic voltage distortion limits for different bus voltages. In the tables 5.2 and 5.3, TDD refers to total demand distortion (Root Sum Squared, RSS). These should be used as system design values for the "worst case" for normal operating conditions (conditions lasting longer than one hour).

For shorter periods, during start-ups or unusual conditions, the limits may be exceeded by 50%. These tables are applicable to six-pulse rectifiers and general distortion situations. However, when phase-shift transformers or converters with pulse numbers (q) higher than six are used, the limits for the characteristic harmonic orders are increased by a factor equal to $\sqrt{q/6}$ provided that the amplitudes of the noncharacteristic harmonic orders are less than 25% of the limits specified in the tables. Table 5.2 lists the harmonic currents based on the size of the load with respect to the size of the power system to which the load is connected. The ratio I_{sc}/I_L is the ratio of the short-circuit current available at the points of common coupling (PCC), to the maximum fundamental load current. It is recommended that the load current, I_L, be calculated as the average current of the maximum demand for the preceding 12 months. All generation, whether connected to the distribution, subtransmission, or transmission, is treated like utility distribution and is therefore held to these recommended practices.

TABLE 5.4
Current Distortion Limits for General Transmission Systems (>161 kV)

Dispersed Generation and Cogeneration						
IEEE Standard 519-1992						
Individual Harmonic Order (Odd Harmonics)						
I_{sc}/I_L	<11	$11 \le h < 17$	$17 \le h < 23$	$23 \le h < 35$	$35 \le h$	TDD
<50*	2.0	1.0	0.75	0.3	0.15	2.5
≥50	3.0	1.5	1.15	0.45	0.22	3.75

Notes:
Even harmonics are limited to 25% of the odd harmonic limits.

Current distortions that result in a dc offset, for example, half-wave converters, are not allowed.

* All power generation equipment is limited to these values of current distortion, regardless of actual I_{sc}/I_L, where
I_{sc} = maximum short-circuit current at PCC,
I_L = maximum demand load current (fundamental frequency component at PCC).

Source: With permission from IEEE Std. 519-1992, IEEE Recommended Practices and Requirements for Harmonic Control in Electric Power Systems.

TABLE 5.5
Voltage Distortion Limits

Bus Voltage at PCC	Individual Voltage Distortion (%)	Total Voltage Distortion THD (%)
69 kV and below	3.0	5.0
69.001 kV through 161 kV	1.5	2.5
161.001 kV and above	1.0	1.5

Note: High-voltage systems can have up to 2.0% THD where the cause is an HVDC terminal that will attenuate by the time it is tapped for a user.

5.6 IEC 61000-SERIES STANDARDS

IEC standards are divided into six parts:[14]

- Part 1: *General*, dealing with fundamental definitions, etc.
- Part 2: *Environmental*, deals with the characteristics of the environment where equipment will be supplied, and its compatibility levels.
- Part 3: *Limits* define the permissible emissions that can be generated by the equipment connected.

- Part 4: *Testing and measurement techniques* provide detailed guidelines for measurement equipment.
- Part 5: *Installation and mitigation guidelines* provide guidelines for cabling of electrical and electronic systems, etc. They also describe protection concepts for civilians against the high-altitude electromagnetic pulse (HEMP) emissions from high-altitude nuclear explosions.
- Part 6: *Miscellaneous*, defining immunity and emission levels required for equipment in general categories or for specific types of equipment.

Because there are so many IEC standards (see the Appendix to this chapter for a list), it is not possible to discuss all of them even briefly. We will restrict our discussion in this section to the IEC 61000-3-6 (1996)[15] standard that deals with the assessment of emission limits for distorting loads in MV and HV power systems. The practicing engineer must study in detail those standards that are applicable to his or her area of practice.

We will discuss in chapter 8, "Monitoring Power Quality," the other two important IEC standards in References 16 and 17, that is, IEC Standard 61000.4.7: Electromagnetic Compatibility (EMC), Part 4: Testing and Measurement Techniques, Section 7: General Guide on Harmonics and Interharmonics Measurements and Instrumentation, for Power Supply Systems and Equipment Connected Thereto; and IEC 61000-4-15: Electromagnetic Compatibility (EMC), Part 4: Testing and Measurement Techniques, Section 15: Flicker Meter—Functional and Design Specifications. The first standard deals with important testing and measurement techniques, and the second deals with the IEC Flicker Meter.

Now we will define the concepts of compatibility level, equipment immunity level, emission limit, and planning level.[15,16]

Compatibility levels are defined as the levels of severity that can exist in the relevant environment, that is, they are based on the 95% probability levels of entire systems using distributions that represent both time and space variations of disturbances.

Figure 5.1 shows an ideal distribution of system disturbances over many points in the power system, and taken over many days. Equipment immunity levels have a similarly shaped distribution. *Immunity test levels* are specified by relevant standards or agreed upon between manufacturers and users. In the whole power system (see Figure 5.1) interference inevitably occurs on some occasions and, therefore, there is significant overlapping between the distributions of disturbance and immunity levels. *Planning levels* are generally equal to or lower than the compatibility level; they are specified by the owner of the network. Planning levels may differ in different parts of the system; for example, they may be allowed to be larger in rural areas where there is less equipment that may be affected. The standard suggests typical values of planning levels as summarized in Table 5.5.

At each (inter)harmonic frequency, the *emission level* from a distorting load is the (inter)harmonic voltage (or current) that would be caused by the load into the power system if no other distorting load is present.

5.6.1 Assessment Procedure (Harmonic Limits)*[15]

The basic standard to be used for harmonic and interharmonic measurement is IEC 61000-4-7. To compare the actual harmonic levels with the planning levels, the

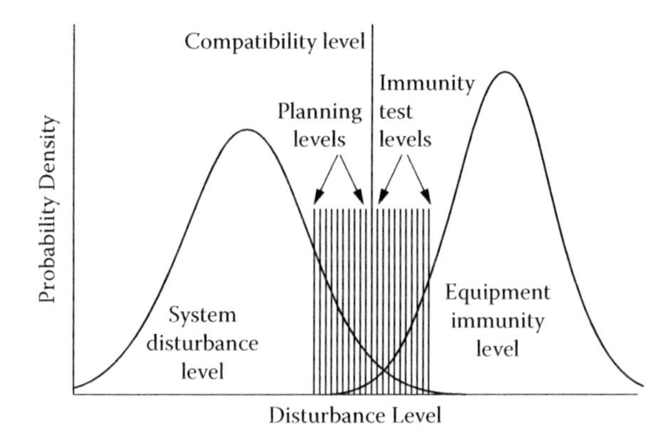

FIGURE 5.1 Illustration of basic voltage quality concepts with time/location statistics covering the whole system. (From IEC 61000-3-6 (1996), Electromagnetic Compatibility (EMC). Part 3.6: Limits-Assessment of emission limits for distorting loads in MV and HV power systems. With permission from IEC.)

TABLE 5.6
Indicative Values of Planning Levels for Harmonic Voltage (as Percentage of the Nominal Value) in MV, HV, and EHV Power Systems

Odd Harmonics (Nonmultiples of 3)			Odd Harmonics (Multiples of 3)			Even Harmonics		
	Harmonic Voltage %			Harmonic voltage %			Harmonic voltage %*	
Order h	**MV**	**HV–EHV**	**Order h**	**MV**	**HV–EHV**	**Order h**	**MV**	**HV–EHV**
5	5	2	3	4	2	2	1.6	1.5*
7	4	2	9	1.2	1	4	1	1
11	3	1.5	15	0.3	0.3	6	0.5	0.5
13	2.5	1.5	21	0.2	0.2	8	0.4	0.4
17	1.6	1	>21	0.2	0.2	10	0.4	0.4
19	1.2	1				12	0.2	0.2
23	1.2	0.7				>12	0.2	0.2
25	1.2	0.7						
>25	0.2+ $0.5\left(\frac{25}{h}\right)$	0.2+ $0.5\left(\frac{25}{h}\right)$						

Notes: Total harmonic distortion (THD): 6.5% in MV networks; 3% in HV networks.

*A value of 1.5% for the second harmonic voltage may seem rather high for HV systems, but such values may be encountered, and it is worth noting that the second harmonic is not always associated with a dc component.

Source: IEC 61000-3-6 (1996). Electromagnetic Compatibility (EMC), Part 3.6: Limits—Assessment of emission limits for distorting loads in MV and HV power systems. (With permission from IEC.)

minimum period should be one week. Table 5.6 furnishes the indicative values of planning levels for harmonic voltage (as a percentage of the nominal value) in *MV*, *HV*, and EHV power systems).

The limits for U_h (rms value of individual harmonic components) for different durations are listed below.

- The greatest 95% probability daily value of $U_{h,\,vs}$ (rms value of individual harmonic components over "very short" 3-s periods) should not exceed the planning level.
- The maximum weekly value of $U_{h,sh}$ (rms value of individual harmonic components over "short" 10-min periods) should not exceed the planning level.
- The 99.9% weekly value of $U_{h,\,vs}$ should not exceed 1.5 to 2 times the planning level.

Note: Harmonics are generally measured up to $h = 40$. In most cases, this is adequate for the evaluation of the distortion effects of power disturbances. However, higher-order harmonics up to the 100th order can be a matter of important concern in some cases. Examples include

- Large converters with voltage notching
- Large installations with converters of high pulse numbers (e.g., aluminum plants)
- Newer types of power electronics equipment with PWM converters interfacing with power system

Such cases can result in induced noise interference in neighboring sensitive appliances (e.g., sensors, communication systems, etc.). It is generally found that higher-order harmonics vary more with location and with time than lower-order harmonics. In many cases, higher-order harmonics are produced by a single consumer, often in combination with power system resonance. (There may be a need for more extensive evaluations when higher-order harmonics are a concern).

Usually, this will be the case when large aluminum smelters are operating using a 48-pulse rectifier supply. In such situations, the characteristic harmonics of the smelter are 47th and 49th. In actual practice, lower-order harmonics, particularly 5, 7, 11, and 13, are also present because of imbalances in the system or when one of the rectiformers is taken out either for maintenance reasons, or because of troubles in a part of the smelter. Hence, filters are usually provided even for these lower-order harmonics.

In addition to harmonics, flicker is observed with arc furnace loads. The topic of flicker and its measurement[17] are discussed in chapter 8, Monitoring Power Quality.

In this standard, the system voltage levels are classified as follows. Let V_L be the line-to-line voltage of the system. Then,

- Low voltage (LV) refers to $V_L \leq 1$ kV.
- Medium voltage (MV) refers to 1 kV $< V_L \leq$ 35 kV.
- High voltage (HV) refers to 35 kV $< V_L \leq$ 230 kV.
- Extra high voltage (EHV) refers to 230 kV $< V_L$.

When a new distorting load applies to a utility for supply connection, an assessment regarding its suitability for connection to the power system is made in three stages.[15,18]

Stage 1: In this stage, one of the three tests can be applied. Test 1 consists of checking for the condition if $S_i/S_{sc} < 0.1\%$ is met, where S_i is the i-th load, and S_{sc} is the short-circuit MVA at the PCC where the load is connected. If this condition is satisfied, it s assumed that the distorting load can be connected without any further investigation. This test is very unfair to large installations that might have a small fraction of distorting load. Tests 2 and 3 are intended to cope with this situation.

In Test 2, the distorting load MVA is summed with weighting factors of 0.7–2.5 as given in Table 5.6 of the standard, depending on the converter type, to give the weighted distorting power S_{DWi}. This test is passed if the $S_{DW\,I}/S_I < 0.1\%$ condition is met.

Test 3 can be used if the current harmonic distortion is known for the installation for each harmonic order of interest.

Stage 2: If a load does not meet stage 1 criteria, the specific characteristics of the harmonic-generating equipment should be evaluated, together with the absorption capacity of the system. This would require detailed system modeling, harmonic analysis, and procedures to apportion the planning levels to individual customers. Detailed system modeling and harmonic analysis are discussed in Chapter 6.

Stage 3: Under special circumstances, higher emission levels than in stages 1 and 2 from certain customers may have to be accepted. In such a situation, the consumer and the utility may agree on special conditions for the connection of the distorting load. These will have to be determined by a careful study of the actual and future system characteristics.

5.6.2 Summation Laws for Combining Harmonics[15,18]

The actual harmonic voltage (or current) at any point of a distribution system is the result of phasor summation of the individual components of each source. Two summation laws are commonly used.

First Summation Law

This is a simple linear law making use of diversity factors k_{hj}:

$$U_h = U_{h0} + \sum_j k_{hj}\, U_{hj} \tag{5.1}$$

where U_{h0} = the background harmonic voltage of the supply network (the harmonic voltage present in the supply network with the j loads disconnected).

For indicative values of k_{hj}, see Tables 3 and 4 in the Standard.

Second Summation Law

The following general summation law can be adopted for both harmonic voltage and current:

$$U_h = \sqrt[\alpha]{\sum_i U_{hi}^{\ \alpha}} \tag{5.2}$$

where

U_h = magnitude of the resulting harmonic voltage of order h for the considered aggregation of sources (probabilistic value)

U_{hi} = magnitude of the i-th individual emission level of order h

α = exponent. The value of this exponent depends mainly upon two factors:

1. The chosen value of the probability for the actual value not to exceed the calculated value
2. The degree to which individual harmonic voltages vary randomly in terms of magnitude and phase

The calculation of the emission limits in MV and HV systems using the stage 2 procedure is described in the Standard using the first and second approximations. This requires rather detailed knowledge of loads and other factors such as

T_{hHM}, the transfer coefficient from the upstream HV system to the MV system at harmonic order h

F_{ML}, the coincidence factor between the two distorting (aggregate) loads of the MV and LV distribution systems

F_{HV}, the coincidence factor for HV loads distorting simultaneously (F_{HV} value, depends on the loads and system characteristics; typical values are between 0.4 and 1)

For reasons of space we will not discuss these topics further. It will be fair to say that the application of the IEC Standards would require a more detailed knowledge of the loads in the system, and it is more complex using the second summation law compared with the application of IEEE 519-1992.

It must be kept in mind that all these standards provide guidelines only, and ultimately, it is up to the consumer and utility to determine the special conditions under which the distorted load could be connected.

5.7 GENERAL COMMENTS ON THE STANDARDS

5.7.1 Allocation of Harmonic Voltage or Current or Both Limits to the Customers

IEEE Std 519-1992 specifies both harmonic voltage and current limits. The current limits are specified in this standard with the implicit assumptions that (1) the system background harmonic voltage is zero and the harmonic current is only due to the customer load that is being considered for connection to the utility, (2) no resonance effects are expected due to this new connection.

IEC Std 61000-3-6 primarily sets limits to the harmonic voltage only for each individual customer, even though indicative values for relative harmonic currents under stage 1 for the total load of a consumer are provided. The following discussion will help to understand the interaction between a power system and a nonlinear load of a customer. Consider the system in Figure 5.2 where the power system is

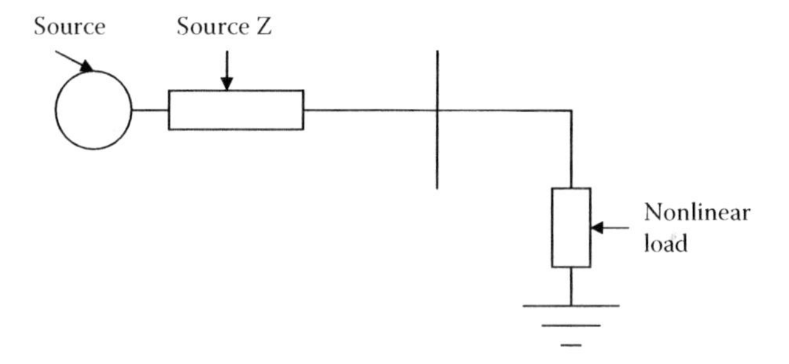

FIGURE 5.2 Power system with a nonlinear load.

represented by a source, Thevenin impedance, and a nonlinear load. First, let us consider an ideal system with no nonlinear loads inside the system. Then, the moment you connect a nonlinear load at the PCC, a harmonic voltage appears at the PCC with harmonic currents flowing into the system. Let us consider a pure resistive load connected to a power system with some background harmonic voltages due to the nonlinear loads within the power system. The moment the resistive load is connected, harmonic currents will flow into the pure resistive load even though it is not a harmonic source. Further, if a capacitive load is connected, under resonance conditions, we can obtain high harmonic voltages at the PCC due to a small harmonic source in the system even though a capacitor bank is not a harmonic source. Finally, if we have harmonic sources in the power system and the load, if the net harmonic currents are exactly 180° out of phase, there will be zero harmonic voltage at the PCC by their mutual cancellation. From this discussion, it should be clear that the harmonic voltage and current either into the power system or load are determined both by the power system, load, and system conditions such as resonance. Hence, whenever one encounters harmonic problems in a power system, tracing the source of the problem will require a rather detailed investigation in some cases. In summary, what one can easily measure are the harmonic voltages at the PCC before and after connecting a nonlinear load. Further, we can easily measure the magnitude of the harmonic current at the PCC. Precise measurement of the phase angle of the harmonic voltages or currents is more difficult. Sometimes, fundamental and harmonic powers are measured to determine the direction of the harmonic current as it is always certain that fundamental power flows into a nonlinear load, and then the relative signs of fundamental power and relevant harmonic power will indicate the direction of the net harmonic current at that location.

5.7.2 Empirical Nature of the Standards

The following statement from the book by J. Arrillaga and Watson[19] is quoted here to emphasize the empirical nature in the development of the different harmonic standards:

> It must be recognized that no standard relating to system harmonic content can be regarded as permanent, but as the current interpretation of system requirements. As understanding improves through the application of improved measurement and analysis techniques, so must the standards change.

5.7.3 Legal Responsibility for Damages due to Harmonic Problems

If damage is caused to either the customer or utility equipment, one of the parties may sue the other, particularly in view of the privatization of the electricity supply industry in many parts of the world. In this context, it is best to remember that all the harmonic standards are only recommended practices, and they do not have the force of a legally binding contract between two parties. Under these conditions the supply agreements will take precedence in the courts. Hence, it is very important to specify the conditions under which harmonic measurements are made, methods proposed to be used, and the limits (harmonic voltage, current, THD, etc.) to be observed,[20] Note that

1. After a disturbing load has been commissioned, system conditions can change, such as new capacitor banks in the system or new transmission lines, etc. Under these new conditions, due to excessive background harmonic voltage at the PCC in the system, harmonic currents can flow into a customer load, even if the customer has a pure resistive load.
2. Due to unforeseen resonance effects of the capacitors or reactors in the system or consumer load, excessive harmonic voltages or currents can be observed.
3. In some cases it may be very difficult to identify the precise cause of the harmonic problems as a single harmonic source or several sources or a particular system configuration. In practice, a particular combination of these may be responsible for the harmonic problems.

5.8 APPLICATION OF THE STANDARDS

Application of the existing harmonic standards to a utility system involves many practical issues. They Include

1. Designing an efficient procedure to evaluate the acceptability of an installation,
2. Specifying rational conditions to perform evaluation, and
3. Defining equitable limits to check acceptability.

Some of the practical difficulties in dealing with these issues, and the assumptions and approximations made to find a workable solution, are discussed as follows. In any practical power system,

1. Before a new harmonic-producing load is connected, some background harmonic voltages will be existing at the points of common coupling. These background harmonic voltages will be varying with time as the loads and other system conditions vary.
2. Once some loads are commissioned, it is not possible to disconnect them for making harmonic measurements with a view to estimating the additional harmonic (current or voltage) contribution due to these loads. One such load is the aluminum smelter load. IEEE Std 519-1992 specifies both

harmonic voltage and current distortion limits for the harmonic-producing loads at the PCC. This current distortion due to a harmonic-producing load such as an aluminum smelter can only be calculated at the planning stage assuming zero background harmonic voltages in the system. Once the aluminum smelter load is connected, it is not possible by measurements to estimate its harmonic current contribution because of the interaction between the power system source and the load. One solution to this problem is to specify the harmonic voltages (which are measurable) at the PCC before and after commissioning of the aluminum smelter load. Limits for harmonic currents from the smelter may only be used at the design stage of the project to ensure that suitable filters are provided so that the resultant harmonic voltages at the PCC are within acceptable limits.

3. If after commissioning of the aluminum smelter load, excessive harmonic voltages are observed at the PCC then detailed harmonic measurements are necessary to identify one of the following likely causes, for example, resonance due to (a) smelter filter banks or (b) capacitor banks in the system, and (c) excessive harmonic currents from (1) the smelter or (2) other sources in the power system.
4. The measurement of the phase angle of harmonic voltages and currents (particularly higher harmonics like 47 and 49) is rather difficult. This aspect will be discussed further in the chapter on harmonic measurements. In view of this, it is sometimes difficult to identify the source of certain harmonics, that is, which nonlinear load is contributing certain harmonics.
5. The following choices exist to define equitable harmonic limits for different consumers, though none of them look ideal from equity considerations:
   a. Absolute harmonic current limits for individual consumers large or small
   b. Harmonic current limits are made proportional to the ratio of the consumer load/total load at the substation
   c. Consumers are permitted to add a certain harmonic voltage level at the PCC, taking the existing background level into account.

Method (a) is disadvantageous to large consumers. Method (b) can become very conservative if a large consumer has little or no disturbing load. Method (c) benefits those consumers connected to a strong PCC. Further, if the power system is weak at the PCC and has high harmonic voltages there, new consumers may be forced to install expensive additional circuitry before connection is permitted.

5.8.1 Application of Standards—B.C. Hydro's Approach

Some salient features of the B.C. Hydro's Harmonic Guide are as follows.[20]

The guide deals with harmonic-producing installations in two categories. Category 1 (small) installations are accepted without performing detailed analysis in the plant design stage. Category 2 (large) installations are required to perform harmonic studies at the design stage. The study shall demonstrate that the harmonic design limits are met.

Criteria for Category Current 1 Installations

With the help of the IEEE 519 Std voltage limits, IEEE 519 Std current limits, and CIGRE stage 1 limits, a harmonic chart is constructed for each voltage level. On the x-axis of the chart, the ratio of the system fault MVA to plant demand MVA, and on the y-axis, percentage harmonic load MVA in a plant are plotted. Figure 2 of Reference 20 shows the three curves for the three voltage levels 69 kV, 138 kV, and 230 kV.

A customer installation is considered as category 1 if two conditions are satisfied. The first is that the ratio of total harmonic load MVA in the plant with respect to the total plant load MVA is below the curves, as shown in Figure 2 of Reference 20. If this condition is satisfied, the harmonics due to this installation are unlikely to violate either the voltage or current limits. The total harmonic loads MVA are estimated according to the following formula:

$$\text{Total harmonic load MVA} = 0.85 \times \text{MVA}_{12} + 1.00 \times \text{MVA}_6 \tag{5.3}$$

MVA_{12} is the total demand of harmonic loads configured in 12 or higher pulses. MVA_6 is the total demand of other harmonic loads including six-pulse drivers. The second condition is that the customer capacitors should not cause harmonic resonances; it means that the following equation is satisfied for every harmonic number h:

$$\begin{aligned} |h_r - h| &> 0.35 \quad h = 5,7,11,13,17\ldots\ldots \text{ odd} \\ |h_r - h| &> 0.10 \quad h = 2,4,6,8,10\ldots\ldots. \text{ even} \\ |h_r - h| &> 0.15 \quad h = 3,9,15.21.27\ldots\ldots.\text{triplen} \end{aligned} \tag{5.4}$$

In the foregoing equation, h_r is the parallel resonance frequency in multiples of the fundamental frequency. Equation 5.4 needs to be checked only for two harmonics adjacent to h_r. The frequency h_r is normally obtained with a frequency scan analysis of the plant. If all plant capacitors are connected to the same bus, the frequency may be estimated according to Equation 5.5,

$$h_r = \surd(\text{MVA}_{sys}/\text{MVA}_{cap}) \tag{5.5}$$

where MVA_{sys} is the system fault MVA at the capacitor bus. It includes the contribution of nonharmonic-producing loads such as motors in the plant. MVA_{cap} is the installed capacity MVA calculated at normal operating conditions of the supply system or the plant. The limits of Equation 5.4 shall be satisfied for all conditions.

The limits specified in Equation 5.4 define a set of "comfort zones" centered at each harmonic resonance. Because odd harmonics in a power system have higher magnitudes than even harmonics, the comfort zones around them have been made wider.

Standardization of the Conditions for the Verification of Limit Compliance

The following resonance conditions are possible: (a) If there is a series resonance within the supply system for a particular harmonic frequency, then the supply system impedance seen from the PCC will be zero. (b) There could be either series or parallel resonance between the system impedance and the customer capacitor banks. If it is series resonance, a large harmonic current can flow into the customer capacitor bank because of other sources within the system. If it is parallel resonance, then the supply system impedance seen from the PCC will be rather high. Under these conditions, large harmonic currents can flow into the system, and the customer capacitor bank with high harmonic voltages at the PCC. To account for these conditions B.C. Hydro specifies that harmonic currents must be determined using two sets of impedance data: (a) The supply system is modeled as a 60-Hz sinusoidal voltage source. Because such an ideal voltage source presents zero impedance for all the harmonic frequencies, this simulates the first resonance condition. This will ensure that customer plants restrain their own harmonic currents, and the harmonic currents escaping into the supply system are minimized. (b) To account for the second case, the actual supply system harmonic impedances based on B.C. Hydro's best estimate under different operating conditions are used. The limits apply to each phase current individually at the PCC.

Harmonic Voltage Limits

B.C. Hydro adopts the IEEE 519 harmonic voltage limits, taking into account system background harmonic voltages. A new customer installation is limited to add a certain level of harmonic voltage distortion at the PCC such that the combined harmonic voltage due to background and customer is within limits:

$$V_{hc} + V_{hb} \le V_{hL} \tag{5.6}$$

$$\sqrt{\sum_{k=2}^{40} (V_{khc} + V_{khb})^2} \le \mathrm{THD}_L \tag{5.7}$$

where V_{hc} and V_{hb} stand for the h-th harmonic voltage distortion due to customer and system background. V_{hL} and THD_L stand for the limits on individual (h-th) harmonic voltage distortion and total harmonic voltage distortion.

Telephone Interference Due to Harmonics

The B.C. Hydro harmonic guide does not apply $I \cdot T$ product limits but imposes a limit on the total harmonic current in amperes in addition to the percentage limits. The limit values for total harmonic current are shown in Table 5.7.

The limits on triplen harmonics are reduced to one-third of the limits for odd-order harmonics in IEEE Std 519 because the triplen harmonics generally have more zero-sequence current.

Harmonic Limits for Measurement Purposes

Harmonic measurements are necessary to ensure compliance with the limits set in supply agreements and the limits recommended in the relevant national standards at the PCC. If there are any problems due to the harmonics, they are required to identify the cause of the problems and to recommend remedial measures.

TABLE 5.7
Total Harmonic Currents

System voltage, kV	69	138	230/287
Total harmonic current (A)	20	10	6

Most harmonics in power systems are of a time-varying nature. In most recent instruments dealing with the measurements of the harmonics, Fast Fourier Transform and some window function will be used. Due to the limitations of existing instruments, B.C. Hydro does not specify the window size. If disputes arise over the accuracy of an instrument, Canadian Standards Association (CSA) standard C22.2,[21] which follows IEC 555, will be used to solve them.

The problem of harmonic bursts is not addressed by the majority of the standards. Harmonic bursts should be of more concern because most harmonic problems are related to them.

B.C. Hydro's limits on measured harmonics are shown in Figure 7 of Reference 20.

5.9 EXAMPLES OF THE HARMONIC STUDIES

In the Std IEEE 519-1992 and in the appendices of the IEC 61000-3-6 (1996) standard, there are examples illustrating the applications of the standards. A simple example of an adjustable speed drive has been reported by S. Mark Halpin and Reuben F. Burch in Reference 22.

Experimental and analytical studies of the effects of voltage and current harmonics on induction machines, transformers, appliances, underfrequency and overcurrent relays are reported in Reference 23 by Ewald F. Fuchs et al. Based on these studies, they are of the opinion that the present IEEE and IEC standard limits are rather conservative. In the revision of these standards, they recommend relaxation of some of the limits. Further, they suggest very restrictive amplitude limits for interharmonics and subharmonics due to their detrimental effects on underfrequency relays (malfunctioning), lighting equipment (flicker), and rotating machines (harmonic torques).

5.10 CONCLUSIONS

In this chapter we first discussed the undesirable effects of the harmonics. Later, salient features of the standards IEEE 519-1992 and IEC 61000-3-6 (1996) were discussed, together with differences in their philosophy in setting harmonic voltage and current limits. In IEEE 519-1992, a series of harmonic voltage limits are specified at different buses at different voltage levels where multiple nonlinear loads are connected. Also, limits on harmonic currents are specified based on the ratio I_{SC}/I_L. The current limits specified in IEEE 519 assume that there is no background harmonic voltage in the system and there are no resonances in the system. The IEC 61000 series deals with the broad area of low-frequency electromagnetic phenomena, including conduction and radiation aspects. It considers harmonics and interharmonics as a part of this whole field. It uses the concepts of compatibility levels, emission limits, and planning levels using probability in the estimation of the harmonic levels. It specifies harmonic levels for the greatest 95% probability daily value and 99.9% weekly value. The IEC procedure for the assessment of the suitability of connecting a disturbing load consists of three stages. In the second stage of the assessment, two different summation laws are suggested for studying the combined effects of the harmonics. The calculation of the emission limits in MV and HV systems using the stage 2 procedure requires rather detailed knowledge of loads and other factors such as T_{hHM}, the transfer coefficient from the upstream HV system to the MV system at harmonic order F_{ML}, and the coincidence factor between the two distorting (aggregate) loads of the MV and LV distribution systems.

F_{HV} is the coincidence factor for HV loads distorting simultaneously. Hence, the procedure in the IEC 61000-3-6 standard is rather more complex than the application of the procedure in IEEE Standard 519-1992. Later, we pointed out the interaction between power system and disturbing loads, and the difficulties in identifying harmonic problems, namely, multiple harmonic sources of equal strength, difficulty in precise measurement of phase angles of harmonic voltages and currents, and resonances in the system. We also discussed the practical issues in the application of the existing standards. These require us to specify

(a) the details of the harmonic studies that should be conducted at the planning stage for major loads and empirical rules to apply for small and medium loads,
(b) standard conditions for the analysis and measurements to determine harmonic levels, and
(c) acceptable limits for harmonic voltages, currents, total harmonic distortion, etc., which are equitable to different consumers.

Problems dealing with harmonic limits specification and monitoring for loads like those from an aluminum smelter, which cannot be disconnected once they are commissioned, are also considered. Finally, we reviewed the salient features of B.C. Hydro's harmonic guide developed in 1995.

REFERENCES

1. Kimbark, E.W. (1971). *Direct Current Transmission*, John Wiley, New York.
2. Gonzalez, D.A. and Mccall, J.C. (1987). Design of Filters to reduce Harmonic Distortion in Industrial Power Systems, *IEEE Trans. Ind. Appl., IA 23,* (3), 504–511.
3. IEEE Std. 519-1992, IEEE Recommended Practices and Requirements for Harmonic Control in Electric Power Systems. (With corrections incorporated on June 15, 2004.)
4. IEEE Std. 18-1992, *IEEE Standard for Shunt Power Capacitors.*
5. *Engineering Reports of the Joint Subcommittee on Development and Research of the Edison Electric Institute and the Bell Telephone System,* New York, 5 Volumes, July 1926 to January 1943.
6. *Directives concerning the Protection of Telecommunication Lines against Harmful Effects from Electricity Lines*, International Telegraph and Telephone Consultative Committee (CCITT), published by the International Communications Union, Geneva, 1963.
7. Arrillaga, J., Bradley, D., and Bodger, P. (1985). *Power System Harmonics*, John Wiley, Chichester, U.K.
8. ANSI/IEEE C57-1200-1987 *General Requirements for Liquid-Immersed Distribution, Power, and Regulating Transformers* (ANSI).
9. Jost, F.A., Menzies, D.F., and Sachdev, M.S., "Effect of System Harmonics on Power System Relays," CEA, System Planning and Operation Section, Power System Protection Committee, Spring Meeting, March 1974.

10. Crevier, D. and Mercer, A., "Estimation of high frequency network impedances by harmonic analysis of natural waveforms," *IEEE Trans, PAS-97, 1978,* pp. 424–431.
11. Barnes, H. and Kearley, S.J., "The measurement of impedance presented to harmonic currents by power distribution networks," *IEE Conf. Publ. 197, 1981,* pp. 71–75.
12. Sherman, W.G., "Summation of harmonics with random phase angles," *Proc. IEE, Vol 119, 1972,* pp. 1643–1648.
13. Rowe, N.B., "The summation of randomly varying phasors or vectors with particular reference to harmonic levels," *IEE Conf. Publ. 110, 1974,* pp. 177–181.
14. Dugan, R.C., McGranaghan, M.F., Santoso, S., and Beaty, H.W. 2002. *Electrical Power Systems Quality.* McGraw-Hill, New York.
15. EC 61000-3-6 (1996). Electromagnetic Compatibility (EMC), Part 3.6: Limits assessment of emission limits for distorting loads in MV and HV power systems.
16. IEC Standard 61000.4.7: Electromagnetic compatibility (EMC), Part 4: Testing and measurement techniques, Section 7: General guide on harmonics and interharmonics measurements and instrumentation, for power supply systems and equipment connected thereto.
17. IEC 61000-4-15, Electromagnetic Compatibility (EMC), Part 4: Testing and Measuring Techniques, Section 15: Flickermeter—Functional and Design Specifications.
18. Gosbell, V.J., Muttik, P., and Geddey, D.K. 2000. *A review of the new Australian Harmonics Standard AS/NZS 61000.3.6, Journal of Electrical and Electronic Engineering, Australia, 20 (1),* 57–64.
19. Arrillaga, J. and Watson, N.R., *Power System Harmonics*, John Wiley, 2003.
20. Xu, W. et al. (1995). Developing utility harmonic regulations based on IEEE STD 519—B.C. Hydro's approach, *IEEE Trans. on Power Delivery,* 10(3) July, 1423–1431.
21. Canadian Standards Association (CSA), 1992 *Measurement of Harmonic Currents, (CSA)* standard C22.2 No.0.16-M92.
22. Halpin, S.M. and Burch, IV, R.F., *Harmonic Limit Compliance Evaluations using IEEE 519-1992.* Chapter 9 of the IEEE Tutorial at the website www.powerit.vt.edu
23. Fuchs, E.F. et al., 2004. Are harmonic recommendations according to IEEE and IEC too restrictive? *IEEE Trans. on Power Delivery,* 19(4) 1775–1786.

APPENDIX—IEC 61000-SERIES STANDARDS

- IEC 61000 Electromagnetic compatibility (EMC)
- IEC 61000-2-1: Guide to electromagnetic environment for low-frequency conducted disturbances and signaling in public power supply systems
- IEC 61000-2-7: Environment—Low-frequency magnetic fields in various environments
- IEC 61000-2-9: Environment—Description of HEMP environment. Radiated disturbance. Basic EMC publication
- IEC 61000-2-10: Environment—Description of HEMP environment. Conducted disturbance
- IEC 61000-3-2: Limits—Limits for harmonic current emissions (equipment input current up to and including 16 A per phase)
- IEC 61000-3-3: Limits—Limitation of voltage changes, voltage fluctuations, and flicker in public low-voltage supply systems, for equipment with rated current ≤ 16 A per phase and not subject to conditional connection
- Power supply systems for equipment with rated current greater than 16 A IEC 61000-3-4: Limits—Limitation of emission of harmonic currents in low voltage

IEC/TR 61000-3-6 - Ed. 1.0 Electromagnetic compatibility (EMC)—Part 3: Limits, Section 6: Assessment of emission limits for distorting loads in MV and HV power systems. Basic EMC publication

- IEC 61000-3-7: Limits—Assessment of emission limits for fluctuating loads in MV and HV power systems
- IEC 61000-3-8: Limits—Guide to signaling on low-voltage electrical installations. Emission levels, frequency bands, and electromagnetic disturbance levels
- IEC 61000-3-11: Limits—Limitation of voltage changes, voltage fluctuations, and flicker in public low-voltage supply systems. Equipment with rated voltage current ≤ 75 A and subject to conditional connection
- IEC 61000-4-1: Testing and measurement techniques—Overview of IEC 61000-4 series
- IEC 61000-4-2: Testing and measurement techniques—Electrostatic discharge immunity tests. Basic EMC publication
- IEC 61000-4-3: Testing and measurement techniques—Radiated, radio-frequency, electromagnetic field immunity test
- IEC 61000-4-4: Testing and measurement techniques—Electrical fast transient/burst immunity test. Basic EMC publication
- IEC 61000-4-5: Testing and measurement techniques—Surge immunity test
- IEC 61000-4-6: Testing and measurement techniques—Immunity to conducted disturbances, induced by radio-frequency fields
- IEC Standard 61000.4.7: Electromagnetic compatibility (EMC), Part4: Testing and measurement techniques—Section 7: General guide on harmonics and interharmonics measurements and instrumentation, for power supply systems and equipment connected thereto
- IEC 61000-4-8: Testing and measurement techniques—Power frequency magnetic field immunity test. Basic EMC publication
- IEC 61000-4-9: Testing and measurement techniques—Pulse magnetic field immunity test. Basic EMC publication
- IEC 61000-4-10: Testing and measurement techniques—Damped oscillatory magnetic field immunity test. Basic EMC publication
- IEC 61000-4-12: Testing and measurement techniques—Oscillatory waves immunity test. Basic EMC publication
- IEC 61000-4-13: Testing and measurement techniques—Harmonics and interharmonics, including mains signaling at ac power port, low-frequency immunity tests
- IEC 61000-4-14: Testing and measurement techniques—Voltage fluctuation immunity test
- IEC 61000-4-15: Testing and measurement techniques—Flickermeter. Functional and design specifications. Basic EMC publication
- IEC 61000-4-16: Testing and measurement techniques—Test for immunity to conducted, common-mode disturbances in the frequency range 0 Hz–150 kHz

- IEC 61000-4-17: Testing and measurement techniques—Ripple on dc input power port immunity test
- IEC 61000-4-23: Testing and measurement techniques—Test methods for protective devices for HEMP and other radiated disturbances
- IEC 61000-4-24: Testing and measurement techniques—Test methods for protective devices for HEMP-conducted disturbance. Basic EMC publication
- IEC 61000-4-25: Testing and measurement techniques—HEMP immunity test methods for equipment and systems
- IEC 61000-4-27: Testing and measurement techniques—Unbalance, immunity test
- IEC 61000-4-28: Testing and measurement techniques—Variation of power frequency, immunity test
- IEC 61000-4-29: Testing and measurement techniques—Testing and measurement techniques. Voltage dips, short interruptions, and voltage variations on dc input power port immunity tests
- IEC 61000-5-1: Installation and mitigation guidelines—General considerations. Basic EMC publication
- IEC 61000-5-2: Installation and mitigation guidelines—Earthing and cabling
- IEC 61000-5-5: Installation and mitigation guidelines—Specification of protective devices for HEMP-conducted disturbance. Basic EMC publication
- IEC 61000-6-6: Generic standards—HEMP immunity for indoor equipment

6 Harmonic Filters

6.1 INTRODUCTION

This chapter discusses the undesirable effects of harmonics occurring in the power system and examines shunt filters (tuned and damped) used to reduce these harmonic levels to the limits recommended in the relevant standards. As it is necessary to estimate the system harmonic impedance at the point of common coupling (PCC) for the design of filter banks, different models for network components are also discussed.

6.2 UNDESIRABLE EFFECTS OF HARMONICS

The following are the undesirable effects of harmonics:

1. Capacitors may draw excessive current and prematurely fail from increased dielectric loss and heating.[1,2]
2. Harmonics may interfere with telecommunication systems, especially noise on telephone lines.
3. Transformers, motors, and switchgear may experience increased losses and excessive heating.
4. Induction motors may refuse to start (cogging) or may run at subsynchronous speeds.
5. Circuit breakers may fail to interrupt currents owing to improper operation of blowout coils.
6. The time–current characteristics of fuses can be altered and protective relays may experience erratic behavior. In particular, maloperation of the relays associated with ripple control systems can occur.

6.3 HARMONIC SOURCES

The following loads are the major harmonic sources in the power system:[3]

1. Three-phase static power converters in high-voltage direct current (HVDC) system.
2. Static var compensators (SVC).
3. Single-phase static power converters in power supplies in computers, television sets, small adjustable speed drives, and other equipment.
4. High-frequency sources, e.g., the electronic ballast for fluorescent lighting. In these devices, distortion is created at the switching frequency, which is generally above 20 kHz. Because of its high frequency, the distortion cannot generally penetrate far into the system.

5. Transformer saturation is due to the nonlinear voltage–current relationship represented by the magnetization characteristic. Arc furnaces also have a nonlinear voltage–current relationship.
6. There are also some other devices which produce harmonic distortion at frequencies that are not integer multiples of the fundamental frequency, e.g., cycloconverters, doubly fed machine drives, and adjustable speed drives.

6.4 TYPES OF FILTERS

Harmonic filters can broadly be classified into two types:[2,4]

1. Series filters
2. Shunt filters

If series filters with high series impedance to block the relevant harmonics are to be used, they must carry full-load current and be insulated for full-line voltage. In contrast, the shunt filters divert the relevant harmonic currents to the ground by providing a low-impedance shunt path. Hence, the shunt filters carry only a fraction of the current that a series filter must carry, making the cost of the shunt filter much less than the series filter. Further, the shunt filters supply the capacitive reactive power to the power system at the fundamental frequency, which is needed to provide voltage support in heavily loaded systems. In view of this, shunt filters are invariably used in power systems to reduce harmonic voltages to acceptable levels at the points of common coupling.

The shunt filters can be classified into two types:[4]

1. Tuned filters
2. Damped filters; these are also referred to as high-pass filters.

6.4.1 Types of Damped Filters

Four types of damped (high-pass) filters are shown in Figure 10.7 in References 4 and 5.

The first-order filter, which consists of a capacitor in series with a resistor characterized by large power losses at the fundamental frequency, is rarely used.

The second-order high-pass filter consists of an inductor in parallel with a resistor. This parallel combination is connected to the line in series with a capacitor. This filter has lower fundamental frequency losses and provides good filtering action.

The third-order filter consists of an inductor in parallel with a capacitor C_2 and resistor in series. This parallel combination is connected to the line in series with a capacitor C_1. Although loss performance of this third-order filter is superior to that of the second-order filter, it is less sensitive in its filtering action.

In the "C" type of filter, the location of C_2 is shifted to be in series with the inductor. In this filter, R is in parallel with series L and C_2. This parallel combination is in series with C_1. The filtering performance of the newly introduced "C" type lies in between second-order and third-order filters. Its main advantage is a considerable reduction in fundamental frequency loss because C_2 and L are series-tuned at that frequency. This filter is more susceptible to fundamental frequency deviations and component value drifts.[4]

6.5 AC NETWORK IMPEDANCE

In most cases, such as SVCs, the ac side of HVDC converters, etc., the harmonics generated can be considered current sources. As the ac network impedance will be in parallel with the harmonic current source and shunt filter, the harmonic voltage at the PCC and the harmonic currents entering into the power system will be dependent on the ac network impedance.[3,4] Hence, it is essential to know the value of the ac network impedance to properly design filters.

The ac network impedance can be determined either by measurement (for an existing system) or by calculation (both existing and future systems). Both methods involve certain difficulties. Harmonic measurements will be discussed in Chapter 7.

Table 6.1 gives some guidance regarding the details of the network that must be represented for harmonic studies.

This section summarizes the typical representations of common network components for the calculation of the ac network impedance.

6.5.1 OVERHEAD LINES

The modeling of transmission lines and transformers over a wide range of frequencies is relatively well documented in the literature.[6–9]

Typical overhead lines can be modeled by a multiphase coupled equivalent π circuit as shown in Figure 6.1. For balanced conditions, the model can be further simplified into a single-phase π circuit determined from the positive-sequence impedance of the line.

The main concerns for modeling overhead lines are[3,4] as follows:

- The frequency dependency of the unit-length series impedance. Major causes of the frequency dependency are the earth return effect and the conductor skin effect.
- The distributed-parameter nature (long-line effects) of the unit-length series impedance and shunt capacitance.

TABLE 6.1
System Representation for Harmonic Studies

System Details	Equivalent Representation
Industrial facility	Represent the utility source by its short-circuit impedance
Two systems are coupled at a single point	Represent the external network (a) by its short-circuit impedance at the point of connection or (b) with the help of frequency scan use different impedances at different frequencies
Complex network	Represent the external network at least five buses and two transformations from the point of interest

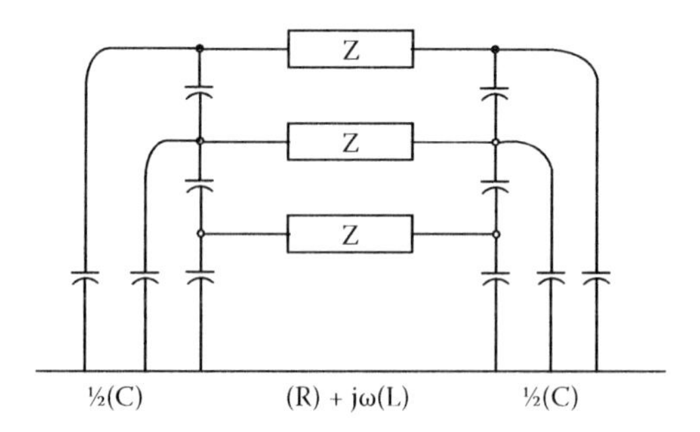

FIGURE 6.1 Overhead line model.

Some guidance can be given regarding the use of the line models. Inclusion of the frequency-dependent effects requires the calculation of unit-length line constants for each harmonic frequency. This requires much input data for harmonic analysis programs. The earth return effects mainly affect the zero-sequence harmonic components. The conductor skin effects mainly affect the line resistance, which in turn affects the damping level at resonance frequencies. Therefore, if zero-sequence harmonic penetration and damping at resonance frequencies are not of significant concern, the frequency-dependent effects may be neglected. In this case, a single unit-length Z matrix computed at the dominant resonance frequency is adequate.

Whether or not to include the long-line effects depends on the length of the line being modeled and the harmonics of interest. An estimate of critical line lengths where the long-line effects should be represented is $150/n$ miles, where n is the harmonic number. Finally, Reference 10 suggests that for line lengths of 250 km for the third harmonic and 150 km for the fifth, transpositions are ineffective and can aggravate unbalance. For such cases, a three-phase line model should be considered.

References 11 to 15 are useful for the development of digital computer programs for the calculation of the line constants incorporating the following features:

- Frequency dependence of transmission lines
- Skin effect
- Bundling of conductors

6.5.1.1 Line Constants

Almost all the electric utilities and consulting companies use a line constants program, because line constants are essential to perform any studies, power flow, short circuit, dynamic simulation, switching transients, etc. The line constants section of the program, EMTP, is versatile and very popular. It uses the usual input data, conductor geometry, both phase and earth conductor details, as well as earth resistivity, and calculates the line constants at any required frequency. Some typical line parameters from Reference 16 are shown in Table 6.2.

TABLE 6.2
Typical Line Parameters

Parameter	66 kV	115 kV	138 kV	230 kV	345 kV	500 kV	750 kV
R_1, ohm/mile	0.340	0.224	0.194	0.107	0.064	0.020	0.020
X_1, ohm/mile	0.783	0.759	0.771	0.785	0.509	0.338	0.528
R_0, ohm/mile	1.220	0.755	0.586	0.576	0.416	0.275	0.500
X_0, ohm/mile	2.370	2.300	2.480	2.235	1.624	1.050	1.584
X_0/R_0	1.950	3.050	4.230	4.080	3.490	3.800	3.170
C_1, MFD/mile	0.014	0.015	0.014	0.014	0.019	0.013	0.020
C_0, MFD/mile	0.009	0.008	0.009	0.009	0.012	0.009	0.013

Source: From Natarajan, R., *Computer-Aided Power System Analysis*, Marcel Dekker, New York, 2002, 16. With permission.

6.5.2 Underground Cables

Underground cable models are very similar to overhead line models.[9] The difficulty for cable modeling is the determination of unit-length parameters for a cable. Dommel[14] provides detailed descriptions on the calculation of cable parameters. Cables have more shunt capacitance than overhead lines. Therefore, long-line effects are more significant. An estimate of critical cable lengths where the long-line effects should be represented is $90/n$ miles.

6.5.3 Transformers

Figure 6.2 shows one relatively general model for a multiwinding transformer that is adequate for harmonic analysis.[3] The following points may be noted regarding this transformer model:

1. Harmonics above the 50th harmonic (3.0 kHz) are usually so small that they do not cause problems in power systems. For most transformers, the effect of stray capacitance becomes noticeable only for frequencies higher than 4 kHz; hence, they can be neglected.
2. Even though the different parameter resistances (R) and inductances (L) in Figure 6.2 are frequency-dependent, treating them as constant is generally acceptable for typical harmonic studies. This is because the frequency-dependent effects are not significant for the harmonic frequencies of common interest.
3. The current source in Figure 6.2 is used to represent the harmonic-generating effects of the magnetizing branch. The value of the current source should be determined from the flux-current curve and the supply voltage.
4. Inclusion of the saturation characteristics is important only when the harmonics generated by the transformer are of primary concern. Exact replication of saturation is virtually impossible owing to the complexities of local phenomena in the magnetic core and minor loop hysteresis.

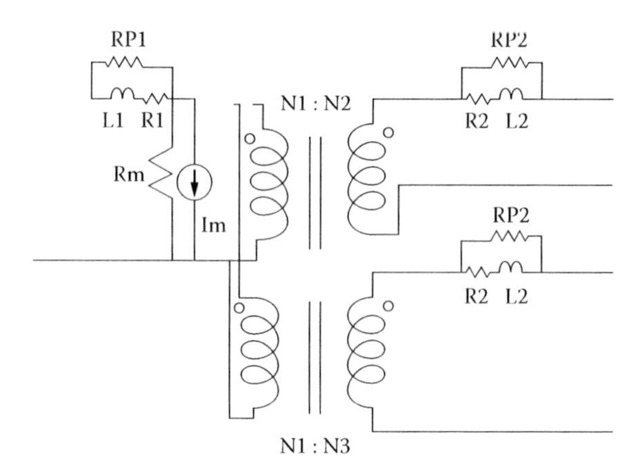

FIGURE 6.2 Transformer model. (From Task Force on Harmonics Modelling and Simulation, *IEEE Trans. Power Delivery*, 11(1), 452–465, 1996. With permission.)

5. If a transformer is subject to a dc current injection, harmonics generated from the magnetizing branch can be significant.
6. Transformers can impart a ±30° phase shift to harmonic voltages and currents, depending on the harmonic order, the sequence, and transformer connections. Modeling phase-shifting effects is essential if there is more than one harmonic source in the system. Three-phase representations automatically include phase-shifting effects. Single-phase approaches should use a phase-shifter model to represent the effects.

Other three-phase transformer models are also reported by Arrillaga and Watson[5] and in other publications in the technical literature. Also refer to papers by Laughton (References 16 and 17 in Chapter 2).

6.5.4 Rotating Machines

In view of the higher speed of the rotating magnetic field of the harmonics compared with that of the rotor in synchronous and induction machines, the harmonic impedance approaches the negative-sequence impedance.[3] In the case of synchronous machines, the inductance is usually taken to be either the negative-sequence impedance or the average of direct and quadature axis subtransient impedances. For induction machines, the inductance is taken to be locked rotor inductance.

For salient-pole synchronous machines, a negative-sequence fundamental frequency current in the stator winding induces a second harmonic current in the field winding. The interaction between stator and field windings produce even harmonics in the field winding and odd harmonics in the stator winding. This harmonic conversion mechanism causes a salient-pole synchronous machine to generate harmonic currents. A more accurate machine model has been proposed to take such effects into account.[17]

Reference 5 deduces a simple induction motor model from the analysis of detailed induction motor models.

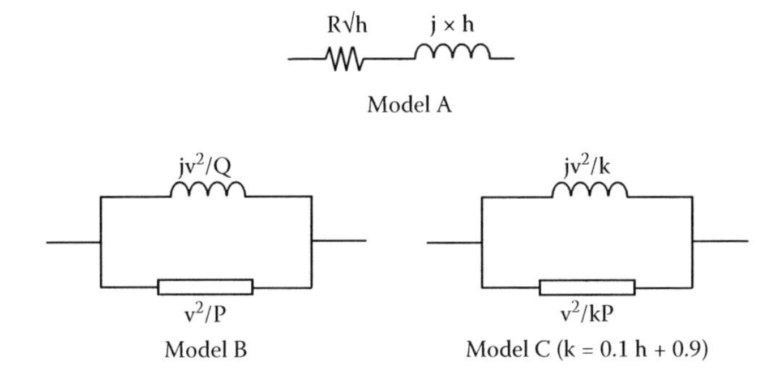

FIGURE 6.3 Load models of harmonic penetration studies.

6.5.5 Passive Loads

Unfortunately, there are no general models to represent different loads. One requires more details of the composition of the load to develop better load models. Utilities generally have some idea of the composition of the load in different areas. They can use this information to develop their load models.

Different load models are shown in Figure 6.3 to determine ac network impedance and for harmonic penetration studies.[5,18]

All these models have limitations. In model A, harmonic impedance increases continuously with frequency and hence gives very pessimistic results at high frequencies. Thus, this model can only be used to represent domestic loads with small motors, etc.

In power flow studies, loads are represented as P, Q loads. Hence when dealing with HV systems, loads are represented by a resistance in parallel with inductance. In model B total reactive load is assigned to an inductor L. Because a majority of reactive power corresponds to induction motors, this model also cannot give accurate results. There are many variations of this parallel form of load representation. Model C shows one form of such representation.

6.5.5.1 Electronic Loads

Representation of power electronic loads is very difficult, as they are harmonic sources and cannot be represented by linear R, L, and C elements. When more details are not available, the power electronic loads are left open-circuited when calculating harmonic impedances. If loads such as arc furnaces, aluminum smelters, etc., with high power rating are present, their effective harmonic impedances must be considered.[5]

6.5.6 Norton Equivalents of Residential Loads

If nonlinear appliances such as personal computers, television sets, and fluorescent lighting are present in the loads, the PSpice program is used (see Reference 19) to derive equivalent components for these loads as well as the low-voltage distribution network of a radial feeder. In this approach, the instantaneous voltages and currents of the network are calculated and the Fourier transform of the wave shapes obtained. The linear load is represented as a single branch with a lumped circuit derived at the PCC.[20]

6.6 DESIGN OF SINGLE-TUNED FILTERS

A filter is not always tuned to the frequency of the harmonic that it is intended to suppress. If a filter is tuned exactly to a particular harmonic frequency, the loss of some capacitor units will cause an upward shift in the tuned harmonic frequency. This will result in a sharp increase in impedance as seen by the harmonic. Should the resonance peak coincide with the harmonic of concern, the resulting voltage amplification may be disastrous (see Figure 6.4). Hence, it is usual practice to tune a filter 3–10% below the desired frequency.

The detuning of the filter can occur for the following reasons:[1,2,4]

1. System frequency variations
2. Manufacturing tolerances in both the capacitor units and the tuning reactor
3. Temperature variations
4. Capacitor fuse blowing, which lowers the total capacitance and thereby raises the frequency at which the filter is tuned

Let f_n = nominal harmonic resonance frequency

f = actual harmonic frequency

$\omega_n = 2\pi f_n$

$\omega = 2\pi f$

$\delta = (\omega - \omega_n)/\omega_n$ = deviation (per unit) of frequency from tuned frequency

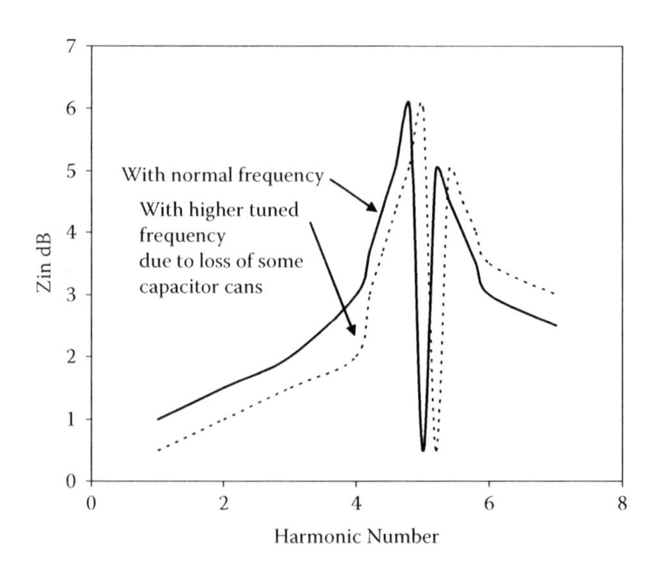

FIGURE 6.4 Typical response of harmonic filter and power system.

The resonance frequency f_n of a tuned filter consisting of a series inductance L and capacitance C is given by

$$f_n = \frac{1}{2\pi\sqrt{LC}} \tag{6.1}$$

For small deviations of Δf, ΔL, and ΔC, the total per unit deviation in system frequency is given by

$$\delta = \Delta f/f + \frac{1}{2}(\Delta L/L + \Delta C/C) \tag{6.2}$$

For filter design purposes, δ is assumed to be wholly attributable to Δf.

Quality factor Q of a filter is defined as

$$Q = X_{Lr}/R = X_{Cr}/R = \sqrt{(L/C)}/R \tag{6.3}$$

where the reactances at the resonance frequency are given by X_{Lr} and X_{Cr}.

It has been shown in References 1, 5, and 6 that in an ac network with a limited impedance angle $+\varphi_a$ that there is an optimum value for Q of a single-tuned filter which will give the lowest harmonic voltage. This value is given by

$$Q = (1 + \cos\varphi_a)/(2\,\delta \sin\varphi_a) \tag{6.4}$$

In many practical situations, to reduce the value of the filter Q to this value, a series resistance will have to be added to the filter inductance and capacitance. This will increase the filter losses considerably and hence will be uneconomical if we take the cost of the energy losses over the total life of the filter.

As can be appreciated from the following discussion, in view of the uncertainties associated with ac network impedance under normal and abnormal operating conditions (existing and future), certain assumptions will have to be made to design suitable filters to reduce the excessive harmonic levels in the system.

The following data are needed for filter design:

1. Preexisting harmonic levels in the power system must be determined by field measurements.
2. Criteria adopted for adequate filtering:
   a. Maximum voltage of an individual harmonic
   b. Maximum total harmonic distortion (THD)
   c. Less commonly used I.T or KV.T products making use of telephone influence factors (TIFs)

   These will have to be decided with the help of the relevant national and international standards and the engineering practices of the customer.

   Items (1) and (2) will assist the designer, in cooperation with the customer, to decide the maximum harmonic voltage contribution due to the new harmonic-producing load at the PCC.
3. Variation of the ac bus voltage under peak and light load conditions and at different times in the year (winter or summer)
4. Maximum power frequency capacitive reactive power that can be connected to the network at which the filter is to be installed.

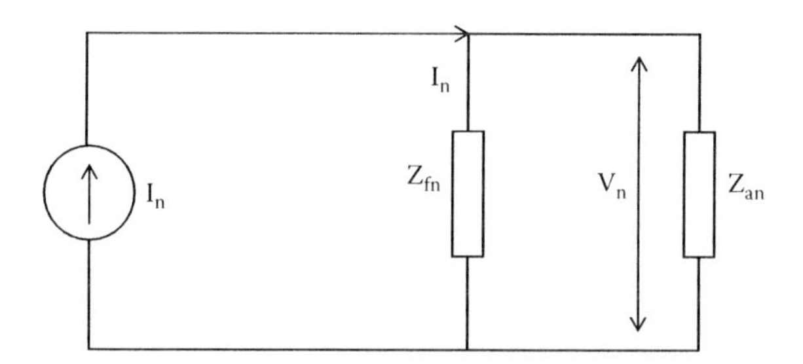

FIGURE 6.5 Equivalent circuit of a filter and AC system per phase.

Items (3) and (4) decide the maximum size of the capacitor bank rating in MVA that we can use in the filter.

In Figure 6.5 let *In* be the n-th harmonic current from the harmonic source, say an SVC or HVDC converter, or an adjustable speed drive. Let Z_{fn} be the single-tuned filter impedance for the n-th harmonic. Let Z_{an} be the ac network impedance for the n-th harmonic. If V_n is the maximum n-th harmonic voltage permitted at the bus, then

$$V_n = I_n\,(Z_{fn}\,Z_{an})/(Z_{fn} + Z_{an}) \tag{6.5}$$

In Equation 6.5, except Z_{fn} all the other quantities are known. Hence, Z_{fn} can be determined from Equation 6.5. Further,

$$Z_{fn} = R + j(\omega L - 1/\omega C) \tag{6.6}$$

At the tuned frequency f_n

$$f_n = \frac{1}{2\pi\sqrt{LC}} \tag{6.7a}$$

$$L = 1/(\omega_n^2\,C) \tag{6.7b}$$

Note: System actual harmonic angular frequency is ω rad/s, even though the filter is tuned to the nominal harmonic angular frequency ω_n rad/s. $\omega = \omega_n(1 + \delta)$

Substituting for L from Equation 6.7b in Equation 6.6 and ignoring R, which is usually small, we can obtain the following:

$$Z_{fn} = j\,[\omega/(\omega_n^2\,C) - 1/(\omega\,C)] \tag{6.8a}$$

In the case of filters with low values of Q (e.g., large filters in HVDC systems), if R is not negligible, estimate the value of R assuming a typical value for Q in the range 50–150. For such cases

$$\omega_n\,L = 1/\omega_n\,C;$$

$$\omega_n\,L/R = 1/(C\omega_n R) = Q$$

$$R = \frac{1}{C\omega_n Q}$$

Instead of Equation 6.8a, which neglects R, use the following equation to solve for C:

$$Z_{fn} = 1/(C\omega_n Q) + j[\omega/(\omega_n^2\, C) - 1/(\omega\, C)]$$

$$Z_{fn} = 1/C\{1/(\omega_n\, Q) + j[\omega/(\omega_n^2\,) - 1/(\omega)]\} \quad (6.8b)$$

From Equation 6.8a or 6.8b, the value of C can be obtained.

From Equation 6.7b, the value of L can be obtained.

Typically, the value of R consists only of the resistance in the inductor. In this case, Q of the filter is equal to R times the X/R ratio of the tuning reactor. This usually results in a very large value of Q and a very sharp filtering action. It has been reported[2] that the response for values of Q above 25 is essentially indistinguishable from the $Q = 100$ plot, except for the magnitude of the peak. The 60-Hz X/R ratio of the filter reactors is usually between 50 and 100.

Reference 20 gives some examples of simulation of the harmonic studies, which will be helpful in modeling some locations with harmonic filters.

6.7 DESIGN OF DOUBLE-TUNED FILTERS

It is possible to design double-tuned filters, for example, 11th and 13th. The equations for the parameters of the double-tuned filters are given in References 4, 5, and 6.

The advantages of a double-tuned filter over two single-tuned filters are the following:[1]

1. Its power loss at fundamental frequency is less.
2. One inductor instead of two is subjected to full impulse voltage.

Such filters are used at both terminals of the Cross Channel Link[1] and in HVDC project in central in India.[5]

6.8 FILTER PERFORMANCE EVALUATION

As the power system operating conditions change, we must ensure that the harmonic levels in the power system do not exceed the levels recommended in the relevant standards.[1,2,6] In other words, there must be adequate filtering of the harmonics under different operating conditions. Hence, ac network impedances should be determined covering the whole range of harmonics of interest under the following conditions:

1. Various load conditions (peak, light, winter, and summer)
2. Different generating conditions (causing maximum and minimum three-phase fault levels)
3. Various outages of lines, transformers, and other critical equipment
4. Split bus operation, etc.

These ac network impedances should be plotted on an $R - X$ plane as shown in Figure 6.6a or 6.6b. These figures show the general shape of the system impedances in

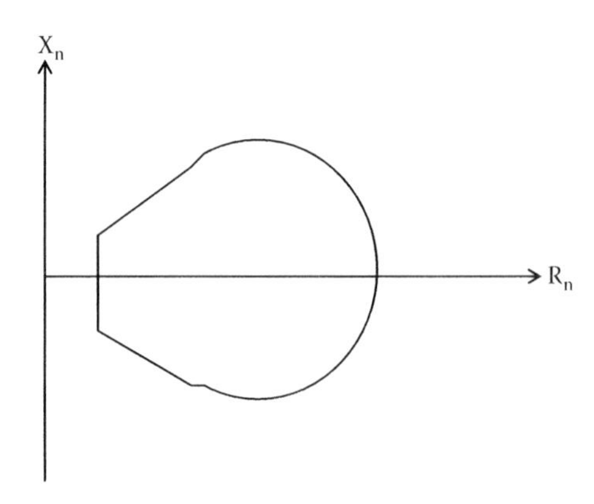

FIGURE 6.6A Bounds of loci impedance to harmonics in an AC network.

a qualitative manner. Figure 6.6b is more appropriate when the network is fully developed and it is possible with reasonable accuracy to establish the system impedances for low-order harmonics. This is the approach used in specifying filters for the British Terminal of the 2000-MW Cross Channel Link.[22] With the assumed filter values, harmonic voltages are calculated at different critical points in the system to cover the different ac network impedances on the boundary of the curves in Figure 6.6a or 6.6b. If more than one filter is needed to improve the system performance, the filter system must be designed for the possibility of having specific filter branches out of service.

By this time it should be obvious to the reader that the filter design involves rather complex and tedious calculations and different trials to optimize performance

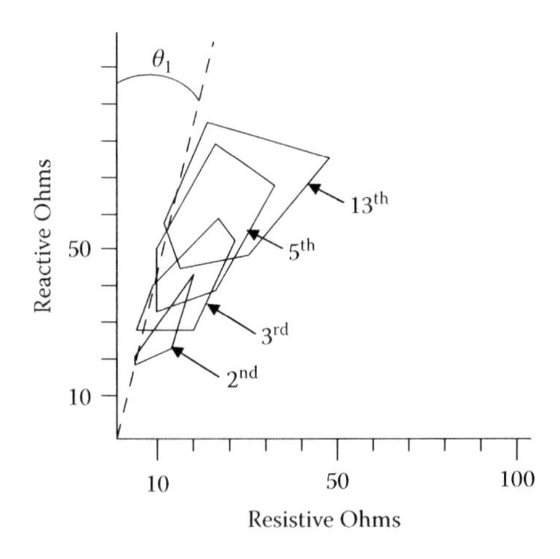

FIGURE 6.6B System impedance representation of 2nd, 3rd, 5th, and 13th harmonics.

and cost. Hence, a digital computer and computer programs to perform frequency scans are necessary.

In general, a frequency scan should be made at each problem node in the system, with harmonic excitation at each point where harmonic sources are connected. This allows the easy evaluation of the effects of system changes on the resonant conditions. The cause of any problem that may arise can be discovered quickly by examination of these plots.

If a parallel resonance point (peak) exists at the sixth harmonic for maximum load and at the 4.8th for 50% load, then at some loading between these points the resonance will exactly occur at the fifth harmonic, a location at which three- and six-pulse variable-speed drives produce significant harmonic current. This will require either a redesign of the filter or the implementation of special operating rules for the system that will minimize the effect of this resonance point.

If the converter is connected to a dc motor with a high-performance regulator, instability may occur if the resonant peak is the fourth harmonic, even though this is an even harmonic.[23]

6.9 DESIGN OF DAMPED FILTERS

In HVDC systems, tuned filters are used for the lower harmonics and one high-pass (damped) filter is used for all the higher harmonics.[1,2,5] The combination of tuned and high-pass arms will be chosen such that the total cost of the filters is minimum still satisfying filtering requirements.

Q of a damped filter is defined as

$$R/X_{Lr} = R/X_{Cr} = R/\sqrt{(LC)} \tag{6.9}$$

Typical values of Q vary from 0.5 to 2.0.

When designing a damped filter, Q is chosen to give the best characteristic over the required frequency band and there is no theoretical optimal Q as with tuned filters.

To assist the designer, some characteristic curves for damped filters have been plotted by Ainsworth[6] with the help of two parameters:

$$f_0 = 1/(2\,\pi\, CR) \tag{6.10}$$

$$m = L/(R^2C) \tag{6.11}$$

where R, L, and C are the parameters used in the second-order damped filter.

Kimbark[1] also plots the scalar impedance of second-order and third-order high-pass filters with the help of dimensionless normalized variables, frequency, etc., to assist the designer.

The performance of high-pass filters is not sensitive to frequency deviations even for harmonics near their resonant frequencies.

Table 6.3 shows the comparison between tuned and damped filters.

TABLE 6.3
Comparison of Tuned and Damped Filters

	Tuned Filters	Damped Filters
Sensitivity to temperature variation, frequency deviation, component manufacturing tolerances and loss of capacitor elements, etc.	Performance is very sensitive to these changes.	Performance is less sensitive to these changes.
Impedance characteristics	Provides very low impedance magnitude at its tuned frequency f_r rises rapidly on both sides of f_r depending on the Q value of the filter. More than one filter branch is required for SVCs and HVDC systems.	Provides low impedance magnitude for a wide spectrum of harmonics. The minimum impedance in its pass band never achieves a value comparable to that of the single-tuned filter at its tuned frequency.
Parallel resonance problems	More likely between the filter and system admittance at a harmonic order below the lower-tuned filter frequency, or in between tuned filter frequencies.	Less likely
Rating (VA) of the filter at fundamental frequency	Less compared with the rating of the damped filters.	
Losses	Low	To achieve a similar level of filtering performance as that of a tuned filter, the damped filter has to be designed for higher fundamental VA rating.

6.10 FILTER COMPONENT RATINGS

6.10.1 Filter Capacitors

ANSI/IEEE Standard 18-1980 specifies the following continuous capacitor ratings[24]:

- 135% of nameplate kvar
- 110% of rated rms voltage (including harmonics but excluding transients)
- 180% of rated rms current (including fundamental and harmonic currents)
- 120% of peak voltage (including harmonics)

In a high-Q (say about 100) single-tuned filter, the value of the resistance is about 1% of the value of the reactor at fundamental frequency, and hence, neglecting this value will not involve serious error in estimating the voltage across the filter capacitor. Let V_C, V_L, V_S be the voltages across filter capacitor, inductor and system voltage applied to the filter. Let X_C and X_L be the reactance of the capacitor and filter reactor at fundamental frequency, respectively, and I the current through the filter. Then,

$$V_C + V_L = V_S = -jI\,X_C + jI\,X_L = -jI\,(X_C - X_L)$$

$$V_C/V_S = -jI\,X_C/-jI\,(X_C - X_L) = X_C/(X_C - X_L)$$

$$= (X_C/X_L)/[(X_C/X_L) - 1\,)]$$

If the filter is tuned to the n-th harmonic, resonance at both the inductive reactance X_{Lr} and capacitive reactance X_{Cr} is equal.

Hence,

$$X_{Lr} = nX_L = X_{Cr} = X_C/n$$

$$X_C/X_L = n^2$$

$$V_C/V_S = n^2/(n^2 - 1)$$

If $n = 3$, because of the series reactor the capacitor voltage rises by 12.5% above the system voltage.

When designing a filter, the limits on the rms voltage and currents, and the arithmetic sum of the peak voltages on the capacitor bank, should be close to 100% rating for normal conditions. This is done so that the overrating capabilities are available to cover system overvoltage and bank unbalance conditions. The harmonic components may increase significantly under unbalanced conditions.[25]

The current limit, although 180% by standards, may be lower because individual capacitor units are usually fused at 125%–165% of their current rating.

A very low temperature coefficient of capacitance is desirable for tuned filters to avoid detuning caused by change of capacitance with ambient temperature or with self-heating of the capacitors. This property is unimportant for damped filters or power capacitors.[2,4]

The required reactive-power rating of the capacitor is the sum of the reactive powers at each of the frequencies to which it is subjected.

6.10.2 Tuning Reactors

The reactors used for filter applications are usually built with an air core, and ±5% tolerance. Air core reactors have linear characteristics owing to the absence of saturation with different harmonic currents flowing in them. The 60 Hz X/R ratio is usually between 50 and 150. In those applications where a lower value of Q is required, this may be achieved by using a series resistor.

Filter reactor ratings depend mainly on the maximum rms current (taking into account both 60 Hz and harmonic currents). They are designed to have the same basic insulation level (BIL) as power transformers connected at that same voltage level. Normally, the R and L form the ground side of a tuned filter.

6.11 OUTLINE OF FILTER DESIGN IN TWO QUEENSLAND (AUSTRALIA) PROJECTS

6.11.1 Transmission SVC Project

We will discuss in this section the general outline of filter design in two Queensland projects involving SVCs. Except for brief details of the specification of the equipment, which are already available in the public domain, other design details of these projects will not be discussed here, for the following two reasons: (1) such material is proprietary to Powerlink Queensland; and (2) even if Powerlink Queensland permission could be sought, limitations of space in the book prevents such detailed discussion.

This compensator is designed to act as a reactive power reserve under contingency conditions to control the voltage in the substation and to control the negative-sequence voltages caused by the single-phase railway loads in the vicinity.

This transmission SVC is designed with the following specifications:

Static var compensator (SVC) rating	260 MVAr capacitive
	80 Mvar inductive
SVC transformer rating	180 MVA, 275/17.7 kV, wye–delta connection

All the following component ratings are given at a bus voltage of 17.7 kV, where they are connected:

Thyristor-controlled reactor (TCR) rating	170 MVAr
Thyristor-switched capacitor (TSC) rating	170 MVAr
Harmonic filters: Third	58 MVAr
Fifth	28 MVAr
Seventh	7 MVAr

Initially, several power flow studies and dynamic simulation (transient stability studies) were conducted to determine the dynamic reactive compensation required at the 275-kV bus of the substation concerned, considering different operating conditions such as different system loads (winter, summer, peak and light loads, etc.), faults, and transformer and line outages. The dynamic simulation studies helped to determine the range of the SVC (both capacitive and inductive) and the capacity of the TSC. The TCR rating was chosen equal to the rating of the TSC. The inductive reactive range of the SVC was determined by the system light load conditions. With partial conduction of the TCR, maximum third, fifth, and seventh harmonics generated by the TCR are estimated for a chosen size of the TCR. The maximum values of these harmonic currents determine the respective filter ratings. Thus, all the ratings of the TSC and TCR, as well as third, fifth, and seventh harmonic filter ratings are determined. Later, several

TABLE 6.4
Reactive Power Range of the SVC in Different Modes

TCR170 MVAr mode	TSC170 MVAr mode	Total Capacity of 3rd, 5th, and 7th Filters 90 MVAr (Always Connected)	Reactive Power Supplied by the SVC
Fully conducting	Switched off	Connected	80 MVAr inductive
Fully blocked	Switched off	Connected	90 MVAr capacitive
Partially conducting	Switched off	Connected	Variable range from 80 MVAr inductive to 90 MVAr capacitive
Fully conducting	Switched on	Connected	90 MVAr capacitive
Fully blocked	Switched on	Connected	260 MVAr capacitive
Partially conducting	Switched on	Connected	Variable range from 90 MVAr capacitive to 260 MVAr capacitive

system studies confirmed that the capacitive rating of the SVC is less than 30% of the three-phase short-circuit level at the substation (1000 MVA) and that it has enough range to cover the following:

- negative-sequence voltages due to the single-phase railway loads in the vicinity of the substation;
- power swings due to drag line loads in that area.

The individual harmonic filter ratings (third, fifth, and seventh) of the transmission SVC are determined primarily by the harmonic currents generated by the TCR of the SVC. In view of the SVC transformer presenting a high impedance, other harmonics (i.e., QR loads, etc.) cannot flow into these filters and hence cannot significantly affect their rating.

The TSC capacitors are connected in delta. The filter banks are connected in a so-called double wye. The usual unbalance protection is provided by connecting a current transformer between the wyes. Table 6.4 shows the reactive range supplied by the SVC with the different modes of TCR and TSC, and it can be seen from this table that a smooth variation of reactive power can be obtained from 80 MVAr inductive to 260 MVAr capacitive by proper choice of the TSC switched on or switched off and TCR in one of its modes, that is, fully conducting, fully blocked, or partially conducting.

6.11.2 Queensland Railway (QR) Project

A brief description of this project is already given in Chapter 2.[26,27] The SVCs are required to perform the following three functions in order of priority:

- Load balancing and power factor correction
- Reduction of negative-sequence voltage
- Regulation of positive-sequence voltage

We will discuss in this section the general outline of the design of the filters in this project, the information that had to be obtained from the client QR to enable filter design, and the harmonic limits applied at the points of common coupling with the QR loads, that is, at the 132-kV buses supplying QR loads.

At most Queensland Railway substations, there are two 30-MVA, 132-kV/50-kV single-phase transformers supplying the traction loads. In addition, there is a wye–delta SVC transformer to provide supply to the SVC. Although the primary voltage of the SVC transformer is always 132 kV, the secondary had different voltages, 10.4 kV, and so on, depending on the number of thyristors used in series at that location. As can be seen from Figure 2.20, three sets of the filters are required in most substations, one for the SVC to filter the harmonics produced by the thyristor-controlled reactor (TCR) and the other two on each 50-kV railway feeder.

Powerlink Queensland had to first know how many fully loaded trains and how many empty trains would be on any section, when maximum harmonic currents are generated by the railway locomotives. Also, QR had supplied the harmonics generated by locomotives. Once the number of locomotives and maximum harmonics generated by the locomotives are known, then by applying suitable diversity factors, the maximum amount of harmonic current that should be absorbed by each filter third, fifth, and seventh were estimated. In addition to this, the filters have to absorb harmonics from the supply system also. This is estimated assuming that the 132-kV bus has a maximum of 1% of each harmonic and simultaneously 0.7% of the adjacent harmonics. This will ensure that the total harmonic distortion will not exceed 1.5% stipulated in the relevant Australian harmonic standards at that time. Once the harmonic currents from the system and the railway loads are known, tuned third, fifth, and seventh harmonic filters are designed, initially assuming each filter absorbs only one harmonic current, and finally, the complete design was checked to allow for other harmonic currents in each filter and the ratings of each filter capacitor and reactor are not exceeded. In other words, when third, fifth and seventh harmonic filters are in parallel, in a fifth harmonic filter in addition to fifth harmonic current, some third and seventh harmonic currents flow.

Further, the following limits are applied to the harmonic voltages' contribution due to the QR loads only at the PCC to allow for background distortion and for future connection of other disturbing loads. The limits applied to the distortion from the railway system are in general 70% of the limits specified in the Australian Standard AS 2279 at that time. These limits are[27] as follows:

a. No individual odd harmonic shall exceed 0.7% of nominal system voltage except that for the 5th, 7th, and 21st harmonics, a limit of 0.35% shall apply because of the existence of tone injection control systems.
b. No individual even harmonic shall exceed 0.35% of the nominal system voltage.
c. The total harmonic voltage distortion, defined as the square root of the sum of the squares of all individual harmonics, shall not exceed 1% of nominal voltage.

The above harmonic limits apply to a normal system and also to any system outage condition. For design purposes, these outage conditions are defined within the tender documentation.

In addition to these traction filters, SVCs had their own filters designed to absorb the harmonics generated by the TCR.

After all the designs of all the filters at 13 QR substations were completed, the entire design was checked by conducting several harmonic studies with the help of a commercial computer package, the VHARM program.

More details of this project can be found in References 26 and 27.

By this time the reader would have an idea of the complexity of the project, the engineering judgments one must make, and the several planning studies that should be done.

Unlike the transmission SVC project, in the case of the QR project, as several SVCs and railway filters are adjacent, steps must be taken to ensure resonance problems are identified and eliminated to avoid damage to the filter capacitors.

At the end of the project, a single-line-to-fault test was conducted at one of the QR substations, and harmonic voltages were also monitored at different key locations under different system conditions, winter, summer, peak, and light load conditions.

6.12 CONCLUSIONS

We have briefly reviewed the different harmonic sources in the power system, and pointed out the deleterious effects of excessive harmonics on power system equipment such as capacitors, transformers, rotating machines, circuit breakers, protective relays, and so on. Different types of shunt filters (tuned and damped) used to reduce these harmonic levels to those recommended by the relevant standards, their relative merits, design, and some important aspects of the specification of their components (filter capacitors and reactors) are discussed. As ac network impedance is necessary for filter design, models for different network components that can be used in digital computer programs to calculate ac network impedance are also discussed.

Outlines of the filter designs in two Queensland projects, one for transmission SVC and another for SVCs to compensate traction loads, are reported.

REFERENCES

1. Kimbark, E.W., 1971, *Direct Current Transmission, Vol. 1*, John Wiley & Sons, New York.
2. Gonzalez, D.A. and McCall, J.C., 1987, Design of filters to reduce harmonic distortion in industrial power systems, *IEEE Trans. Ind. Appl. A,* **23**(3), 504–511.
3. Report by Task Force on Harmonics Modelling and Simulation, Modelling and Simulation of the Propagation of Harmonics in Electric Power Networks, Part 1, 1996, Concepts, models and simulation techniques, *IEEE Trans. on Power Delivery*, **11**(1), 452–465.
4. Arrillaga, J., Bradley, J.D., and Bodger, P., 1985, *Power System Harmonics,* John Wiley & Sons, Chichester, U.K.
5. Arrillaga, J. and Watson, N.R., 2003, *Power System Harmonics*, John Wiley & Sons, Chichester, U.K.
6. Ainsworth, J.D., 1965, *Filters, damping circuits, and reactive volt-amps in HVDC converters, in High Voltage Direct Converters and Systems*, Cory, B.J., Ed., Macdonald, London, pp. 137–174.
7. CIGRE Working Group 36–05, 1981, Harmonics, characteristic parameters, methods of study, estimates of existing values in the network, *Electra*, **77**(July), 35.

8. Grady, W.M., Heydt, G.T., Mahmoud, A.A., and Schultz, R.D., 1984, *System Response to Harmonics, IEEE Tutorial Course on "Power System Harmonics,"* IEEE publication *84 EHO221-2–PWR*, 10–20.
9. Mahmoud, A.A. and Schultz, R.D., 1982, A method for analyzing harmonic power distribution in AC power systems, *IEEE Trans. PAS,* **101**(6), 1815–1824.
10. Arrillaga, J., Acha, E., Densem, T.J., and Bodger, P.S., 1986, Ineffectiveness of transmission line transpositions at harmonic frequencies, *Proc. IEE,* 133C(2), 99–104.
11. Carson, J.R., 1926, Wave propagation in overhead wires with ground return, *Bell System Tech. J.,* 5, 539–556.
12. Deri, A., Tevan, G., Semlyen, A., and Castanheira, A., 1981, The complex ground return plane: a simplified model for homogeneous and multilayer earth return, *IEEE Trans. PAS,* 100, 3686–3693.
13. Semlyen, A. and Deri, A., 1985, Time domain modelling of frequency dependent three-phase transmission line impedance, *IEEE Trans. PAS,* 104, 1549–1555.
14. Dommel, H.W., 1986, *Electromagnetic Transients Program Reference Manual (EMTP Theory Book),* Prepared for Bonneville Power Administration, Dept. of Electrical Engineering, University of British Columbia.
15. Hesse, M.H., 1963, Electromagnetic and electrostatic transmission-line parameters by digital computer, *IEEE Trans. PAS,* 82, 282–291.
16. Natarajan, R., 2002, *Computer-Aided Power System Analysis,* Marcel Dekker, New York.
17. Semlyen, A., Eggleston, J.F., and Arrillaga, J., 1987, Admittance matrix model of a synchronous machine for harmonic analysis, *IEEE Trans. on Power Systems,* 2, 833–840.
18. Ranade, S.J. and Xu, W., 1998, An overview of harmonics modeling and simulation, *IEEE Tutorial,* http://www.powerit.vt.edu/AA/c5.htm
19. Capasso, A., Lamedica, R., and Prudenzi, A., 1998, Estimation of net harmonic currents due to dispersed non-linear loads within residential areas, *International Conference on Harmonics and Quality of Power* (ICHPQ1998), Athens, 700–705.
20. Lamedica, R., Prudenzi, A., Tironi, F., and Zanineelli, D., 1997, A model for large loads for harmonic studies in distribution networks, *IEEE Trans. on Power Delivery,* 12(1), 418–425.
21. Report by Task Force on Harmonics Modeling and Simulation, 1996, Modeling and Simulation of the propagation of harmonics in electric power networks, Part II, Sample systems and examples, *IEEE Trans. on Power Delivery,* 11(1), 466–474.
22. Anderson, B.R. and Macleod, N.M., 1996 Filter performance for the British Terminal of the 2000 MW Cross Channel Link, *Second International Conference on Harmonics in Power Systems.*
23. Steeper, D.E. and Stratford, R.P., 1976, Reactive compensation and harmonic suppression for industrial power systems using thyristor converters, *IEEE Trans. Ind. Appl. A,* 12(3), 232–254.
24. Dugan, R.C., McGranaghan, M.F., and Beaty, H.W., 1996, *Electrical Power Systems Quality,* McGraw-Hill, New York.
25. Vilcheck, W.S. and Gonzalez, D.A., 1985, Guidelines for applying shunt capacitors on industrial power systems, Presented at the Industrial and Commercial Power Systems Conference, Denver, CO, May 13–16.
26. Sastry, V.R., Hill, D.V., and Lee, C.J., 1986, Static var compensators for balancing the single-phase traction loads in Central Queensland, *J. Electr. Electron. Eng. Aust.,* 6(3),184–190.
27. Wright, P.O., 1986, Planning of electricity supply to A.C. electrified railways from a weak network in Central Queensland, *J. Electr. Electron. Eng. Aust.,* 6(2), 148–155.

7 Computational Tools and Programs for the Design and Analysis of Static Var Compensators and Filters

7.1 INTRODUCTION

Electric utilities usually conduct some system studies for the purpose of reactive power planning. In the initial stages, several power flow studies are conducted to determine the reactive power (capacitive or inductive) required at different locations in the power system under investigation. Then it is necessary to identify whether static (capacitors or reactors) or dynamic compensation (static var compensators [SVCs] or synchronous compensators) are required. This is usually accomplished by dynamic simulation (i.e., transient stability studies). In cases involving single-phase traction loads, one must ensure that negative-sequence voltages at the points of common coupling (PCC) are within acceptable limits. Three-phase power flow programs are used to estimate these negative-sequence voltages. For filter design studies, special-purpose harmonic analysis programs such as VHARM or SuperHarm are used, whereas for switching transient studies, Electromagnetic Thansients Program (EMTP), Alternative Transients Program (ATP), and Electromagnetic Transients Program for Direct Current (EMTDC) are used. In the locations close to arc furnace loads or rapidly fluctuating loads with adjustable speed drives, flicker due to voltage fluctuations should be within the limits prescribed by the supply authority. At the design stage, some empirical rules are used and, after commissioning, field measurements are made using an IEC flicker meter.

The purpose of this chapter is to briefly review what computational tools are available to do these studies and to understand their relative merits and limitations.

7.2 COMPUTATIONAL TOOLS

The following computational tools are used in general to solve different electrical network problems:

a. Digital computers
b. Analog computers
c. Transient electrical network analyzers
d. Special-purpose simulators such as the high-voltage direct current (HVDC) simulator

As indicated earlier, the types of studies usually conducted are as follows:

a. Power flow studies
b. Dynamic simulation (i.e., transient stability studies)
c. Control system parameter optimization studies
d. Harmonic studies
e. Switching transient studies

We will briefly discuss the nature of the equations that have to be solved to conduct each of these studies, and later, the computation tools commonly used in performing these studies.

- Power flow studies require the solution of nonlinear algebraic equations.
- Dynamic solution requires the solution of nonlinear differential equations.
- Control system parameter optimization studies usually involve a set of linearized differential equations around an operating point and to optimize certain parameters such as voltage regulator or governor gains and some time constants in the voltage regulator.
- Harmonic studies require the solution of generally linear algebraic equations. On occasion, when dealing with transformer saturation, etc., nonlinear characteristics have to be suitably modeled and some iterative methods of solution have to be adopted.
- Switching transient studies require the solution of ordinary differential equations. When it includes long transmission lines, it requires the solution of partial differential equations.

Now we can discuss tools that are more appropriate for the solution of these problems and their relative merits.

Let the mathematical model of the system be represented by the usual state-space model, that is,

$$[X]=[A][X]+[B][U] \tag{7.1a}$$

$$[Y]=[C][X]+[D][U] \tag{7.1b}$$

In Equations 7.1a and 7.1b, the vector $[X]$ comprises the system state variables, that is, a set of system variables (voltages, currents, torques, flux linkages, etc.) in terms of which all other variables may be calculated. The choice of $[X]$ is not unique, but certain combinations of variables may prove computationally more efficient than others. $[U]$ is the vector of independent variables, and $[Y]$ is the vector of variables of interest other than the state variables.

7.3 DIGITAL COMPUTERS

Digital computers are the most versatile and can be used to solve all the earlier-mentioned problems, although in particular cases and depending on the facilities available, other methods can be more advantageous and economical. As very large and fast digital computers are available today, invariably all large problems are solved using digital computers with commercial software packages or locally developed special-purpose computer programs. Considerable effort is required in originally developing the program and initially testing some sample cases to validate the algorithms used.

The first step in solving the problems using either digital or analog computers is to obtain the mathematical model of the system described by linear or nonlinear differential equations, algebraic equations, and other relationships involving limiters (e.g., in voltage regulator or governor blocks, transformer saturation, etc). Afterward, in the digital computers, the problem is solved serially, i.e., the state of the system is computed after each chosen time.

7.4 ANALOG COMPUTERS

In analog computers the problem is solved continuously in a parallel manner. The analog computer consists of electronic integrators, amplifiers, summers, multipliers, and switches, which are interconnected in a manner providing a solution to the set of describing system equations. All the state variables appear as the output of the integrators in the analog computer, that is, voltages to a datum. Then they can be recorded with the help of suitable recorders. Before one can set up a problem on an analog computer, one has to choose amplitude and time scales and manually make the connections on a patch panel. This requires considerable effort. Further, any human errors in digital computers can be traced back easily from the input data; it is somewhat difficult in the case of analog computers. This also somewhat limits the size of the problems that can be solved using analog computers. However, where the problems are of medium size with several parameters to be optimized such as voltage regulator or governor gains, time constants, etc., the analog technique is ideally suited and solutions can be obtained in much less time than by digital simulation. Sometimes digital computers are employed to solve the problems formulated as analog computer blocks.

7.5 SPECIAL PROBLEMS IN THE SIMULATION OF POWER ELECTRONIC CIRCUITS*

In power electronic circuits, switching causes the network topology or structure to change; hence, the coefficient matrices in Equations 7.1a and 7.1b change after each switching instant. There are two philosophically different explanations for switching. The first models the network in topological time invariance and will be referred to as the "constant topology" approach. The second explicitly represents the time-varying structure and will be called the "varying topology" approach.[1]

7.5.1 CONSTANT TOPOLOGY

The constant topology simulation is conceptually simple. Switching devices such as transistors, thyristors, and diodes are modeled by time-varying impedances—high-

* The material in Sections 7.5, 7.5.1, and 7.5.2 is taken from Kassakian, J.G., *Proc. IEEE*, 67(10), 1428–1441, 1979. With permission.

impedance device is off, low-impedance device is on. The result is a matrix [A] whose elements vary with time but that never becomes singular. Only single-circuit analysis is needed to determine [A] for all time, instead of 2^n analyses in the general case. Also, as the current through each switched network branch is always defined, only these branch currents are needed to determine the state of the switching device. In the case of the diode, for instance, the modeling impedance reverts from a high value to a low value when it is sensed that the respective branch current has changed sign from negative to positive, indicating a positive anode–cathode voltage. For a thyristor, one would control the impedance by a logical combination of branch current and gate signal. At each step in the solution, the state of each switching device must be checked. The ability to do this by inspection of only element current rather than by both current and voltage improves the efficiency of the numerical computation.

If the system eigenvalues are known, it is a simple matter to select an appropriate integration step size. Unfortunately, interesting systems are seldom simple. The result is that either numerical means must be employed to determine the eigenvalues or a heuristic approach to their determination must be adopted. In the latter case, the integration algorithm calculates an error associated with the current step and step length. If the error exceeds a certain threshold, the step size is reduced and the calculation repeated. Although it is implicit in the step size, the maximum eigenvalue is never explicitly calculated. If the system exhibits many states with widely differing eigenvalues, this iterative procedure can become computationally expensive. In this case, a constant step size computed on the basis of calculated eigenvalues may be more efficient.

7.5.2 Varying Topology

If the switching devices are actually represented as switches, that is, zero current when "off" and zero voltage when "on," then the network will exhibit a time-varying structure. Each state is then represented by a distinct [A]. If the system is simple enough and the number of possible states small, the representative state of *A* matrices may be calculated manually. The great advantage of explicitly accommodating the time-varying network structure is that spurious eigenvalues are not introduced into the mathematical model. Also, the order of the matrices is a minimum. As the complexity of the numerical techniques for matrix manipulations increases rapidly as the order of the matrix increases, this last point is significant.

In its simplest form, a topologically time-varying algorithm simply selects the appropriate [A] and [B] from the set calculated *a priori* and provides them to the simulation program as input data.

In its most sophisticated form, the time-dependent topology technique uses graph theory to derive the normal form of the state equations (Equation 7.1a) and (Equation 7.1b) for each state of the network.

In the input data, on states of the switches in the electrical network are represented by 1 or Boolean logical variable "True". Off states of the switches in the electrical network are represented by zero or Boolean logical variable "False".

All switching element voltages and currents must be checked at the conclusion of each calculation step to determine when state changes occur. When a state change occurs, the network incidence matrix is changed accordingly. Through the use of existing algorithms and with a series of matrix manipulations involving this incidence matrix, the appropriate set of state equations may be obtained. Although this technique produces an efficient solution of the state equations, there is clearly a considerable expense associated with the derivation in the general case. As the simulated time interval becomes longer and the number of times a state is entered becomes greater, the fraction of the computation cost attributable to the derivation of the state equations becomes less. Although the question of whether to determine the integration step size heuristically or analytically is the same as in the previous discussion of constant topology approaches, the analytical determination of eigenvalues is much simpler in this case because the orders of the *matrices* are smaller.

When computer generation of network equations is desired, a nodal rather than state formulation is sometimes advantageous.

Further, it must be appreciated that in the system being simulated that there should not be large differences (several orders of magnitude) between the different time constants or eigenvalues. This will result in digital computers with very small steps, making the solution uneconomical and inaccurate due to rounding-off errors. In analog computers one will meet with very small and very large coefficients in the scaled equations, making the computer setup very inaccurate.

However, the nodal formulation may not result in a minimal number of equations.

7.6 TRANSIENT ELECTRICAL NETWORK ANALYZERS

Before the arrival of digital computers, sinusoidal steady calculations such as power flows were calculated using AC calculating boards or network analyzers. These were basically scaled-down versions of model power systems. Transient problems were similarly solved using transient network analyzers. It is fair to say they have gone out of fashion and the use of the existing installations has considerably declined.

7.7 SPECIAL-PURPOSE SIMULATORS

The different control schemes of many HVDC and SVC schemes are usually designed and tested with the help of these special-purpose simulators. One of the attractions of these simulators is the actual control systems can be connected to these simulators with suitable changes in signal levels and their performance assessed. These can be more economical than the digital computers; also, some phenomena for which precise mathematical models are not available can be incorporated here with scaled-down versions of the components.

7.8 COMPUTER PROGRAMS

One of the authors found the following programs (or earlier versions or similar programs) to be useful in his work:

PSS TM E Version 30	For power flow, short-circuit, and dynamic simulation studies.
EMTP/ATP/PSCAD/EMTDC	For switching transient studies, insulation coordination studies, surge arrester rating studies.
SuperHarm	For harmonic studies.
DADiSP	For signal processing in connection with harmonic measurements. This was also used to obtain graphical output from the alternative transients program with some local software development.
Powerlink Three-Phase Powerflow Program	To calculate negative phase sequence (NPS) voltages at the points of common coupling in Queensland Railway project.

We will briefly discuss the details of these programs.

7.8.1 PSS TM E Version 30*

One of the authors has used the earlier versions of this program (before 1997 when it was solely being developed by Power Technologies, Inc., Arkansas). Since then, Siemens and PTI have combined and several improvements have been made to the program. This program provides the following features:

- Power flow.
- Short-circuit studies.
- Dynamic simulation (including long-term, detailed governor models, most standard boiler control strategies, very detailed HVDC models).
- Application of FACTS devices, voltage source converter HVDC light systems.
- Microsoft® Foundation Class (MFC), graphical user interface (GUI). The GUI contains command recording capability to aid the user in building macros that can be used to automate repetitive calculations.
- PV analysis: to investigate the relationship between power transfer and voltages.

To **find** the maximum power transfer (or load supply) level before voltage collapse:

- QV analysis: To **investigate** the relationship between reactive power demand and voltage at a bus.

To **find** the maximum amount of reactive power required to avoid voltage collapse:

- Python scripting language: PSS TM E version 30 introduced this language for program automation processing.
- Wind Energy: Siemens have permission to make public the following wind farm models—GE 1.5 MW, Vestas V80, GE 3.6 MW, Vestas V47.

* The material in Section 7.8.1 is taken from the Web site www.usa.siemens.com/energy. With permission.

- Optimal power flow (OPF): to minimize costs of operating the system. It can also be used in the following applications:
  1. Voltage-limited transfer analysis
  2. Reactive power margins
  3. Remedial action under a contingency
  4. Marginal cost analysis
  5. Loss minimization
  6. Congestion analysis
- Accommodates user-written dynamic models.

For further details please see the Web sites http://www.pti.us.com and http://www.siemens.com/power-technologies.

7.8.2 EMTP*

The Electromagnetic Transients Program[2] (EMTP) has been used widely internationally to solve electrical network transient problems and capacitor energization transients. It is also used for insulation coordination studies, surge arrestor ratings determination, and to obtain the frequency response of the network with harmonic current injection at a node (i.e., considering a harmonic-producing load as a current source).

The development of the algorithms of the EMTP described in the following text follows closely that presented by Dommel.[2]

EMTP uses the trapezoidal rule and derives equivalent circuits for lumped inductances and capacitances. These equivalent circuits consist of pure resistances in parallel with current sources whose values depend upon the past history, that is, the values of currents and voltages at the earlier instant of time or time step of integration $t - \Delta t$. With the help of Bergeron's method, equivalent circuits can be derived for lossless lines consisting of pure resistances and current sources, which again depend upon past history. In this case, the terminals are not topologically connected (see Table 7.1); the conditions at the other end are only seen indirectly and with a time delay τ through the equivalent current sources.

7.8.2.1 Inductance

Let the time step of integration $\Delta t = h$.

For the inductance L of a branch k, m (see Table 7.1) we have

$$e_k - e_m = L(d\, i_{km}/d_t) \tag{7.2}$$

$$i_{km} = \frac{1}{L}\int (e_k - e_m)dt \tag{7.3}$$

---

* The material in Section 7.8.2 is taken from Dommel, H.W., *IEEE Trans. PAS*, 88(4), 388–399, 1969. With permission.

Professor Dommel has kindly brought to the author's attention a mistake on p. 391 of his original paper in the algorithm for approximating the resistance of a transmission line with $R/4$ at each end and $R/2$ in the middle. Subsequently, the necessary correction has been reported in Section 4.2.2.5 of the EMTP theory book dealing with single- and multiphase lines with lumped resistances. The interested reader is advised to refer to that section.

which must be integrated from the known state at $t-h$ to the unknown state at t?

$$i_{km}(t)=I_{km}(t-h)+(1/L)\int_{t-h}^{t}(e_k-e_m)dt \tag{7.4}$$

Using the trapezoidal rule of integration yields the branch equation

$$i_{km}(t)=(h/2L)\,[e_k(t)-e_m\,(t)]+I_{km}\,(t-h) \tag{7.5}$$

where the equivalent current source I_{km} is known from past history:

$$I_{km}\,(t-h)=i_{km}\,(t-h)+(h/2L)\,[e_k\,(t-h)-e_m\,(t-h)] \tag{7.6}$$

7.8.2.2 Capacitance

For the capacitance C of a branch k, m, by similar reasoning we can obtain the equivalent circuit in Table 7.1. It may be observed from Table 7.1 that the equivalent circuits for L and C are similar. For L the resistance value in the equivalent circuit is $2L/h$, whereas for C it is $h/2C$. Both have parallel current sources whose values are determined by past history.

7.8.2.3 Lossless Line

Consider a lossless line with inductance L' and capacitance C' per unit length. Then, at a point x along the line voltage and current are related by

$$-\partial e/\partial x=L'(\partial i/\partial t) \tag{7.7}$$

$$-\partial i/\partial x=C'\,(\partial e/\partial t) \tag{7.8}$$

The general solution, first given by d′ Alembert, is

$$i(x,t\,)=f_1(x-vt)+f_2(x+vt) \tag{7.9}$$

$$e(x,t)=Z\bullet f_1(x-vt)-Z\bullet f_2(x+vt) \tag{7.10}$$

with $f_1(x-vt)$ and $f_2(x+vt)$ being arbitrary functions of the variables $(x-vt)$ and $(x+vt)$. The physical interpretation of $f_1(x-vt)$ is a wave traveling at a velocity v in a forward direction and of $f_2\,(x+vt)$, a wave traveling in a backward direction. Z in Equation 7.10 is the surge impedance, and v the phase velocity, given by

$$Z=\sqrt{(L'/C')} \tag{7.11}$$

$$v=1/\sqrt{(L'C')} \tag{7.12}$$

Multiplying Equation 7.9 by Z and adding it to and subtracting it from Equation 7.10 gives

$$e(x,t)+Z\bullet i(x,t)=2Z\bullet f_1\,(x-vt) \tag{7.13}$$

$$e(x,t)-Z\bullet i(x,t\,)=-\,2Z\bullet f_2\,(x+vt) \tag{7.14}$$

Note that in Equation 7.13 the expression $(e+Zi)$ is constant when $(x-vt\,)$ is constant and in Equation 7.14 $(e-Zi)$ is constant when $(x+vt)$ is constant. The expressions

TABLE 7.1
Equivalent Circuits for Network Elements

Network Elements	Equivalent Circuit (Representation in EMTP)
Inductance  k m L	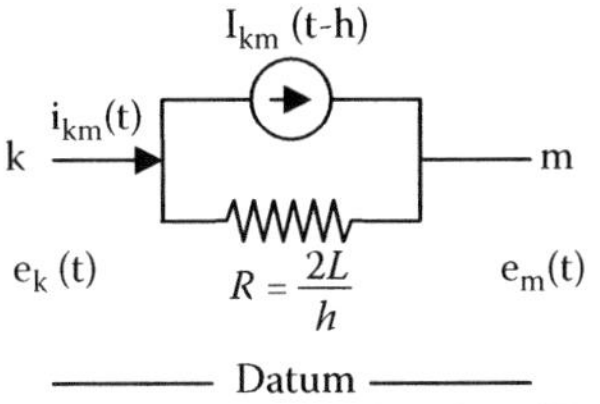 I_{km} (t-h) $i_{km}(t)$ k m e_k (t) $R = \frac{2L}{h}$ $e_m(t)$ Datum
Capacitance 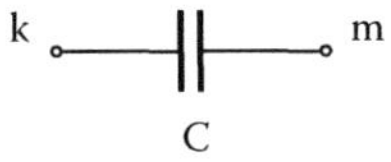 k m C	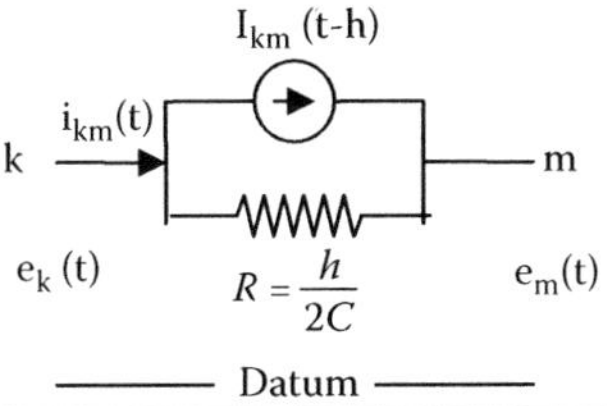 I_{km} (t-h) $i_{km}(t)$ k m e_k (t) $R = \frac{h}{2C}$ $e_m(t)$ Datum
Resistance 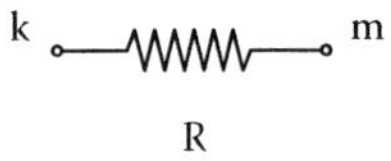 k m R	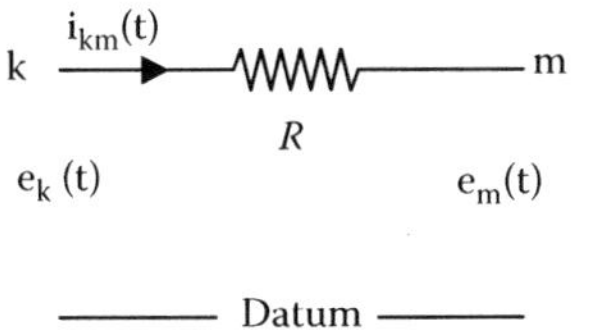 $i_{km}(t)$ k m R e_k (t) $e_m(t)$ Datum
Lossless line 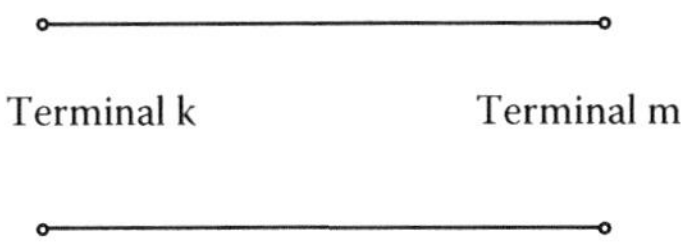 Terminal k Terminal m	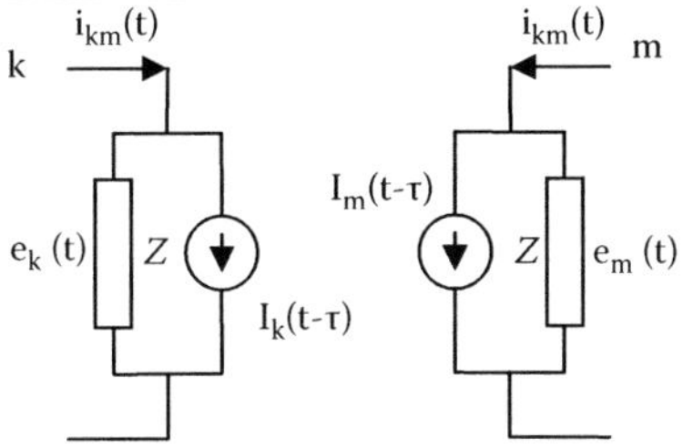 $i_{km}(t)$ $i_{km}(t)$ k m $I_m(t-\tau)$ e_k (t) Z Z e_m (t) $I_k(t-\tau)$

Source: From Dommel, H.W., *IEEE Trans. PAS*, 88(4), 388–399, 1969. With permission. (This table was prepared from the material in the paper.)

$(x - vt)$ = constant and $(x + vt)$ = constant are called the characteristics of the differential equations.

The significance of Equation 7.13 may be visualized in the following way: let a fictitious observer travel along the line in a forward direction at velocity v. Then $(x - vt)$, and consequently, $(e + Zi)$ along the line will be constant for him or her. If the travel time τ to get from one end of the line to the other is

$$\tau = d/v = d\sqrt{(L'C')} \tag{7.15}$$

(where d is the length of the line), then the expression $(e + Zi)$ encountered by the observer on leaving node m at time $t - \tau$ must still be the same on arrival at node k at time t, that is,

$$e_m(t - \tau) + Zi_{mk}(t - \tau) = e_k(t) + Z[-i_{km}(t)] \quad (7.16)$$

(i_{mk} and i_{km} are current directions as in Table 7.1). From this equation follows the simple two-port equation for i_{km}

$$i_{km}(t) = (1/Z)\, e_k(t) + I_k(t - \tau) \quad (7.17)$$

and analogous

$$i_{mk}(t) = (1/Z)\, e_m(t) + I_m(t - \tau) \quad (7.18)$$

with equivalent current sources I_k and I_m which are known at state t from the past history at time $t - \tau$,

$$I_k(t - \tau) = -(1/Z)\, e_m(t - \tau) - i_{mk}(t - \tau) \quad (7.19)$$

$$I_m(t - \tau) = -(1/Z)\, e_k(t - \tau) - i_{km}(t - \tau) \quad (7.20)$$

Table 7.1 shows the corresponding equivalent circuit, which fully describes the lossless line at its terminals.

7.8.2.4 Nodal Equations

With all network elements replaced by the equivalent circuits in Table 7.1, it is very simple to establish the nodal equations for any arbitrary system. The result is a system of linear algebraic equations that describes the state of the system at time t:

$$[Y]\,[e(t)] = [i(t)] - [I] \quad (7.21)$$

with

- $[Y]$ nodal conductance matrix
- $[e(t)]$ column vector of node voltages at time t
- $[i(t)]$ column vector of injected node currents at time t (specified current sources from datum to node)
- $[I]$ known column vector, which is made up of known equivalent current sources I

The EMTP algorithm solves Equation 7.21 in each time step, updating the relevant quantities from the past history.

7.8.2.5 Frequency Scan Facility

It is possible to inject a current (say, 1 A or pu) at a node and obtain the frequency response at different nodes. It is possible to inject a current (say 1A, or 1 pu) and solve the electrical network for all the voltages at different nodes and currents through the different branches. This network solution can be repeated with different frequencies of the injected current. Thus the frequency response (i.e magnitude and phase of the voltage) at each node of the electrical network can be determined. This facility is very convenient for harmonic analysis (by considering a harmonic-producing load as a current source at that node), identification of resonances, filter designs, and so on.

Availability of Other Models

The following are the other models available in EMTP:

a. Transformer models
b. Rotating machine models, synchronous machines, etc.
c. Surge arrester models (zinc oxides, etc.)
d. Nonlinear and time-varying resistances, inductances
e. Switches to simulate circuit breakers, isolators
f. Frequency-dependent transmission line models
g. Line constant calculation facility
h. Transient analog computer simulation (TACS) facility

This facility can be used to simulate HVDC controls, Exciter and Governor models, etc.

Numerical Oscillations[3,4]

The trapezoidal rule of integration for the solution of systems with lumped inductances and capacitances (algorithm used in EMTP) can cause spurious numerical oscillations after switch openings, especially in power systems with converter circuits. Hence, some experience in the use of EMTP solutions is necessary on certain occasions to distinguish between spurious and genuine numerical oscillations in EMTP solutions. To understand why the trapezoidal rule of integration causes these oscillations, let us integrate the terminal equation for inductance,

$$v = L\frac{di}{dt} \tag{7.22}$$

or

$$i(t) = i(t-h) + \frac{1}{L}\int_{t-h}^{t} v(u)du \tag{7.23}$$

where $h = \Delta t$.

The trapezoidal rule of integration yields

$$i(t) = i(t-h) + \frac{h}{2L}[v(t) + v(t-h)] \tag{7.24}$$

If we assume the switch opens at the exact instant at which $i(t-h) = 0$ after the switch opening, $i(t) = 0$. Hence, from Equation 7.24, because $h/2L$ is not equal to zero,

$$v(t) = -v(t-h) \tag{7.25}$$

As this equation holds good for all subsequent time steps as well,

$$v(2) = -v(1), \quad v(3) = -v(2), \quad v(4) = -v(3), \text{ and so on.}$$

TABLE 7.2
Comparison of Backward Euler Method and Trapezoidal Rule

	Backward Euler Method	Trapezoidal Rule
Equivalent resistance	L/h	$2L/h$
Parallel current source	$i(t-h)$	$i(t-h)+(h/2L)\,v(t-h)$

Source: From Dommel, H.W. and Cogo, J.R., *Proceedings of the Tenth Power System Computation Conference*, Graz, pp. 19–24, 1990. (This table was prepared from the material in the paper.)

That is, the voltage oscillates around the correct value $v = 0$ after switch opening. Mathematically, the problem is caused by incorrect initial conditions for integration from time $(t-h)$ to t. State-space formulations can circumvent this problem by using state variables that are always continuous (e.g., flux linkages as state variables for inductances, etc.), but such state-space formulations are less general than nodal formulations.

After a switch opening, reinitialization with the backward Euler method[5] eliminates the spurious numerical oscillations. The backward Euler method converts Equation 7.22 into

$$v(t) = L\frac{[i(t)-i(t-h)]}{h} \tag{7.26}$$

Note that $v(t-h)$ does not appear in this equation, and hence, its initial conditions after switch opening cannot affect computational results. Table 7.2 shows the values of equivalent resistances and parallel current sources that represent a lumped inductance.

The nodal equation formulations known for the trapezoidal rule are, therefore, directly applicable to the backward Euler method.

The backward Euler method is not as accurate as the trapezoidal rule. It is thus best to use it only to restart the solution after switch opening and then revert to the trapezoidal rule of integration. The nodal conductance matrix needed for the restart solution becomes identical with the one needed after reverting to the trapezoidal rule, if the restart solution is integrated over half-step size $h/2$. The principle of this method has been implemented in the Siemens transients program NETOMAC.[5]

The following three other methods for the elimination of spurious numerical oscillations are discussed in Reference 3:

a. The method of averaging.
b. The critical damping adjustment scheme. These two are also related to the backward Euler integration method.
c. Parallel damping resistances have also been used in networks with occasional switch openings and closings. They are less useful for converter circuits in which thyristor valves open and close repeatedly with relatively

short time intervals in between. For reasons of space these methods are not further discussed here and the interested reader can find more details of these methods in Reference 3.

7.8.3 ATP

This computer program also uses the basic algorithm for its network solution similar to EMTP. The computer program is available royalty-free for noncommercial use through national ATP groups.

The ATPDraw program, which provides graphical user input to this program, was initiated in Trondheim, Norway, as a university project for educational purposes by Dr. Hans Kristian Heidalen at the Norwegian University of Science and Technology. Further development was carried out at SINTEF Energy Research financed by Bonneville Power Administration.

It can also solve all the problems that EMTP can solve with more user-friendly graphical input.

7.8.4 PSCAD/EMTDC

This is a transients program developed by the Manitoba HVDC Research Centre, Canada. Now Manitoba HVDC Research Centre and RIKEI Corporation in Japan appear to be jointly promoting this product. This program also uses the basic algorithm for its network solution similar to EMTP. It has sophisticated graphical input and output facilities. While in the older EMTP the input data was provided with the help of fixed format FORTRAN statements; in the EMTDC and ATPdraw, we use graphical input i.e first a single line diagram is produced to reflect the topology of the electrical network. Then, for each element in the electrical network like generator, transformer, transmission line, capacitor etc., data are provided by typing the data in a free format using a drop down menu on the screen.

In the EMTP the output was obtained as tables of different variables and other auxiliary programs like TPPLOT were used to obtain graphical output. In the EMTDC and ATPdraw, PSCAD program facilities were used to obtain graphical output. Further the output results can be obtained as tables also. The input of data is easier than for older EMTP but similar to ATPDraw. Further, we can observe the solution on the screen while it is running, and the parameters of the problem can be changed interactively. HVDC control problems can be set up more easily than by using TACS (Transient Analogs Computers Simulation) facility in EMTP.

For more details please see the following Web site: http://www.rikei.co.jp/.

7.8.5 SuperHarm*

SuperHarm[6] evaluates harmonics on electric power systems. SuperHarm enables one to develop a computer model of a power system to explore variations on system loads

* The material in Section 7.8.5 is taken from the Web site www.pqsoft.com and reprinted here. With the Permission of Electrotek Concepts Inc.

and configurations, along with the resulting impact on system frequency response and distortion levels.

7.8.5.1 SuperHarm Models

SuperHarm includes a wide variety of device and source models as follows:

- Generic harmonic voltage and current source models
- Three-phase equivalent models
- Simple single- and three-phase RL branch models
- Capacitor models
- Long-line corrected PI models
- Mutually coupled, multiphase transmission models
- Induction and synchronous machine models
- Advanced load models

SuperHarm can solve both balanced and unbalanced three-phase systems. This is accomplished by using phase domain nodal admittance matrix techniques rather than sequence component solution methods.

7.8.5.2 SuperHarm Solution Procedure

The solution engine reads the user-created text file that describes the system to be simulated. SuperHarm circuit description language (CDL) consists of the keywords representing device models as well as control commands. The devices are "connected" together by specifying alphanumeric names for the power system buses.

The solution engine reads the CDL file and converts it into a binary form. After checking for errors, the solver calculates the nodal admittance matrix of the system for the first frequency to be solved. The resulting matrix is factored using LDU decomposition, and a voltage solution factor is obtained by forward and backward substitution using the driving current vector. The solution vector is saved to the output file, and the process is repeated for all requested frequencies.

The CDL syntax allows you to develop libraries of device and subsystem models. These models can have calling parameters and internal preprocessing formulas to facilitate the entry of nameplate data. The CDL, coupled with the SuperHarm front end, provides facilities for batch solutions of many cases with varying parameters.

The harmonic-producing load models allow you to enter either typical worst-case or actual measured data for the harmonic current injections. SuperHarm automatically scales these data to match the normal system conditions at the fundamental frequency. SuperHarm utilizes TOP, the output processor, to visualize the simulation result. TOP takes advantage of the Microsoft Windows graphical user interface and the clipboard to allow simple transfer of data to programs such as Microsoft Excel and Microsoft Word. TOP reads the resulting voltage solution vector file and manipulates, visualizes, and prints the results. Outputs available from TOP include the following:

- Waveform and spectrum plots
- Frequency response plots

- Summary tables (including IEEE Standard 519 application)
- Summary bar/column charts
- Cumulative probability plots
- Probability density charts

SuperHarm uses state-of-the-art modeling and solution methods to ensure efficient operation on personal computers. Sparse matrix techniques are utilized to minimize solution times and storage requirements.

7.8.5.3 Support Options

Electrotek concepts provide support for SuperHarm through PQ Soft Support Service, a comprehensive support offering for power system simulation and analytical tools. Features include upgrades, specialized models, and online resources. To learn more, send an e-mail to pqsoft@electrotek.com or call 865-470-9222.

7.8.5.4 Pricing and Availability

Publishing, pricing, and availability information regarding SuperHarm can be obtained by contacting Electrotex Concepts directly.

7.8.6 Three-Phase Power Flow Programs

7.8.6.1 Powerlink Program[7,8]

When single-phase traction loads are supplied by an electric utility, the negative phase sequence (NPS) voltage should not exceed the limit specified by the relevant national standard at the point of common coupling (PCC). As per the Australian Standard AS1359.31, the limit at the rotating machine terminals is 1% continuous and 1.5% for a few minutes if the zero-sequence voltage does not exceed 1%. Hence, at the design stage, three-phase power flow studies have to be conducted to calculate the NPS voltages at the PCCs.

At Powerlink Queensland, a three-phase power flow program was developed on the basis of the method reported by Laughton.[7,8] One of the important steps in the development of a three-phase power flow program is the development of models for wye–delta wye–wye–delta transformers. These models have been adequately documented by Laughton[7,8] and Arrillaga et al.[9] The following point regarding load presentation should be noted while solving a system using a three-phased power flow program and the Gauss–Seidel method. Loads should be represented as constant impedances Z rather than as a constant P, Q load, which is the normal practice in power flow studies with a positive-sequence network. A system can have multiple power flow solutions with several of them having unacceptable operating conditions (e.g., with very low bus voltages). When constant P, Q loads and the Gauss–Seidel method are used, convergence of the three-phase power flow study is difficult. Even in those cases when convergence is obtained, the solution will yield unrealistic bus voltages and NPS voltages. Hence, constant Z representation must be used to obtain more realistic bus voltages.

Another important difference is in the modeling of transformer taps in three-phase power flow programs and the positive-sequence power flow programs. In the three-phase transformer models, the magnitude of one line voltage controls the taps on the other two line voltages also; that is, taps move uniformly on all the three phases. In the positive-sequence power flow, one has to consider only one primary (say, HV) and secondary voltage (say, LV), and the taps control only one of these voltages.

7.8.6.2 HARMFLO Program

Xia and Heydt[10] and Grady[11] developed the HARMFLO (three-phase harmonic power flow) program with the support of EPRI at Purdue University and based it on Newton–Raphson power flow techniques. The program was the first to adjust the harmonic current output of the load for the harmonic voltage distortion. Heydt[12] points out in his book that the harmonic power flow study is not a replacement for conventional power flow study, because it does not usually contain features for tap changing under load transformers (TCULs), tap limits, reactive power limits at PV buses, and phase shifter applications.

Further harmonic power flow programs are usually not competitive with time domain software (EMTP), because the capabilities of the software are different; that is, EMTP gives a full-time domain solution and HARMFLO can only give voltages for the periodic steady state. EMTP requires more detailed system data than HARMFLO.

7.9 DADISP[a]

DADiSP (Data Analysis and Display) is a graphical software tool for displaying and analyzing data from virtually any source. The DADiSP Worksheet applies key concepts of spreadsheets to the often-complex task of displaying and analyzing entire data series, matrices, data tables, and graphic images.

Because it is often easier to interpret data visually rather than as a column or table of numeric values, DADiSP's default presentation mode is usually a graph within a Worksheet Window. A Worksheet Window in DADiSP serves two purposes at once: it provides a place to store data and serves as a tool for viewing data.

DADiSP offers over 1000 analysis and display functions within an easy-to-use, menu-driven environment.

With DADiSP one can

- Acquire and import data using DADiSP's easy-to-follow pop-up menus;
- Reduce and edit your data and save the intermediate results;
- Display your data through a range of interpretive 2-D, 3-D, and 4-D graphical views;
- Transform your data using a variety of matrix and other mathematical operations;
- Analyze your data with statistical techniques;

[a] This material in section 7.9 is taken from the web site www.dadisp.com. With permission from DSP Development Corporation.

- Analyze digital signals using DADiSP's extensive signal processing tools;
- View, store, edit, and analyze digital images;
- Produce quality output using a variety of annotation features.

Furthermore, DADiSP is completely customizable, allowing you to create your own menus, macros, functions, and command files that meet your data analysis and display needs.

DADiSP can be ordered by contacting the following people in the United States:
DSP Development Corporation
3 Bridge Street
Newton, MA 02458
Toll Free: 800-424-3131
Phone: 617-969-0185
Fax: 617-969-0446
E-mail: info@dadisp.com
Web: www.dadisp.com
In the United Kingdom, you can contact the following:
Adept Scientific
Amor Way
Letchworth Garden City
Herts, SG6 1ZA l
Tel: +44 (0)1462 480055

***Note*:** The aforementioned programs were discussed to give the reader some idea of the computer programs that are useful in the design and analysis of SVCs and filters. The list is not meant to be comprehensive, and hence, there is no implication that these are the only programs available for this purpose. In fact, most electric utilities, large equipment manufacturers such as ABB, Siemens, GE, Westinghouse, and consulting engineers will have several programs to perform similar tasks. They will be based on their particular requirements and to be compatible with their internal databases of their systems and equipment.

7.10 CONCLUSIONS

In this chapter, we have discussed the different computational tools and computer programs available for the design and analysis of SVCs and filters. Digital and analog computers and their relative merits were briefly discussed. All the problems, particularly large and complex ones, are usually solved using digital computers in view of the large number of digital computer installations available. For medium-size problems, personal computers and analog computers are used. Analog computers are useful where several parametric studies have to be completed because of their speed and interactive capabilities.

We pointed out the special problems involved in analyzing power electronic circuits. In the constant topology approach, very large spurious eigenvalues are introduced in certain states, leading to numerical instability or uneconomically short steps. In the varying-topology approach, derivation of the state equations in the general case involves considerable expense. We also mentioned transient network analyzer and special-purpose simulators used in HVDC and SVC control system and design studies.

We also briefly discussed the following digital computer programs and some of the applications where they can be used:

PSS TM E version 30, EMTP, ATP, PSCAD/EMTDC, SuperHarm, Powerlink three-phase power flow program, HARMFLO, and DADiSP.

REFERENCES

1. Kassakian, J.G., 1979, Simulating power electronic systems: a new approach, *Proc. IEEE*, 67(10), 1428–1441.
2. Dommel, H.W., 1969, Digital computer solution of electromagnetic transients in single- and multiphase networks, *IEEE Trans. PAS*, 88(4), 388–399.
3. Dommel, H.W. and Cogo, J.R., 1990, Simulation of transients in power systems with converters and static compensators, *Proceedings of the Tenth Power System Computation Conference*, Graz, pp. 19–24.
4. Campos Barros, J.D. and Rangel, R.D., 1985, Computer simulation of modern power systems: the elimination of numerical oscillations caused by valve actions, *Fourth International Conference on AC and DC Transmission*, September 23rd to 26th, London, England, IEEE publication 255, 254–259.
5. Kulicke, B., 1981, Simulations program NETOMAC: difference conduction method for continuous and discontinuous systems, *Siemens Research and Development Reports,* 10(5), 299–302.
6. PQ Soft Web site, http://www.pqsoft.com/index.htm
7. Laughton, M.A., 1968, Analysis of unbalanced polyphase networks by the method of phase coordinates, Pt. 1: System representation in phase frame of reference, *Proc. IEE*, 115(8), 1163–1172.
8. Laughton, M.A., 1969, Analysis of unbalanced polyphase networks by the method of phase coordinates, Pt. 2: Fault analysis, *Proc. IEE*, 116(5), 857–865.
9. Arrillaga, J., Bradley, T., and Bodger, P., 1985, *Power System Harmonics*, John Wiley & Sons, Chichester, U.K.
10. Xia, D. and Heydt, G.T., 1982, Harmonic power flow studies, Pt. 1: Formulation and solution, *IEEE Trans. PAS*, June, 1257–1265.
11. Grady, W.M., 1983, Harmonic power flow studies, Ph.D. thesis, Purdue University.
12. Heydt, G.T., 1991, *Electric power quality*, Stars in a Circle Publications, West Lafayette, IN.

8 Monitoring Power Quality

8.1 INTRODUCTION

With the deregulation of the electricity supply industry, increased attention is being focused on power supply reliability and power quality.[1,2] Although utilities spend great effort to prevent power interruptions, there is no way to completely control disturbances on the supply system. Many disturbances are due to normal operations such as switching loads and capacitors or faults and the opening of circuit breakers to clear faults. Faults are usually caused by events outside the utility's control. These events include acts of nature such as lightning, birds flying close to power lines and getting electrocuted, and accidental acts such as trees or equipment contacting power lines.

Not all disturbances are caused by the utility. Disturbances generated by customer-owned equipment and plant operations are also beyond the utility's control. In industrial and commercial facilities, disturbances may be caused by the operation of arc welders and the switching of power factor capacitors and inductive loads such as motors, transformers, and lighting ballast solenoids. Moreover, fluorescent lamps, and other devices that use power electronics such as switch-mode power supplies, television sets, light dimmers, and adjustable-speed drives can also inject harmonics into the power system.

Utilities and customers alike are concerned about reliable power, whether the focus is on interruptions and disturbances or extended outages. One of the most critical elements in ensuring reliability is monitoring power system performance. Monitoring can provide information about power flow and demand and help identify the cause of power system disturbances. It can even help identify problem conditions on a power system before they cause interruptions or disturbances.

Effective monitoring programs are important for power reliability assurance for both utilities and customers, because most customer power quality problems originate within the customer facility. Monitoring power quality ensures optimal power system performance and effective energy management.

The key to an effective monitoring program is flexibility, powerful data processing, and easy access to information. The emergence of the Internet and intracompany intranets has made this possible, allowing quick viewing of data and assuring effective decision making and fast response time. Event notification and program scalability are also critical for addressing the ever-changing environment of the energy business.

Before embarking on any power quality program one should clearly define the monitoring objectives, as they determine the choice of monitoring equipment. The monitoring objectives can be as follows:

- Characterizing system performance
- Characterizing specific problems

- Enhancing power quality service
- Predictive or just-in-time maintenance

8.2 SITE SURVEYS

We have already listed and categorized the different types of disturbances that occur in a power system in Table 1.1 in Chapter 1. To monitor these disturbances we will have to measure voltage, or current, or both. Depending on the phenomena being investigated, different instruments are necessary.[3]

Some power quality problems can be easily solved without extensive and expensive monitoring just by a simple site survey and talking to the customer and asking him or her the right questions about their grounding practices, possible sources of disturbance such as capacitor switching, motor starting, etc.

The majority of power quality problems reported by customers are caused by problems with wiring or grounding within the facility. Wiring and grounding testers are very useful to detect isolated ground shorts, open grounds, open neutral, etc.

Overloading of circuits, under- and overvoltage problems, and unbalances between circuits can be detected using multimeters. However, one has to be careful when using a multimeter; the method of calculation used in the meter has to be understood. Reference 3 describes the details of the following different methods:

1. Peak method
2. Averaging method
3. True root-mean-square (rms) method

Reference 3 shows the comparison of meter readings for various waveforms when different methods are used (see Table 8.1 in Reference 3). Hence, in interpreting the measurements obtained by a multimeter, considerable attention must be paid to the method of calculation used in the meter, for example, PC current waveform that has two peaked pulses when measuring by different meters. Using true rms method, peak method, and average responding method yield results 100, 184, and 60%, respectively. Similarly, adjustable-speed drives (ASD) waveform current yields 100, 127, and 86%, respectively. All the measurements are normalized taking the reading from the true rms method as 100%.

Oscilloscopes[3,5] are useful when performing real-life tests. They have sampling rates far higher than transient-disturbance analyzers. Digital storage oscilloscopes can usually be obtained with a communications cable so that waveform data can be uploaded to a PC for additional analysis with a software package. Using multichannel digital storage oscilloscopes, more than one electrical parameter may be viewed and stored. Passive voltage probes can be used depending on their rating in circuits of 300–1000 V ac. Active voltage probes use field-effect transistors and have high-input impedance to measure low-level signals. The high-frequency current probe can be used to measure stray noise and ground loop currents in the ground grid.

Data loggers and chart recorders are used sometimes to record voltage, current, demand, and temperature data in electrical power systems.

Harmonic analyzers[5] are simple instruments for measuring and recording harmonic distortion data. They comprise a meter with a waveform display screen, voltage leads, and current probes. Some of the analyzers are hand-held devices, and others are intended for tabletop use. Some instruments provide a snapshot of the waveform and harmonic distortion at a particular instant, whereas others are capable of recording snapshots as well as a continuous record of harmonic distortion over time. The nature of a particular power quality problem will determine which instrument one should use.

When measuring currents one should make sure that the probe has sufficient current rating and that it is suitable to monitor high frequencies. One can obtain current probes that are useful between the frequencies 5 Hz and 10 kHz for a maximum current rating of 500 A rms. At higher frequencies, currents and distortions are considerably lower than at lower frequencies. Hence, some loss of accuracy might not be all that important. Typically, a 5.0% loss in accuracy can be expected if the waveform contains significant levels of higher-order harmonics.

8.2.1 Spectrum Analyzers[4,5]

As spectrum analyzers are designed for a broader market, they are reasonably priced. Because they are not designed specifically for sampling power frequency waveforms, care must be taken when using them, to ensure accurate harmonic analysis.

8.2.2 Special-Purpose Power System Harmonic Analyzers[4,5]

These are based on the FFT with sampling rates specifically designed for determining harmonic components in power signals. They can generally be left in the field with include communications capability for remote monitoring.

8.2.3 Transient-Disturbance Analyzers[4,5]

These instruments are advanced data acquisition devices for capturing, storing, and presenting short-duration, subcycle power system disturbances. The sampling rates for these instruments are high. Some of the transient-disturbance recorders have sampling rates in the range of 2 million to 4 million samples per second. Higher sampling rates provide greater accuracy in describing transient events in terms of their amplitude and frequency content. The amplitude of the waveform provides information about the potential for damage to the affected equipment. The frequency content enables one to identify those events which generate high frequency components. Electromagnetic interference problems are usually caused by the coupling of these high frequency components with the neighbouring sensitive circuits. Once these events which generate these high frequency components are identified, then suitable mitigation techniques can be adopted to avoid this interference problem. Typically, lead lengths of 6 ft or less should not introduce significant errors in the measurement of fast transients.

8.2.4 Combination Disturbance and Harmonic Analyzers[4]

Some instruments combine harmonic sampling and energy-monitoring functions with complete disturbance-monitoring functions as well. The output is graphical, and the data are remotely gathered over phone lines into a central database. Statistical analysis can then be performed on the data. The data are also available for input and manipulation into other programs such as spreadsheets and other graphical output processors.

8.2.5 Flicker Meters

These instruments are used to measure flicker, especially that induced by arc furnace loads in the power system. These will be discussed in the section dealing with flicker measurement.

Reference 4 (see Table 8.2 in Reference 4) gives a summary of monitoring requirements for different types of power quality variations.

8.3 TRANSDUCERS

Monitoring of power quality on power systems often requires transducers to obtain acceptable voltage and current signals. In many situations, particularly at HV and EHV levels, direct connections are not possible.

In many practical situations that measurements to the 25th harmonics are sufficient to indicate the makeup of the waveform. Also their data are adequate to calculate THD, individual maximum voltage, and current harmonics. These data help to identify the local harmonic problems. However, there are other times when harmonic measurements must be made up to the 50th order, for example, if there is an aluminum smelter load with 48-pulse operation having resonance at the 49th harmonic. Hence, the current or voltage transformers used in such locations should have an acceptable frequency response, that is, up to the 50th harmonic (i.e., 3.0 kHz in 60-Hz power systems or 2.5 kHz in 50-Hz power systems). Hence, the accurate measurement of frequency response is important. One of the methods used at Georgia Power Company will be discussed in the next section.

According to the IEC 61000-4-7-2002 standard, accuracy requirements for current, voltage, and power measurements are furnished in Table 8.1.

Two classes of accuracy are suggested for instrumentation measuring harmonic components. Class I instruments are recommended for precise measurements, and Class II instruments for general surveys. The maximum allowable errors given in Table 8.1 refer to single-frequency and steady-state signals, in the operating frequency range, applied to the instrument under related operating conditions to be indicated by the manufacturer (temperature range, humidity range, instrument supply voltage, etc.)

Note: When testing appliances according to IEC 61000-3-2, the uncertainty terms are related to the permissible limits (5% of the permissible limits) or to the rated current (*Ir*) of the tested appliance (0, 15% *Ir*), whichever is greater. This should be considered when choosing the proper input current range of the measuring instrument.*

* The above two paragraphs are from the IEC Standard 61000-4-7-2002. With permission.

TABLE 8.1
Accuracy Requirements for Current, Voltage, and Power Measurements

Class	Measurement	Conditions	Maximum Error
I	Voltage	$U_m \geq 1\%\ U_{nom}$	$\pm 5\%\ U_m$
		$U_m < 1\%\ U_{nom}$	$\pm 0.05\%\ U_{nom}$
	Current	$I_m \geq 3\%\ I_{nom}$	$\pm 5\%\ I_m$
		$I_m < 3\%\ I_{nom}$	$\pm 0.15\%\ I_{nom}$
	Power	$P_m \geq 150$ w	$\pm 1\%\ P_{nom}$
		$P_m < 150$ w	± 1.5 w
II	Voltage	$U_m \geq 3\%\ U_{nom}$	$\pm 5\%\ U_m$
		$U_m < 3\%\ U_{nom}$	$\pm 0.15\%\ U_{nom}$
	Current	$I_m \geq 10\%\ I_{nom}$	$\pm 5\%\ I_m$
		$I_m < 10\%\ I_{nom}$	$\pm 0.5\%\ I_{nom}$

I_{nom}: Nominal current range of the measurement instrument.
U_{nom}: Nominal voltage range of the measurement instrument.
U_m *and* I_m: Measured values.

Source: IEC Standard 61000.4.7-2002: Electromagnetic compatibility (EMC), Part 4: Testing and measurement techniques — Section 7: General guide on harmonics and interharmonics measurements and instrumentation, for power supply systems and equipment connected thereto. (Reprinted with permission IEC)

IEEE 519-1992 has similar requirement. Please refer to Section 9.3.1 for accuracy requirements for instrument response.

8.3.1 Measurement of the Frequency Response of Instrument Transformers

References 7 to 13 discuss the results of some of the tests on instrument transformers. All magnetic transformers report responses that exhibit significant variations in transformation ratio with frequency.

In particular, References 12 and 13 discuss the problems associated with the accurate determination of the ratio and phase angle between output and input of the instrument transformers.

Hence, in this section we discuss the technique for laboratory characterization of instrument transformers designed for transmission-level voltage and current measurements developed at Georgia Institute of Technology.[6]

The test technique consists of applying a signal to the input of the transformer. Input and output waveforms are recorded. Then, the transfer function is computed as the ratio of the output waveform Fourier transform over the input waveform Fourier transform. However, this conceptually simple procedure presents a number of practical problems when applied to transmission-level instrument transformers: (a) high voltages and currents must be applied, (b) the input signal must contain energy spread over the frequency range of interest in order to accurately obtain the instrument transformer transfer function, and (c) noise generated by high-voltage sources must be dealt with. These problems were solved by appropriately selecting the input waveforms within feasibility constraints and transferring the captured input and output waveforms to a personal computer for analysis. Analysis of the captured data permitted derivation of the instrument transformer transfer function. In addition, the measured data were utilized to quantify the following performance characteristics:

- Instrument transformer transfer function accuracy using coherence analysis
- Sensitivity of the transfer function with respect to impedance
- Instrument transformer linearity using coherence analysis

For accurate computation of the instrument transformer transfer function, all sources of measurement error must be minimized. Potential sources of significant error are as follows:

- Signal truncation due to the memory limitations of the digitizers
- Signal aliasing due to sampling of the input and output waveforms
- Discretization error due to the finite length (8 to 12 bits) of analog to digital converters
- Noise superimposed on signals by electromagnetic coupling

Signal truncation was completely avoided by using impulse waveforms with fast extinction of input signals. Thus, both input and output waveforms were captured from beginning to end. Signal aliasing was also kept to a minimum by using very fast digitizers (sampling period ranged from 0.0004 to 0.05 ms). The spectrum of a finite duration signal is theoretically not band-limited. However, analysis of the used input waveforms indicated that the spectrum beyond the Nyquist rate of the digitizers is insignificant, when compared to the spectrum in the frequency range of interest. For example, consider the waveform shown in Figure 8.4a. Assuming a double exponential model with a rise time of 0.03 ms and a decay time of 3 ms, the spectrum was computed analytically. The value of the spectrum in the frequency range of interest (0–1500 Hz) was found to be at least 1000 times greater than at the minimum Nyquist rate (100 kHz). This result becomes apparent if one considers that the input pulse of Figure 8.4a is generated by a laboratory switching impulse generator (shown in Figure 8.2) whose frequencies are limited by the electrical RC time constants of the machine. Finally, noise due to sample discretization and external interference was attenuated using the cross-spectrum averaging method. This aspect of the technique is described next.

Let $I_{in}(t)$ and $I_{out}(t)$ represent the exact input and output waveforms of a current transformer under test. Then the transfer function of this device (including the effect of the burden) is

$$H(\omega) = I_{out}(\omega)/I_{in}(\omega) \tag{8.1}$$

where $I_{in}(\omega)$ and $I_{out}(\omega)$ are the Fourier transforms of the input and output waveforms, respectively. In practice, the measured signals (In $\bar{I}_{in}(t)$ and $\bar{I}_{out}(t)$ the symbol — at top denotes measured data) contain added noise so that

$$\bar{I}_{in}(t) = I_{in}(t) + n_i(t) \tag{8.2}$$

$$\bar{I}_{out}(t) = I_{out}(t) + n_o(t) \tag{8.3}$$

where n_i and n_o are uncorrelated noise signals added to the exact input and output waveform functions.

By performing N series of measurements of $\bar{I}_{in}(t)$ and $\bar{I}_{out}(t)$, a good approximation of the instrument transformer transfer function can be obtained as follows:

$$\bar{H}(\omega) = \frac{\sum_{k=1}^{N} \bar{I}_{out}(\omega)\bar{I}_{in}^{*}(\omega)}{\sum_{k=1}^{N} \bar{I}_{in}(\omega)\bar{I}_{in}^{*}(\omega)} \tag{8.4}$$

where the summation index k indicates the measurement number, and * denotes complex conjugation.

It can be easily shown that the foregoing estimate $\bar{H}(\omega)$ approaches the actual response $H(\omega)$ as the number of measurements (N) grows. In practice, it has been found that a total of five sets of measurements ($N = 5$) suffices.

Note: The alternative way to calculate $\bar{H}(\omega)$ is to use the equation

$$|\bar{H}(\omega)|^2 = \frac{\sum_{k=1}^{N} I_{out}(\omega)\, I_{out}^{*}(\omega)}{\sum_{k=1}^{N} I_{in}(\omega) I_{in.}^{*}(\omega)}$$

$$= \frac{\text{Autocorrelation function of the output}}{\text{Autocorrelation function of the input}}$$

This method yields no phase information and, further, is less accurate because of noise in the output. If we rms-average the autocorrelation function of the output, the square of the noise terms exists. If we rms-average the cross-correlation function, the noise in the output and conjugate of the input signal are uncorrelated. Hence, all the noise terms average to zero. For more details, please see References 14 and 15.

An added benefit of multiple measurements is that the validity of the instrument transformer transfer function (as computed using the foregoing procedure) can be checked by computing the coherence function associated with the measured input and output signals to the instrument transformer under test. The coherence function is defined as follows:

$$Coh(\omega) = \frac{\left|\sum_{k=1}^{N} \bar{I}_{out}(\omega)\bar{I}_{in}^{*}(\omega)\right|}{\sqrt{\left(\sum_{k=1}^{N} |\bar{I}_{in}(\omega)|^2 \left|\sum_{k=1}^{N} \bar{I}_{out}(\omega)\right|^2\right)}} \tag{8.5}$$

where the summation index k indicates the measurement number, and * denotes complex conjugation. By definition, the value of the coherence function ranges between 0 and 1. If no noise were contained in the measurements, the coherence function would be equal to 1 for any frequency. Conversely, a low coherence value at a given frequency indicates that those measurements have been corrupted by excessive noise, and therefore, the computed transfer function at that frequency may contain a large error. Low coherence may also indicate that the tested device is nonlinear.

In summary, the test technique provides the transfer function of the instrument transformer, and in addition, quantifies the accuracy of the transfer function.

8.3.2 Description of the Instrument Transformers' Tests

The technique for the laboratory measurement of the instrument transformer frequency response is based on injecting a signal into the input terminals of the instrument transformer under test. The input and output waveforms are recorded.[6]

Then, the frequency response is computed using averaging Fourier transform techniques.

Many types of inputs were considered initially: step functions, impulse functions, swept sine wave, and white noise. The final selection was based on the following criteria:

- Feasibility of producing the waveform.
- Sufficient energy distributed over the frequency range of interest.
- If the signal is a pulse, it must have minimal overshoot and high damping.
- The amplitude level of the input waveform must be near the nominal rating of the tested instrument transformer.

As a best compromise, an impulse-type double-exponential input signal was selected. Various high-voltage and high-current pulse circuits with adjustable waveform rise and fall times were developed to achieve the earlier-mentioned criteria. Input and output signals were recorded using laboratory-grade voltage and current probes connected to digitizers in compliance with ANSI/IEEE Std 4-1978. The digitizer parameters were selected so as to avoid aliasing, to provide sufficient dynamic range, and to sample the entire waveform.

Measurements were performed with various loads connected to the instrument transformer outputs. Several resistive, capacitive, and inductive loads were used. The impedance of the loads was selected to be in the range of typical loads in actual high-voltage substation installations.

For reasons of space the details of instrument transformers and event recorders tested are not included here. They can be found in Reference 6. Only a summary of the conclusions from these tests are reported here.

Figures 8.1 and 8.2 illustrate the laboratory setup for current and voltage instrument transformers characterization, respectively. The major parameters of the

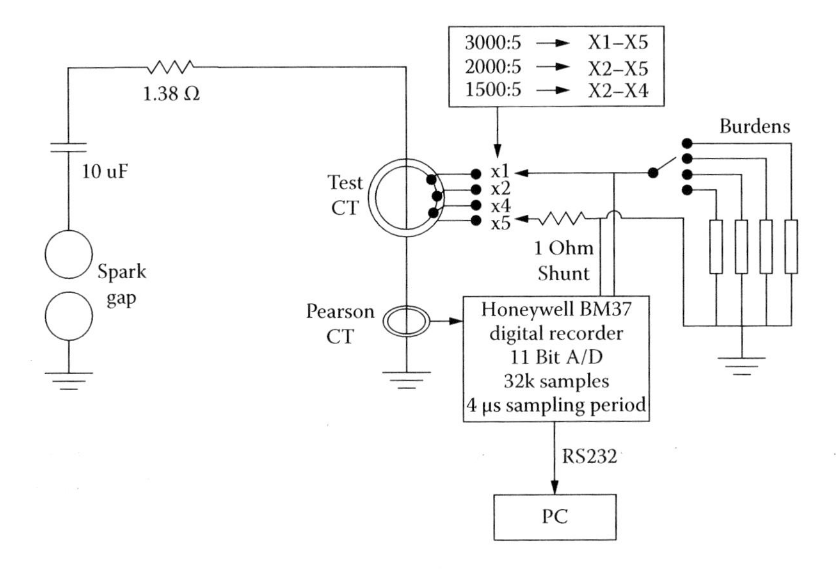

FIGURE 8.1 Typical laboratory setup for current transformer characterization. (From Meliopoulos, A.P.S., Zhang, F., Cokkinides, G.J., Coffeen, L., Burnett, R., McBride, J., Zelingher, S., and Stillman, G., *IEEE Trans. Power Delivery*, 8(3), 1507–1517, 1993. With permission.)

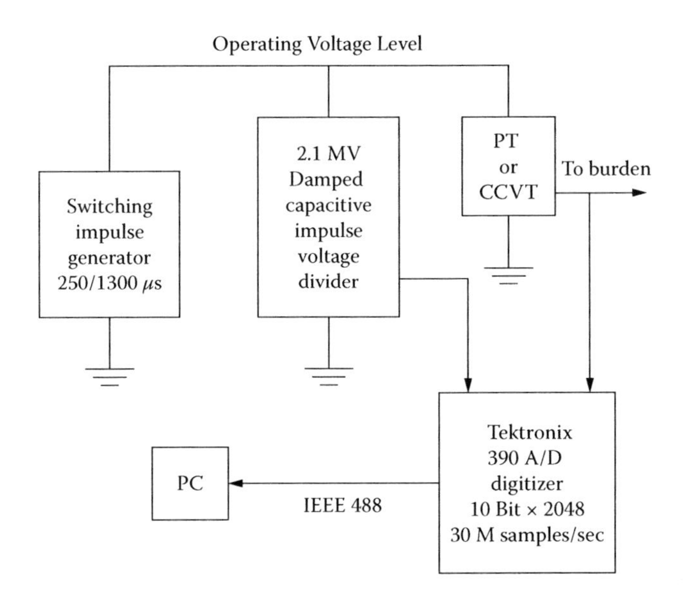

FIGURE 8.2 Typical laboratory setup for PT or CCVT characterization. (From Meliopoulos, A.P.S., Zhang, F., Cokkinides, G.J., Coffeen, L., Burnett, R., McBride, J., Zelingher, S., and Stillman, G., *IEEE Trans. Power Delivery*, 8(3), 1507–1517, 1993. With permission.)

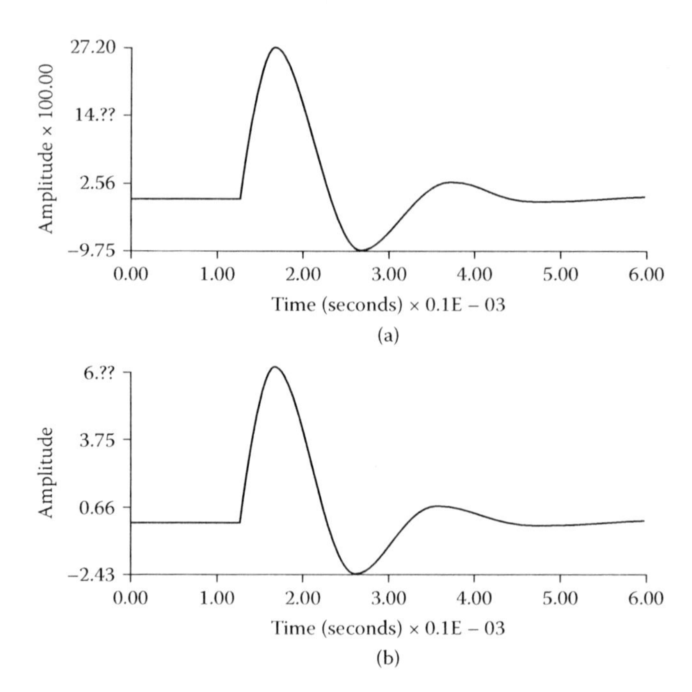

FIGURE 8.3 A sample of input and output waveforms measured during the Nissan Electric CT tests (1-MHz sampling frequency): (a) input waveform; (b) output waveform. (From Meliopoulos, A.P.S., Zhang, F., Cokkinides, G.J., Coffeen, L., Burnett, R., McBride, J., Zelingher, S., and Stillman, G., *IEEE Trans. Power Delivery*, 8(3), 1507–1517, 1993. With permission.)

impulse generators are listed in the figures. Typical input and output waveforms for the setup of Figures 8.1 and 8.2 are illustrated in Figures 8.3 and 8.4, respectively.

8.3.3 Summary of the Conclusions from the Tests

The accuracy of current transformers is sufficient for harmonic measurements of the first 25 harmonics, that is, up to 1.5 kHz.[6]

The tests performed on potential transformers indicate that wound-type PTs are usable for harmonic measurements, provided that transfer function correction algorithms are used. Specifically, the response of the tested potential transformer varies considerably over the frequency range of interest. Sensitivity to burden impedance is small over most of the frequency range except for the regions near the transfer function resonance peaks.

The tests performed on CCVTs (capacitive coupling voltage transformers) indicate that these devices are not suitable for harmonic measurements.[9]

Transient event recorders (TERs) were characterized in the laboratory under steady-state and transient conditions. The main conclusions are given in the following text.

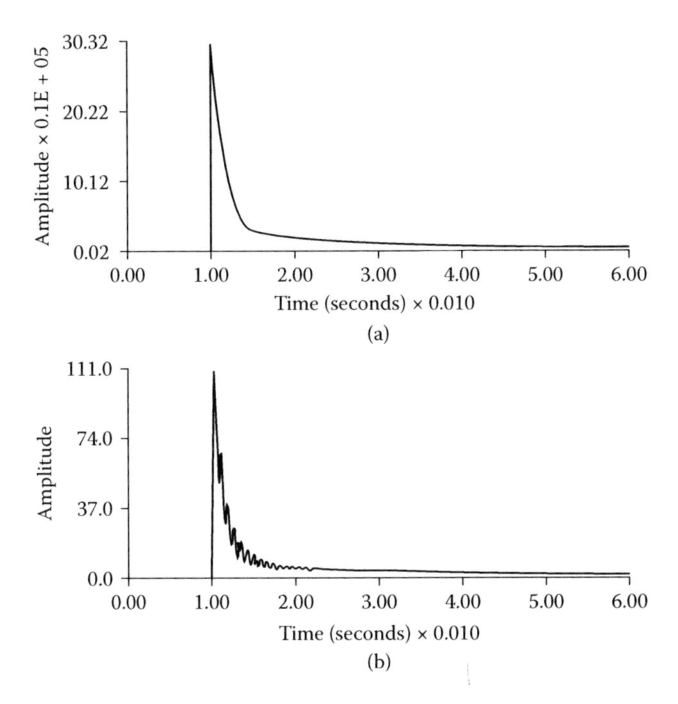

FIGURE 8.4 A sample of input and output waveforms measured during the ABB, MSV 1300, 200 kV/115 V PT tests (PT #1; 20-kHz sampling frequency): (a) input waveform; (b) output waveform. (From Meliopoulos, A.P.S., Zhang, F., Cokkinides, G.J., Coffeen, L., Burnett, R., McBride, J., Zelingher, S., and Stillman, G., *IEEE Trans. Power Delivery*, 8(3), 1507–1517, 1993. With permission.)

The transfer function magnitude of the tested TERs is nearly flat over the frequency range 0–8 kHz with a linear phase shift. Time delays caused by the input isolator board, input filter, and sampling operation are on the order of a few microseconds to tens of microseconds, with substantial uncertainty. This fact impacts phase measurements.

TERs can be used for harmonic measurements provided that means for in-service measurement of individual channel transfer function and dynamic range enhancement are available.

Standard metering-class CTs are generally adequate for frequencies up to 2 kHz (phase error may start to become significant before this). For higher frequencies, window-type CTs with a high turns ratio (doughnut, split core, bar type, and clamp-on) should be used.[16]

1. Additional desirable attributes for CTs include[4,16] large turns ratio, e.g., 2000:5 or greater.
2. Window-type CTs preferred; small remnant flux, e.g., ≤10% of the core saturation value.

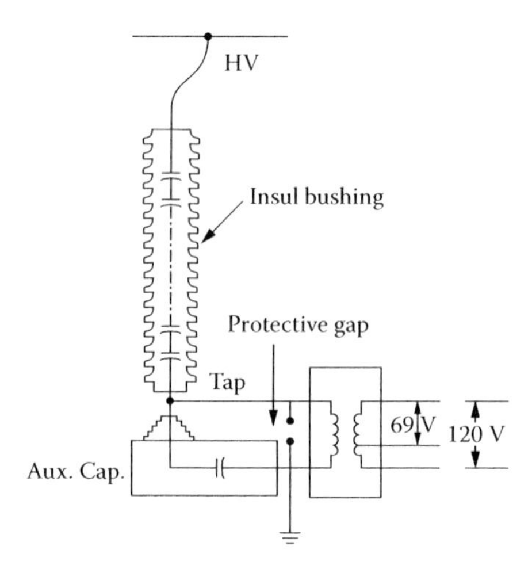

FIGURE 8.5 Capacitive voltage divider. (From IEEE Standard 519-1992, Figure 9.7. With permission.)

3. Large core area, to obtain better frequency response of the CT.
4. Secondary winding resistance and leakage impedance as small as possible.
5. Small burden.

8.3.3.1 Voltage Transformers

Magnetic voltage transformers, generally available, are designed to operate at the fundamental frequency.[17,18] In these transformers, resonance between winding inductance and capacitance can cause large ratio and phase errors. With a high-impedance load, the response is usually adequate to at least 5 kHz and the accuracy is within 3%.

In Reference 12, some frequency response tests on voltage transformers are reported. From these tests, it is concluded that the transformer of a CVT (capacitive voltage transformer) unit should be disconnected and the CVT should be used as a purely capacitive divider. Otherwise, typically, the lowest frequency response peaks appear at frequencies less than 200 Hz.

Figure 8.5 shows a capacitive voltage divider built using an HV bushing and its dielectric loss angle (DLA) tap and another bottom end capacitor. High-input-impedance instrumentation amplifiers must be included in such measurements. The leads from the low-voltage capacitors to the input amplifier should be as short as possible to reduce the errors in phase angle measurements.

The results of transfer ratio measurements on about 40 VTs (voltage range from 6 kV to 40 kV) are shown in Figure 8.6. In view of the small number of CTs tested so far, similar graphs could not be reproduced for CTs.

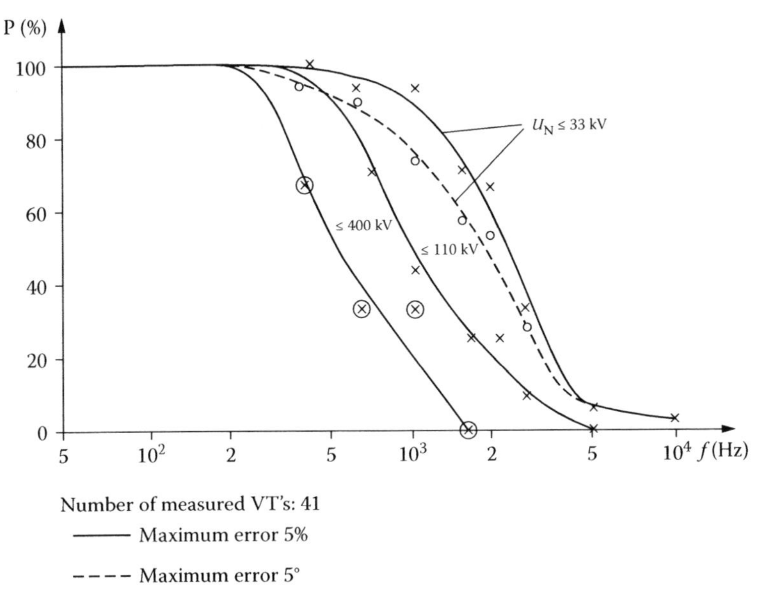

FIGURE 8.6 Percentage p of inductive voltage transformers, the transfer ratio of which has a maximum deviation (from the nominal value) of less than 5% or 5* up to frequency f. (From IEC Standard 61000.4.7-1991. With permission.)

8.4 IEC-RECOMMENDED MEASUREMENT TECHNIQUES FOR HARMONICS

8.4.1 Harmonics

IEC Standard 61000.4.7-2002 deals with harmonics and interharmonics measurements and instrumentation.[17,18]

A joint IEEE/CIGRE/CIRED working group study report on interharmonics[21] identified the main problem associated with measurement of interharmonics so that a waveform consisting of two or more nonharmonically related frequencies may not be periodic.

The type of harmonic measurements and the instruments to be used will depend on the purpose of the assessment, for example,

- Emission measurements
- Compliance with standards
- Resolving of disputes

The standard recommends two classes of instruments. Table 8.1 provides the accuracy requirements for current, voltage, and power measurements for these two classes of instruments.

Note:

1. Class I instruments are recommended when precise measurements are necessary, such as for verifying compliance with the standards, and resolving disputes. Any two instruments that comply with the requirements of Class I, when connected to the same signals, produce matching results within the specified accuracy (or indicate an overload condition).
2. Class I instruments are recommended for emission measurements. Class II is recommended for general surveys, but can also be used for emission measurements if the values are such that even allowing for the increased uncertainty, it is clear that the limits are not exceeded. In practice, this means that the measured values should be lower than 90% of the allowed limits.
3. Additionally, for Class I instruments, the phase shift between individual channels should be smaller than $n \times 1°$. (Table 8.1 and the foregoing three notes are from the IEC Standard 61000.4.7-2002: Electromagnetic compatibility [EMC], Pt. 4: Testing and measurement techniques—Section 7: General guide on harmonics and interharmonics measurements and instrumentation for power supply systems and equipment connected thereto. With permission.)

The general structure of the measuring instrument recommended in the preceding standard is shown in Figure 8.7. The standard recommends fixing the sampling intervals of

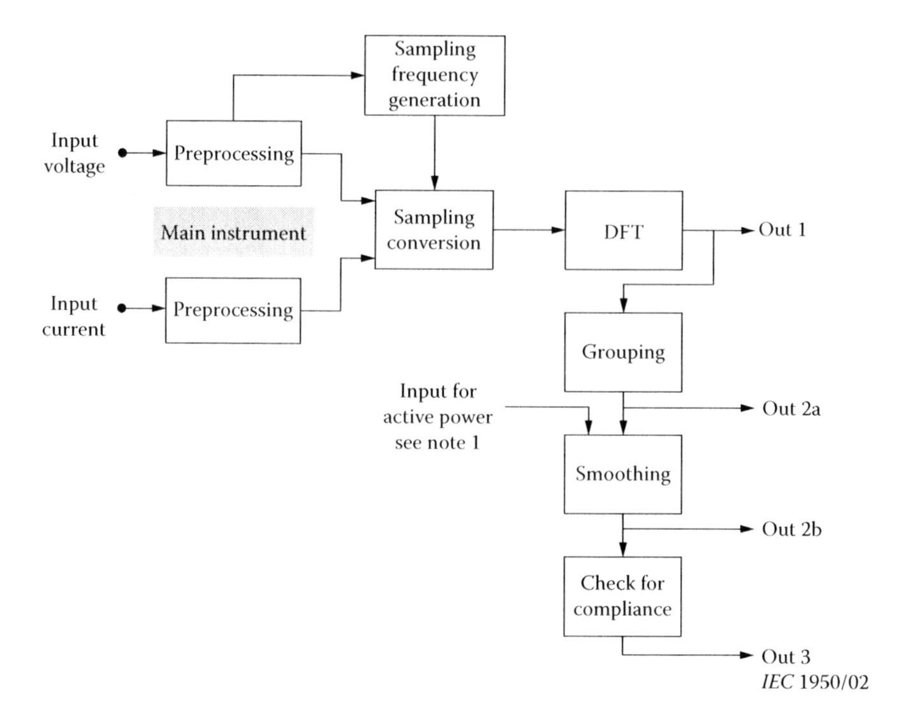

FIGURE 8.7 General structure of a measuring instrument. (From IEC Standard 61000.4.7-2002. With permission.)

the waveform to 10 and 12 cycles for 50- and 60-Hz systems, resulting in a fixed set of spectra with 5-Hz resolution for harmonic and interharmonic evaluation. Instruments can use a phase-locked loop or other means of synchronization. Hanning weighting is allowed only in the case of loss of synchronization. The loss of synchronization shall be indicated on the instrument display, and the data so acquired shall be flagged.

To reduce the errors in the FFT-based harmonic-monitoring equipment, this standard has suggested the following grouping and smoothing procedures. Associated with each harmonic or an interharmonic, there are an rms component value, group value, and subgroup value. We will define these terms in the following text.

The rms value of a harmonic component is the rms value of the spectral component corresponding to an output bin (spectral component) of the DFT.

8.4.1.1 RMS Value of a Harmonic Group: $G_{g,n}$

It is the square root of the sum of the squares of the rms value of a harmonic and the spectral components adjacent to it within the time window (200 msec), thus summing the energy contents of the neighboring lines with that of the harmonic power. The harmonic order is given by the harmonic considered. The harmonic group of order n has the magnitude $G_{g,n}$ (rms value) and is given by the following equations:

$$G^2_{g,n} = \frac{C^2_{k-5}}{2} + \sum_{i=-4}^{4} C^2_{k+i} + \frac{C^2_{k+5}}{2} \text{ \{50-Hz system\}} \tag{8.6a}$$

$$G^2_{g,n} = \frac{C^2_{k-6}}{2} + \sum_{i=-5}^{5} C^2_{k+i} + \frac{C^2_{k+6}}{2} \text{ \{60-Hz system\}} \tag{8.6b}$$

In these equations, $Ck_{+}i$ is the rms value of the spectral component corresponding to an output bin (spectral line) of the DFT.

8.4.1.2 RMS Value of a Harmonic Subgroup: $G_{sg,n}$

It is the square root of the sum of the squares of the rms value of a harmonic and the two spectral components immediately adjacent to it. For the purpose of including the effect of voltage fluctuation during voltage surveys, a subgroup of output components of the DFT is obtained by adding the energy contents of the frequency components directly adjacent to a harmonic to that of the harmonic proper. The harmonic order is given by the harmonic considered.

$G_{sg,n}$ is calculated using the following equation:

$$G^2_{s,g,n} = \sum_{i=-1}^{1} C^2_{k+i} \tag{8.7}$$

8.4.2 Total Harmonic Distortion (THD)

For the definition of total harmonic voltage distortion and total harmonic current distortion, refer to Section 1.7, Chapter 1.

Similar to the above, the IEC standard defines group total harmonic distortion for voltage and current.

8.4.2.1 Group Total Harmonic Distortion (THDG)

THDG is defined as the ratio of the rms value of the harmonic groups (g) to the rms value of the group associated with the fundamental

$$\mathrm{THDG} = \sqrt{\sum_{n=2}^{H} \left(\frac{G_{gn}}{G_{g1}} \right)^{2}} \tag{8.8}$$

8.4.2.2 Subgroup Total Harmonic Distortion (THDS)

THDS is defined similar to THDG by replacing $G_{g,n}$ with $G_{sg,n}$ and G_{g1} with G_{sg1} in Equation 8.8. The subscript g refers to the group and sg to the subgroup.

8.4.2.3 Partial Weighted Harmonic Distortion (PWHD)

PWHD is defined as the ratio of the rms value, weighted with the harmonic order n, of a selected group of higher-order harmonics (from the order $H_{\min}$ to $H_{\max}$) to the rms value of the fundamental

$$\mathrm{PWHD} = \sqrt{\sum_{n=H_{\min}}^{H_{\max}} n \left(\frac{G_n}{G_1} \right)^{2}} \tag{8.9}$$

Note:

1. The concept of partial weighted harmonic distortion is introduced to allow for the possibility of specifying a single limit for the aggregation of higher-order harmonic components. The partial weighted group harmonic distortion can be evaluated by replacing the quantity G_n by the quantity $G_{g,n}$. The partial weighted subgroup harmonic distortion can be evaluated by replacing the quantity G_n by the quantity $G_{sg,n}$.
2. The value of $H_{\min}$ and $H_{\max}$ are defined in each standard concerned with limits (IEC 61000-3-series).
3. PWHD is defined in this standard because it is used in IEC 61000-3-4 and in IEC 61000-3-2. Ed. 2 with amendment 1.

8.4.3 Interharmonics

8.4.3.1 RMS Value of an Interharmonic Component

It is the rms value of a special component of an electrical signal with a frequency between two consecutive harmonic frequencies.

Note:

1. The frequency of the interharmonic component is given by the frequency of the spectral line. This frequency is not an integer multiple of the fundamental frequency.

2. The frequency interval between two consecutive spectral lines is the inverse of the width of the time window, approximately 5 Hz for the purpose of the standard.
3. For the purpose of this standard, the interharmonic component is assumed to be the special component C_k for $k \neq n \times N$.

8.4.3.2 RMS Value of an Interharmonic Group ($C_{ig,n}$)

It is the rms value of all interharmonic components in the interval between two consecutive harmonic frequencies.

Note: For the purposes of this standard, the rms value of the interharmonic group between the harmonic orders n and $n + 1$ designated as $C_{ig,n}$, for example, the group between $n = 5$ and $n = 6$, is designated as $C_{ig,5}$.

$C_{ig,n}$ can be calculated using the following equations:

$$C_{ig,n}^2 = \sum_{i=1}^{9} C_{k+i}^2 \quad \text{(50-Hz power system)} \tag{8.10a}$$

$$C_{ig,n}^2 = \sum_{i=1}^{11} C^2_{k+i} \quad \text{(60-Hz power system)} \tag{8.10b}$$

8.4.3.3 RMS Value of an Interharmonic-Centered Subgroup ($C_{isg,n}$)

$C_{isg,n}$ is the rms value of all interharmonic components in the interval between two consecutive harmonic frequencies, excluding frequency components directly adjacent to the harmonic frequencies.

$C_{isg,n}$ can be calculated using the following equations:

$$C_{isg,n}^2 = \sum_{i=2}^{8} C_{k+i}^2 \quad \text{(50-Hz power system)} \tag{8.11a}$$

$$C_{isg,n}^2 = \sum_{i=2}^{10} C^2_{k+i} \quad \text{(60-Hz power system)} \tag{8.11b}$$

graphical interpretation of the equations 8.6 to 8.11 is provided in Ref. 18.

If instantaneous effects are considered important, the maximum value of each harmonic should be recorded and the cumulative probability (at least 95 and 99%) of these maxima should be calculated. On the other hand, if long-time thermal effects are considered, the maximum of the rms value at each harmonic and its cumulative probabilities (at 1, 10, 50, 90, 95, and 99%) are to be calculated and recorded, that is,

$$C_{n,rms} = \sqrt{\frac{\left(\sum_{k=1}^{M} C_{n,k}^2\right)}{M}} \tag{8.12}$$

where all the single calculated values of C_n shall be determined over the time interval 3 s for selectable individual harmonics[17,20] (preferably up to $n = 50$).

8.4.4 Relative and Absolute Harmonic Phase Angle Measurement

The amplitudes of the harmonic voltages and currents, the relative and absolute phase angles between them, are required for the following purposes[17,20]:

1. Evaluation of the harmonic power flows throughout the power system
2. Detection of the harmonic sources
3. If more than one disturbing load is connected to the same node, to assess summation factors of harmonic currents from those loads
4. At the planning stage, to calculate the effects of new disturbing loads by establishing system-equivalent circuits and also to calculate the effectiveness of remedial measures, such as filters

At a point of common coupling, the direction of the individual harmonic power flows into or out of the different feeders are determined after the measurement of the relative phase angles between harmonic voltages and currents. If the active power flows into the system, the plant is the harmonic source, otherwise it is a sink. The phase lag of harmonic voltage and current in relation to the fundamental (absolute phase angle) need not be known in this case.

The measurement of absolute phase angle is necessary for the following purposes:

1. Measurements at different nodes of the same system or different systems can be compared.
2. If some different harmonic loads have similar phase angles, the effect will be to increase harmonic levels in the system; or if they have opposite phase angles, they cancel each other and decrease harmonic levels in the systems. Hence, a decision can be taken regarding the connection and rearrangement of different systems to ascertain the most favorable arrangement to reduce harmonic levels.
3. Phase angles of disturbing loads, especially from rectifier circuits without firing control, can be measured in order to evaluate their overall disturbing effect or to find countermeasures.

Extra care is needed in measuring phase angles:

1. When there are delta–wye transformers causing phase angle shifts;
2. When the absolute phase angles are being measured, because one needs precise synchronization, preferably to the zero crossing of the fundamental system voltage.

8.5 NECESSITY FOR THE MEASUREMENT OF HARMONIC VOLTAGES AND CURRENTS

1. A knowledge of the background levels is necessary to check them against planned levels or limits specified in the relevant standards. Further, it assists tracking the trends with time or seasonal variations.
2. At the boundaries of a system under study, sometimes it is necessary to develop equivalent networks that are frequency dependent. Further, the driving point impedance at a location enables one to assess the system capability to withstand power quality disturbances.
3. Some equipment may be generating excessive harmonics. Further, some equipment performance may be unacceptable to the utility or customer. The harmonic tests enable one to identify such situations.

8.6 HARMONIC MONITORING SYSTEM

Reference 20 describes the functional details of the following three subsystems of a modern harmonic-monitoring system:

1. Input signal conditioning and acquisition subsystem
2. Digital processing and storage subsystem
3. User interphase subsystem

In a conventional centralized processing architecture, reliance is placed on the outputs of the CTs and VTs being directly routed to the metering room. Although this configuration is normally sufficient for relay operation or metering purposes, the limited bandwidth and EMI susceptibility of the long analog communication links create serious concern over the integrity of the measurements.

In the conceptual distributed processing architecture proposed in Reference 20

1. Fiber-optic links are proposed to reduce the system's susceptibility to electromagnetic noise;
2. A single digital signal processor (DSP) (or CPU) is proposed for every data channel to implement computationally intensive manipulations;
3. A centralized source of sampling channels to synchronize the sampling across all data channels.

GPS-generated timing signals are used to provide synchronization of the samples at different sites.

8.7 CONTINUOUS HARMONIC ANALYSIS IN REAL TIME (CHART)

Miller and Dewe[22] have reported the results of this project, which enables one to measure harmonic levels of as many as 32 power system voltage or current signals on a cycle-by-cycle basis continually in real time.

As it was initially intended to monitor up to the 50th harmonic, if f_0 is the fundamental frequency and f_s is the sampling frequency, then to satisfy the sampling theorem the following relationship is required:

$$f_s > 2 \times 50 f_0 \tag{8.13}$$

This corresponds to exactly 100 samples per cycle. The FFT algorithm requires the number of samples (N) in a record length to be an integral power of 2. The lowest value of N satisfying Equation 8.13 is 128. Hence, to resolve up to the 50th harmonic, 128 samples per cycle of the fundamental are required.

A passive analog filter requires much bulky circuitry and can be very sensitive to component variations. The CHART instrument realizes this filter specification by using a digital finite impulse response (FIR) filter, implemented by each of its front-end digital signal processors (DSPs), which are also used to find the FFT of each channel.

The CHART instrument employs an oversampling rate of $M = 8$, resulting in a sampling frequency of $8 \times 128 f_0 = 1024 f_0$. This enables a simple five-pole Butterworth filter to be used as the front-end antialiasing filter.[22,23] An oversampling rate (M) of 8 was settled on as a compromise between the analog filter being very simple (achieved with M large) and being able to perform a reasonable amount of computation by the DSPs (achieved with M small).

Digital antialiasing FIR filters used in the CHART instrument have the following features suited to this application:

- Linear phase
- Easy design and implementation
- Always stable

For a 50-Hz fundamental frequency, the sampling rate is $1024 \times 50 = 51.2$ kHz. If one uses a 128-tap FIR filter, computing one output from the FIR filter will require $51.2 \times 128 = 6.5$ million instructions per second (MIPs). By breaking the filter up into a polyphase network consisting of eight smaller filters that contribute to the filter output for different time slots, the DSP computations are to be reduced to $(6.5/8) = 0.8$ MIPs in the CHART instrument.

For the FFT implementation details and more detailed reasons for the design choices, see Reference 22.

The effects of sampling rate and analog to digital conversion (ADC) quantization noise on the recovery of harmonic amplitudes and phases in the presence of large fundamental amplitudes were examined by Viktor Kuhlman et al.,[24] to determine the noise limits of instrumentation system design for power systems monitoring and harmonic power flow measurement. Their conclusions are that amplitude and phase errors are independent of the harmonic order; the amplitude error is independent of harmonic amplitude for sufficiently large signal-to-noise ratios; and the phase error is inversely proportional to harmonic amplitude. Both amplitude and phase rms error are inversely proportional to the ADC width and the square root of the FFT length. Hence, the noises may be reduced by increasing the ADC width or the transform length.

8.8 PRESENTATION OF HARMONIC MEASUREMENTS

Harmonic measurements and results from digital simulations can be presented in different forms or as tables.

Figure 8.8 shows the harmonic spectrum using a linear scale.

Figure 8.9 shows the harmonic spectrum using a decibel scale.

Figure 8.10 shows the recording of the fifth harmonic current sampled at 6-min intervals.

Figure 8.11 shows the cumulative probability distribution of the seventh harmonic voltage.

Figure 8.12 shows the histogram of the fifth harmonic current amplitudes.

When one uses oscilloscopes or analog harmonic analyzers for harmonic measurements, the signals are monitored continuously. One should be careful in the choice of sampling intervals when using digital systems. Wide intervals in sampling times will lead to significant errors in interpretation. With digital systems, one can derive cumulative probability graphs (Figure 8.11) and histograms (Figure 8.12).

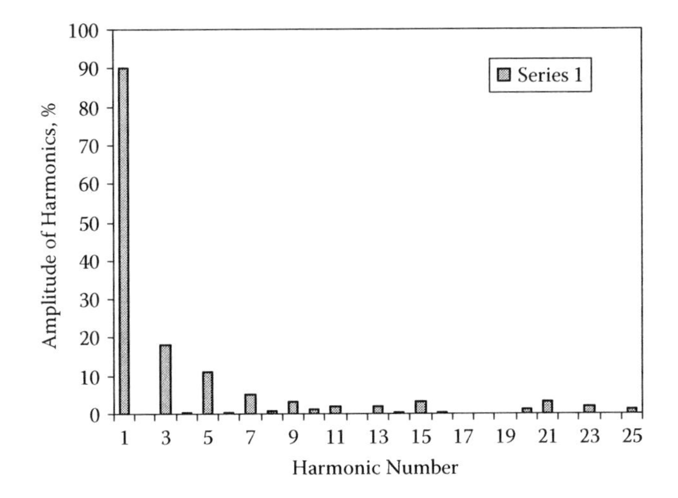

FIGURE 8.8 Harmonic spectrum (linear).

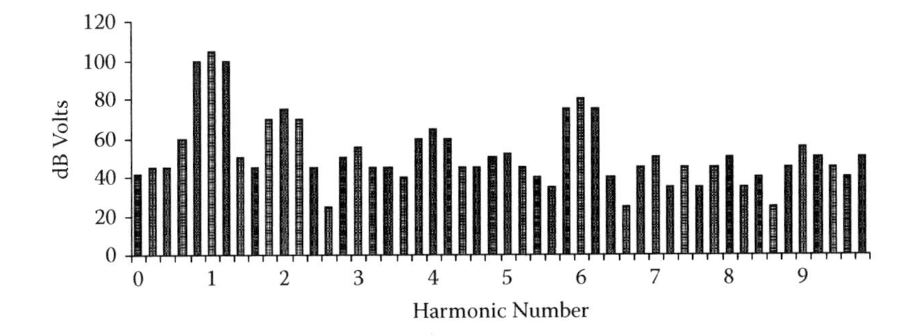

FIGURE 8.9 Harmonic spectrum (decibel scale).

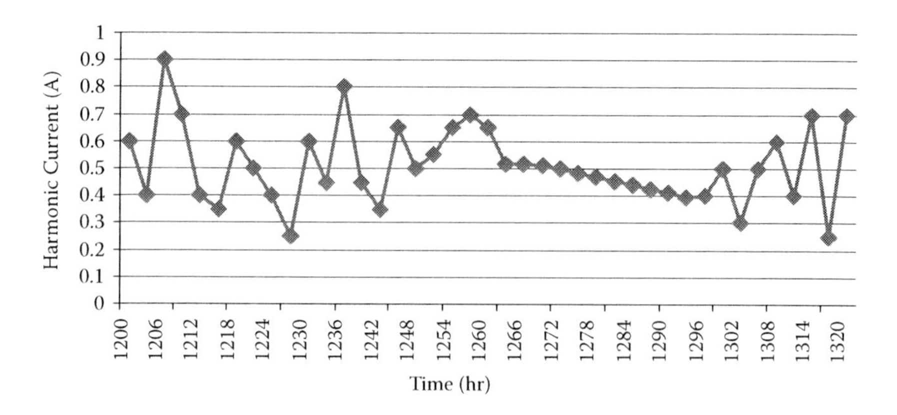

FIGURE 8.10 Fifth harmonic current showing the effect of sampling at 6-min intervals.

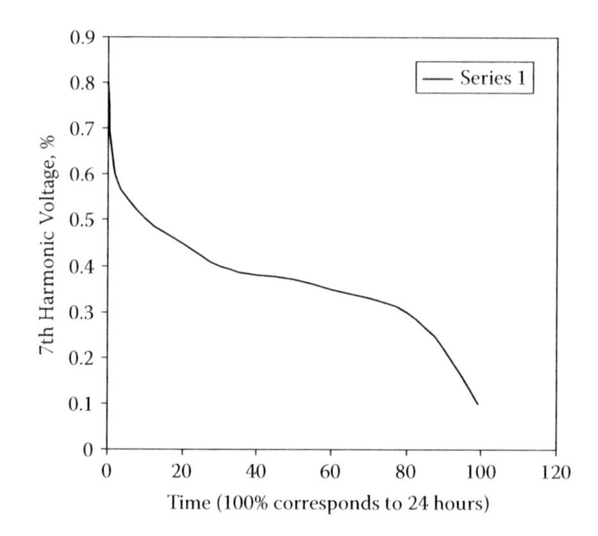

FIGURE 8.11 Cumulative probability distribution of 7th harmonic voltage.

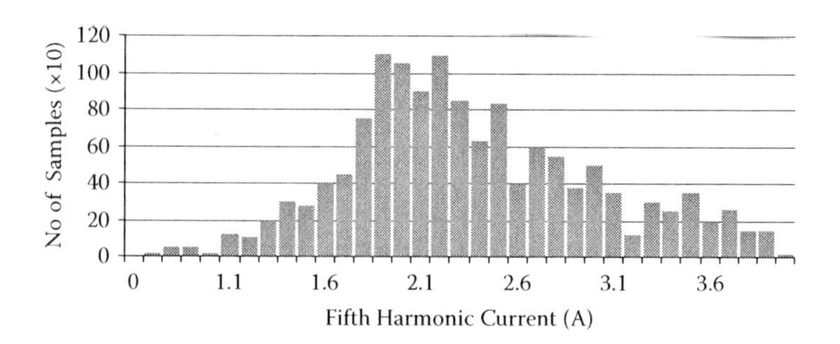

FIGURE 8.12 Histogram of harmonic amplitudes.

8.9 EXAMPLES OF PROJECTS NEEDING HARMONIC MONITORING IN THE POWERLINK QUEENSLAND SYSTEM, AUSTRALIA

Some examples of real-life projects in which one of the authors has participated are reported in this section.

8.9.1 Aluminum Smelter Project

Figure 8.13 shows a somewhat simplified diagram of the layout of the power supply to a large aluminum smelter project. This project was completed in two stages. The first stage of the project required power supply for two potlines approximately 300 MW. This power was supplied by two 132-kV feeders from a nearby power station. The second stage of the project required an additional 400 MW of power for the third potline, which could be supplied through a 500-MVA, 275-kV/132-kV auto-transformer from a separate 275-kV bus. All the three potlines are connected to a smelter 132-kV bus. This 132-kV bus has three-bus tie circuit breakers, giving flexibility to supply any of the potlines either from the power station or from the nearby 275-kV bus. In the initial stages of the project, before all the six units of 275 MW were commissioned at the power station, the smelter load was a big load on the power station compared with its capacity. Hence, there was a concern about the harmonic voltage levels at the point of common coupling at the power station 132-kV bus and

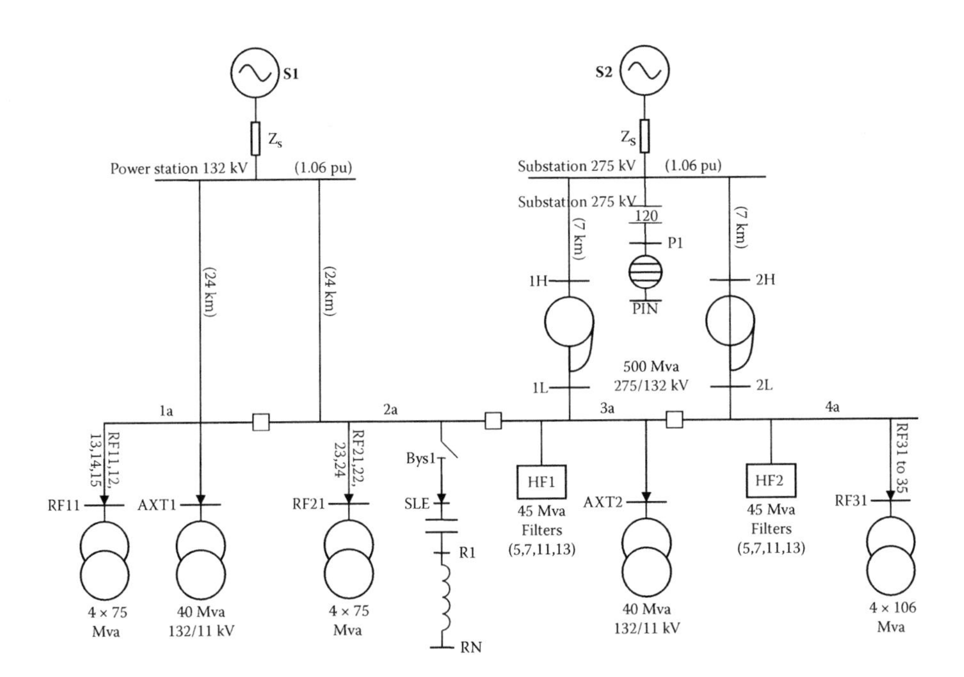

FIGURE 8.13 Layout of the power supply to a large aluminum smelter project.

nearby areas. Very detailed computer simulation studies were performed and the harmonic current levels that the customer can inject at the point of common coupling at the power station 132-kV bus were specified in the power supply agreement. Hence, harmonic voltage levels and current levels at the power station 132-kV bus were measured as the smelter load slowly increased depending upon the production schedule. The aluminum smelter has 48-pulse operation to ensure that harmonic currents injected into the system are at acceptable levels.

In the first stage, two harmonic filters are installed, each of 45 MVA capacity. Each has two double-tuned filters, one tuned for 5th and 7th harmonics, the other tuned for 11th and 13th harmonics. In the second stage, initially, a 60-MVA capacitor bank was connected to the aluminum smelter 132-kV bus. After performing detailed computer simulations to predict harmonic levels and switching transients, this was upgraded to a 95-MVA bank and tuned as a 5th harmonic filter.

8.9.2 Central Queensland Railway Electrification Project

This project was already described in Section 2.7 in Chapter 2 in connection with SVCs. This also required detailed computer simulations initially, to determine harmonic filter ratings and to specify permitted harmonic current levels that the Queensland Railways (QR) can inject into the Powerlink Queensland system at the points of common coupling. Later, several harmonic measurements were taken at different points of common coupling before and after commissioning of the project and at different seasons to account for seasonal load variations in the system winter, summer, etc., and QR loads. Figure 8.14 shows the total harmonic distortion at a

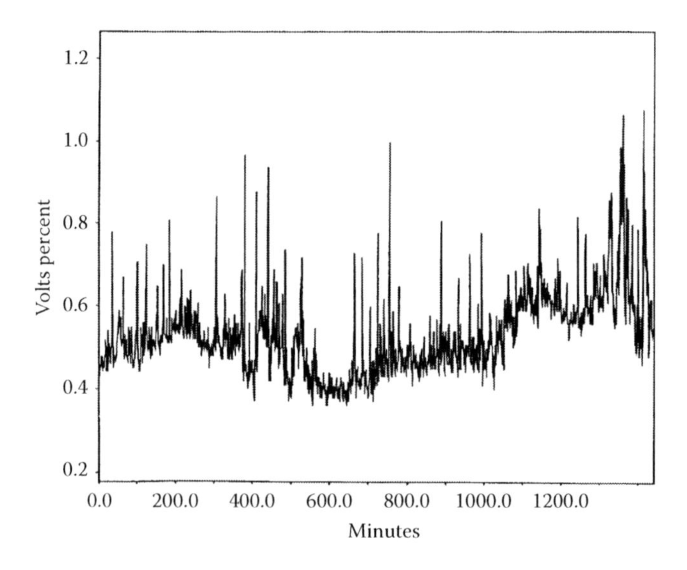

FIGURE 8.14 Power supply to aluminum smelter. (With permission of Queensland Railways.) THD in percent at one of the 132-kV substations supplying QR loads. (With permission of Powerlink Queensland.)

substation 132-kV bus, that is, one of the points of common coupling from which QR loads are supplied.

In this project, in the initial stages, unacceptable sixth harmonic resonance was observed because of the interaction between adjacent SVCs installed to reduce harmonics from the QR loads.

The filter design was later modified to overcome this 6th harmonic resonance problem.

Reference 25 discusses a technique used to determine the harmonic sources in this multiharmonic source environment. This required harmonic power flow determination using the FFT to identify the harmonic sources. Later, the harmonic filters were modified and the problem was solved.

8.10 FLICKER

Flicker[5,26] is defined as the variation of input voltage sufficient in duration to allow visual observation of a change in electric light source intensity. Loads that can exhibit continuous, rapid variations in the load current magnitude can cause voltage variations that are often referred to as flicker. Flicker occurs on systems that are weak relative to the amount of power required by the load, resulting in a low short-circuit ratio. Quantitatively, flicker may be expressed as the change in voltage over nominal expressed as a percentage. For example, if the voltage in a 120-V circuit increases to 125 V and then drops to 119 V, the flicker (f), is calculated as $f = 100 \times (125 - 119)/120 = 5\%$.

Three-phase electric arc furnaces are used to make high-quality steel.[27] These furnaces have two important modes in their operation: melting and refining modes. During the melting period, pieces of steel between the electrodes in these furnaces produce short circuits on the secondary of the transformer to which the electrodes are connected. Therefore, the melting period is characterized by severe fluctuations of current at low power-factor values. The melting process takes around 50–120 min, depending on the type of furnace. In the refining mode, the steel is melted down to a pool, the arc length can be maintained uniformly by regulating the electrodes, and the electrical load is constant with a high power factor. The power is supplied to the furnace during the refining period every 2 h for a period of 10–40 min, depending on the size of the furnace. The severe fluctuations during the melting process are responsible for significant voltage drop in the power system and cause flicker. These voltage fluctuations are not uniform in the different phases and hence increase the negative-phase sequence voltages at the point of common coupling. When the busbar voltage drops, so does the melting power, which increases with the square of the voltage. Further, these voltage fluctuations cause the light from incandescent and fluorescent lamps to flicker unpleasantly.[27]

The human eye is particularly sensitive to periodic fluctuation frequencies between 6 and 10 Hz. Voltage amplitudes of 0.3% are perceptible, whereas at 0.5% they are already objectionable.[28] Arc furnaces mainly produce fluctuations with frequencies from 2–15 Hz, and in some cases they can extend up to 30 Hz. Some studies have shown that voltage fluctuations amounting to 0.5% are already perceptible, with 1–2 half-waves and a repetition frequency of 7 Hz. As the extent and frequency distribution of reactive power fluctuations are random, they cannot be calculated. Industrial experience has shown that the reactions caused by arc furnaces do not lead to complaints if at the connection point the short-circuit power of the network is between 80 and 120 times the rated power, depending on the size of the furnace. Recent investigations have yielded higher values. If the short-circuit power of the network is below these values, flicker will be perceptible to a certain extent, depending on the ratio. It is possible to reduce flicker considerably with the aid of SVCs. Arc furnaces produce a high proportion of harmonic currents during the melt-down phase, and these give rise to disturbances in the network. SVCs can be used to reduce the flicker and unbalance in the system voltages at the point of common coupling, and filters can be used to reduce the effect of harmonics.[28]

To reduce flicker, the mitigation techniques adopted are installing static capacitors, SVCs, and increasing system capacity. Cost is a significant factor in determining the final solution of the flicker problem, whereas the type of load causing the flicker and capacity of the system supplying the load play a part.

If the motor starting is causing a flicker problem, the following modifications might provide a solution:

- Rewinding a motor
- Using a flywheel energy system
- Installing a step starter for the motor

Reference 29 discusses the incandescent and fluorescent lamp flicker problems experienced by a 55-MW arc furnace together with a 65-MVAr SVC. One of their conclusions is that flicker is not only caused by voltage fluctuations below 20 Hz but also by the frequencies above 20 Hz that are not multiples of 60 Hz and can beat with the fundamental 60-Hz frequency or its harmonic components, resulting in fluorescent flicker.

Reference 30 discusses an improved arc furnace model and EMTP simulations to predict the flicker levels in arc furnaces. It has been reported in this paper that although series reactors in electrical plants supplying static power converters reduce voltage distortion,[31] the opposite occurs with electric arc furnaces. In short, series reactors with arc furnaces reduce flicker by reducing the current variations during the melting period and reduce the supply-side harmonic levels.[32] However, they increase harmonic voltage distortion slightly because of the arc furnace at the point of common coupling by increasing the length of the arc.[29] Hence, care is required in the design of series reactors with electric arc furnaces.

8.11 IEC FLICKER METER

Historically, flicker measurements have been carried out using rms meters, load duty cycle, and a flicker curve. Nowadays, it has become a necessity to use a single quantity to characterize flicker severity (P_{st}) for contractual purposes. P_{st} is determined using an IEC flicker meter.[29,33] The design details of this instrument and measuring procedure are comprehensively described in IEC Standard 61000-4-15.

The flicker meter architecture is described by the block diagram of Figure 8.15, and can be divided into two parts:

- Simulation of the response of the lamp–eye–brain chain
- Online statistical analysis of the flicker signal and presentation of the results

The first task is performed by blocks 2, 3, and 4 of Figure 8.15, whereas the second task is accomplished by block 5.

Block 1 is an input voltage adaptor. This consists of a signal generator and a circuit that scales the mean input half-cycle rms values to an internal reference level. Taps on the input transformer establish suitable input voltage ranges to keep the input signal to the voltage adaptor within its permissible range. This allows flicker measurements to be independent of the actual carrier input voltage level and is expressed as a percent ratio.

Block 2 is a square law demodulator. The purpose of this block is to recover the voltage fluctuation by squaring the input voltage scaled to the reference level, thus simulating the behavior of a lamp. The squaring process in the demodulator introduces a dc offset and double mains frequency components. It may be noted that $\sin^2 k\omega t = (1 - \cos 2\, k\omega t)/2$.

Block 3 consists of two filters: one is to eliminate the dc and double mains frequency ripple components of the demodulator output, and the second is a weighting filter block that simulates the frequency response to sinusoidal voltage fluctuations of a coiled-filament gas-filled lamp (60 W – 230 V or 60 W – 120 V) combined with the human visual system. The response function is based on the perceptibility threshold found at each frequency by 50% of the persons tested.

(Note that a reference filament lamp for 100-V systems would have a different frequency response and would require a corresponding adjustment of the weighting filter. The characteristics of discharge lamps are totally different, and substantial modifications of the IEC Standard 61000-4-15 would be necessary if they were taken into account.)

The basic transfer function for the weighting filter is

$$H(s) = \frac{k\omega_1 s}{s^2 + 2\lambda\, s + \omega_1^2} \cdot \frac{1 + s/\omega_2}{(1 + s/\omega_3)\,(1 + s/\omega_4)} \tag{8.14}$$

where s is the Laplace complex variable, and

$$\omega_1 = 2\pi f_1$$
$$\omega_2 = 2\pi f_2$$
$$\omega_3 = 2\pi f_3$$
$$\omega_4 = 2\pi f_4$$

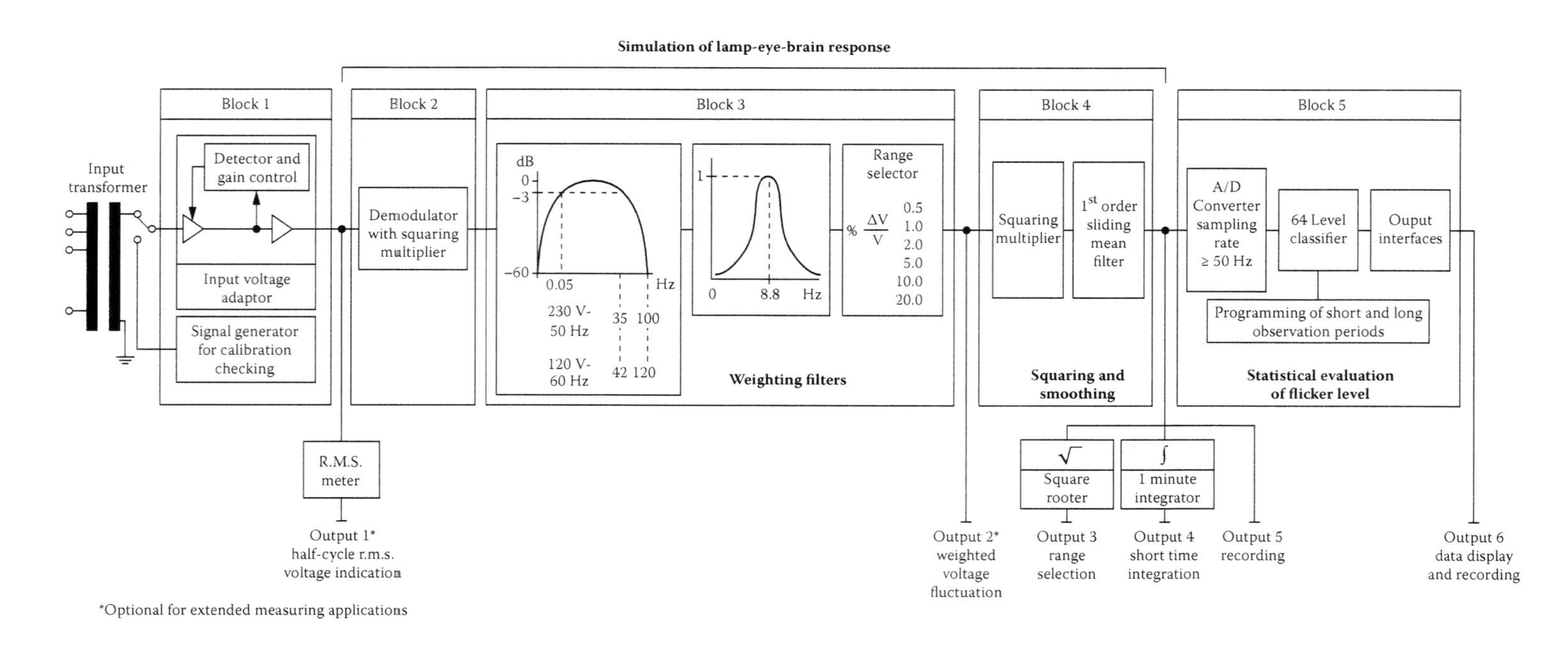

FIGURE 8.15 Functional diagram of IEC flicker meter. (From IEC Standard 61000–4–15 (2003). With permission.)

Indicative values for the different parameters in the foregoing transfer function are given in Table 8.2.

TABLE 8.2
Indicative Values for the Parameters of Lamps

Variable	220-V Lamp 50-Hz System	120-V Lamp 60-Hz System
K	1.74802	1.6357
λ	2π * 4.05981	2π * 4.167375
f_1	9.15494	9.077169
f_2	2.27979	2.939902
f_3	1.22535	1.394468
f_4	21.9	17.31512

Source: IEC 61000-4-15, Electromagnetic Compatibility (EMC). Part 4: Testing and Measuring Techniques, Section 15: Flicker-meter-Functional and Design Specifications. (Reprinted with permission from IEC)

Block 4 consists of a squaring multiplier for the voltage control signal, to simulate the nonlinear eye–brain response. The sliding mean filter in this block averages the signal to simulate the short-term storage effect of the brain. The output of this block is considered to be the instantaneous flicker level. A level of 1 in the output of this block corresponds to perceptible flicker.

Block 5 consists of a statistical analysis of the instantaneous flicker level and creates a histogram by dividing the output of block 4 into suitable classes. A probability density function is created for each class, and from this a cumulative distribution function can be formed.

8.11.1 Short-Term Flicker Evaluation

Short-term flicker evaluation of flicker severity, based on an observation period of 10 min, is denoted PST from the statistical data obtained from block 5.

The following formula is used:

$$PST = \sqrt{}\ (0.0314\ P_{0.1} + 0.0525\ P_{1s} + 0.0657\ P_{3s} + 0.28\ P_{10s} + 0.08\ P_{50s}) \quad (8.15)$$

where the percentiles $P_{0.1}$, P_{1s}, P_{3s}, P_{10s}, and P_{50s} are the flicker levels exceeded for 0.1, 1, 3, 10, and 50%, respectively, of the time during the observation period. The suffix s in the formula indicates that the smoothed value should be used; these are obtained using the following equations:

$$P_{50s} = (P_{30} + P_{50} + P_{80})/3 \quad (8.16)$$

$$P_{10s} = (P_6 + P_8 + P_{10} + P_{13} + P_{17})/5 \quad (8.17)$$

$$P_{3s} = (P_{2.2} + P_3 + P_4)/3 \quad (8.18)$$

$$P_{1s} = (P_{0.7} + P_1 + P_{1.5})/3 \quad (8.19)$$

where the combined effect of several disturbing loads operating randomly (e.g., welder motors) has to be taken into account or when flicker sources with long and variable duty cycles (e.g., arc furnaces have to be considered), it is necessary to provide a criterion for the long-term assessment of the flicker severity. Therefore, the long-term flicker severity *PLT* is derived from *PST* using the equation

$$P_{LT} = 3\sqrt{\left[\sum_{i=1}^{N} P_{sti}^{3}/N\right]} \tag{8.20}$$

where N is the number of P_{sti} (i = 1, 2, 3,...) readings and is determined by the duty cycle of the flicker-producing load. If the duty cycle is unknown, the recommended number of P_{sti} readings is 12 (with a 2-h measurement window).

8.11.2 Flicker Standards

At present the United States does not have a standard for flicker measurement. However, there are IEEE Standards 141-1993[33] and 519-1992[19] that recommend flicker limits. Both these curves are very similar. The limits recommended by the IEEE Standard 141-1993 are shown in Figure 8.16. For the limits recommended in IEEE Standard 519-1992, the reader can find them in Figure 10.3 of that standard. That curve shows four different regions characterized by fluctuations per hour to fluctuations per second.

House pumps have very (1–20) low fluctuations (per hour), whereas arc furnaces and reciprocating compressors have fluctuations 2 to 15 per second.

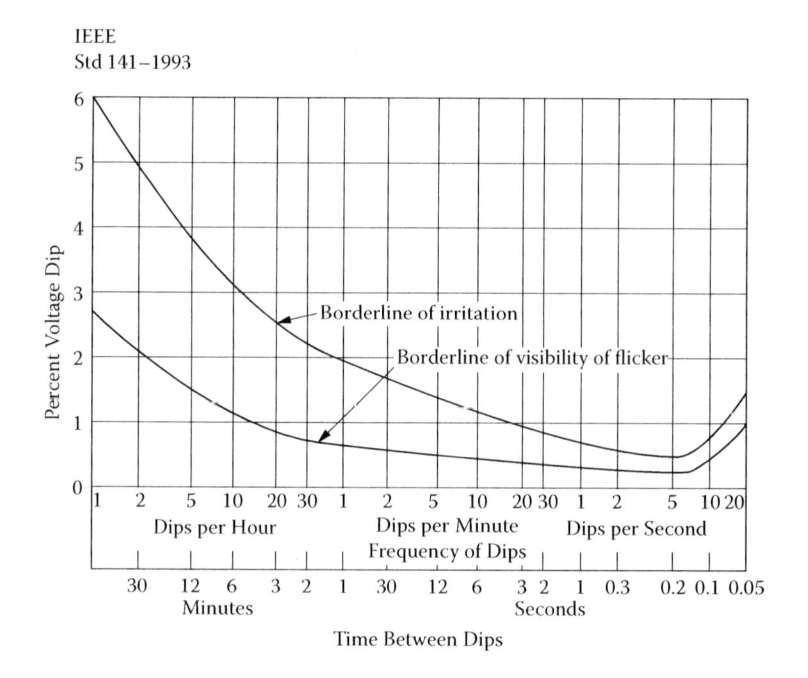

FIGURE 8.16 Range of observable and objectionable voltage flicker versus time. (From IEEE Standard 141-1993. With permission.)

8.12 CONCLUSIONS

In this chapter, we have discussed different instruments, both analog and digital, for the monitoring of power quality. The analog instruments record a continuous waveform. One has to be careful in choosing a suitable time interval to record all harmonics of interest. Otherwise, while using FFT analysis, aliasing errors creep in, yielding very inaccurate results. Further, if the time interval is large in some recording instruments, interpretation of the results is rather difficult, because one does not know how the harmonics have varied between successive recordings with a large time interval.

We have also discussed the necessity for harmonic measurements

- to determine background levels,
- to determine where the specified harmonic levels exceed statutory limits,
- to improve our system models to ensure better prediction of harmonic levels,
- to help determine the harmonic power flow and the sources of harmonics in the power system, by measuring the harmonic phase angle.

We have also discussed the transducer CTs and VTs to reduce high voltage and current levels to levels suitable for instrumentation purposes. CVTs are not suitable for harmonic measurements in view of their inadequate frequency response. For accurate harmonic measurements using the DLA tap on a transformer bushing, a capacitor divider should be used. With a high impedance burden, the response of many magnetic voltage transformers is usually adequate to at least 5 kHz and the accuracy is within 3%.

Standard instrument CTs can be used up to 2 kHz.

IEC Standard 61000-4-7-2002 recommends two classes of instruments. Class I instruments with higher accuracy are recommended for verifying compliance with standards and resolving disputes. Class II instruments are recommended for general surveys, but can also be used for emission measurements if the measured values are lower than 90% of the limits.

The accurate measurement of the frequency response of CTs and VTs is rather difficult. We have discussed a method of measuring frequency response developed by Meliopoulos et al.[6] using autocorrelation and cross-correlation spectrums and coherence function to estimate the accuracy of the measurements.

We have also discussed recommendations of the IEC for digital instruments that use FFT analysis.

The CHART system with 32 channels to determine 32 harmonic variables simultaneously has been described. This system uses multirate digital processing techniques and FFT analysis using the decimation in time algorithm.

The IEC flicker meter has also been described.

REFERENCES

1. ElektroTek Concepts Web site, *Power Monitoring Solutions.*
2. Hatanaka, G., 1988, Measurement and monitoring, *Power Quality Seminar, Perspectives in Ontario*, October 20.

3. Dugan, R.C., McGranaghan, M.F., and Beaty, H.W., 1996, *Electrical Power Systems Quality*, McGraw-Hill, New York.
4. Dugan, R.C., McGranaghan, M.F., Santoso, S., and Beaty, H.W., 2002, *Electrical Power Systems Quality*, McGraw-Hill, New York.
5. Sankaran, C., 2002, *Power Quality*, CRC Press, Boca Raton, FL.
6. Meliopoulos, A.P.S., Zhang, F., Cokkinides, G.J., Coffeen, L., Burnett, R., McBride, J., Zelingher, S., and Stillman, G., 1993, Transmission level instrument transformers and transient event recorders characterization for harmonic measurements, *IEEE Trans. Power Delivery*, 8(3), 1507–1517.
7. Douglass, D.A., 1981, Potential transformer accuracy at 60 Hz voltages above and below rating and at frequencies above 60 Hz, *IEEE Trans. PAS*, 100(3), 1370–1375.
8. Cornfield, G.C., 1981, Record of discussions, CIRED, Brighton, *IEE Conf. Pub.* 197(Pt. 2), 48.
9. CIGRE Working Group 36-05 (Disturbing Loads), 1981, Harmonics, characteristic parameters, methods of study, estimates of existing values in the network, *Electra*, 77, 35–54.
10. Casu, A. and Simoncini, V., 1974, Harmonics due to rectifiers connected to the Sardinian 220 kV AC network, *IEE Conf. Pub.*, 110, 215–255.
11. Malewski, R. and Douville, J., 1976, Measuring properties of voltage and current transformers for the higher harmonic frequencies, *Canadian Communication and Power Conference, Montreal*, pp. 327–329.
12. Bradley, D.A., Bodger, P.S., and Nyland, P.R., 1985, Harmonic response tests on voltage transducers for the New Zealand power system, *IEEE Trans. PAS*, 104(7), 1750–1756.
13. Olivier, G., Bouchard, R.P., Gervais, Y., and Mukhedkar, F.D., 1980, Frequency response of HV test transformers and the associated measurement problems, *IEE Trans. PAS*, 99(1), 141–146.
14. Hewlett-Packard, *Application Notes on Dynamic Signal Analyzer.*
15. Julius, S.B. and Piersol, A.G., 1980, *Engineering Applications of Correlation and Spectral Analysis*, Wiley-Interscience, New York.
16. Douglass, D.A., 1981, Current transformer accuracy with asymmetric and high frequency fault currents, *IEEE Trans. PAS*, 100(3), 1006–1012.
17. IEC Standard 61000.4.7-1991, *Electromagnetic Compatibility* (*EMC*), Pt. 4: Testing and measurement techniques, Section 7: General guide on harmonics and interharmonics measurements and instrumentation for power supply systems and equipment connected thereto.
18. IEC Standard 61000.4.7-2002, *Electromagnetic Compatibility* (*EMC*), Pt. 4: Testing and measurement techniques, Section 7: General guide on harmonics and interharmonics measurements and instrumentation for power supply systems and equipment connected thereto.
19. IEEE Standard 519-1992, *IEEE Recommended Practices and Requirements for Harmonic Control in Electric Power Systems, 1993.*
20. Arrillaga, J. and Watson, N.R., 2003, *Power System Harmonics*, John Wiley & Sons, Chichester, U.K.
21. *IEEE Interharmonic Task Force*, CIGRE 36.05/CIRED 2 CC02, 1997, Voltage Quality Working Group, *Interharmonics in Power Systems.*
22. Miller, A.J.V. and Dewe, M.B., 1993, The application of multi-rate digital signal processing techniques to the measurement of power system harmonic levels, *IEEE Trans. Power Delivery*, 8(2), 531–539.
23. Zverev, A.I., 1967, *Handbook of Filter Synthesis*, John Wiley & Sons, New York.

24. Kuhlman, V., Sinton, A., Dewe, M., and Arnold, C., 2007, Effects of sampling rate and ADC width on the accuracy of amplitude and phase measurements in power-quality monitoring, *IEEE Trans. Power Delivery*, 22(2), 758–764.
25. George, T.A. and Bones, D., 1991, Harmonic power flow determination using the fast Fourier transform, *IEEE Trans. Power Delivery*, 6(2), 530–535.
26. Natarajan, R., 2005, *Power System Capacitors*, CRC Press, Boca Raton, FL.
27. Wanner, E. and Herbst, W., 1977, Static power factor compensators for use with arc furnaces, *Brown Boveri Review*, 2, 124.
28. Bhargava, B., 1993, Arc furnace flicker measurements and control, *IEEE Trans. Power Delivery*, 8(1), 400–410.
29. Montanari, G.C., Loggini, M., Cavallini, A., Pitti, L., and Zaninelli, D., 1994, Arc-furnace model for the study of flicker compensation in electrical networks, *IEEE Trans. Power Delivery*, 9(4), 2026–2036.
30. Montanari, G.C. and Loggini, M., 1987, Voltage-distortion compensation in electrical plants supplying static power converters, *IEEE Trans. Ind. Appl.,* 23(1), 181–188.
31. Mendis, S.R., Bishop, M.T., Day, T.R., and Boyd, D.M., 1995, Evaluation of supplementary series reactors to optimize electric arc furnace operations, *Conference Record*, IEEE IAS Annual Meeting, Fla, Orlando, October 1995, 2154-2161.
32. IEC 61000-4-15, *Electromagnetic Compatibility (EMC)*, Pt. 4: *Testing and Measuring Techniques*, Section 15: Flickermeter-Functional and Design Specifications.
33. IEEE Standard 141-1993, *Recommended Practice for Power Distribution in Industrial Plants.*

9 Reactors

9.1 INTRODUCTION

Prior to the development of thyristors and their use in static var compensators (SVCs), dynamic var compensation was provided by synchronous machines. An overexcited synchronous machine behaves similar to a capacitor, supplying lagging vars to the system. It is usually called a *synchronous condenser.* An underexcited synchronous machine behaves similar to a reactor absorbing lagging vars from the system. Since the advent of the SVCs, they have largely replaced synchronous condensers because of the latter's additional cost of initial purchase and, later, of maintenance because they are rotating machines. However, they still find application in weak high-voltage dc (HVDC) systems to provide additional short-circuit capacity, which SVCs cannot provide.

Passive reactive var compensation is provided by shunt capacitors and shunt reactors. In this chapter, we will discuss shunt reactors. It is fairly well known that, in long, lightly loaded extra-high-voltage (EHV) lines or open-circuited lines, due to the Ferranti effect, the receiving-end voltage is more than the sending-end voltage. The application of shunt reactors to compensate for a portion of the charging current has been recognized by EHV system designers as a practical and economical method of controlling voltage rises.[1–3] Table 9.1 lists some typical charging MVAs for various EHV transmission lines and voltages.[1] In contrast, at lower voltages such as 110, 132, and 275 kV, typical line-charging MVAs are much lower, that is, 7.5, 10.8, and 43.2 MVA per 100 miles, respectively (see Table 9.2).

At the end of those transmission lines, which carry more than their surge impedance loading, generally no shunt reactors are required. Thus, in well-developed high-voltage (HV) systems, shunt capacitors may be required to keep bus voltages within acceptable limits. Many EHV transmission lines are justified by reserve considerations[4]—some for seasonal power interchange, and a few for the connection of mine-mouth plants to systems. These applications often give power flows that are below the surge impedance loading.

9.2 LOSSES IN THE POWER SYSTEM

In long EHV lines with considerable line capacitance, losses will increase when the load power factor is unity or when there are light loads with a lagging power factor. But when there are heavy loads with a lagging power factor, because line-charging currents and the inductive part of lagging load currents are in phase opposition, they cancel each other and line losses are reduced. However, if the shunt reactor capacity is too high under heavily loaded conditions, the inductive part of the load and shunt reactors can overcompensate the line-charging MVA, and losses can increase. It is usual to switch part of these shunt reactors out of service at such times. Voltage-rise

TABLE 9.1
Approximate Charging MVAr for EHV Transmission Lines

Line Voltage (kV)	Phase Separation Feet[a]	Conductors per Bundle[b]	Charging MVA per 100 miles
345	28	2	80
500	40	2	165
500	40	3	180
500	40	4	195
700	50	3	350
700	50	4	375

[a] Flat spacing.
[b] Bundle spacing, 18 in.

Source: From Alexander, G.H., Hopkinson, R.H., and Welch, A.U. (1966). Design and Application of EHV Shunt Reactors. *IEEE Trans. on PAS,* 12, 1247–1258. With permission from IEEE.

conditions may prohibit the switching of all shunt reactors, and a certain part must always be solidly connected to the lines and switched with it.

Reactors for HV and EHV connections may have 10–20% higher losses than low-voltage (LV) reactors. A factor of at least equal importance is the transformer loss. For an approximate method of calculating transformer losses when LV reactors are used, see Reference 5, page 18.

9.3 SWITCHING SURGES

It is fairly well known from traveling-wave theory that the voltage at the receiving end of an open-circuited line reaches twice the applied voltage. Suppose we tripped a line breaker and it retains its charge on the line, and the magnitude of the line is one per unit (pu). If we close the line at the peak of the system voltage with opposite polarity, this voltage will appear at the receiving end of the line with 2.0 pu with

TABLE 9.2
Approximate Charging MVAr for 110- to 275-kV Lines

Line Voltage (kV)	Charging MVA Per100 Miles
110	7.5
132	10.8
275	43.2

opposite polarity. Then the total reflected wave will be 3 pu with respect to ground. The Ferranti effect could increase this even further, and a theoretical value of 4.0 pu is possible.

9.4 SHUNT REACTOR INSTALLATIONS IN EHV LINES

In the following cases, shunt reactors are used in EHV lines:

- For wholly or partly neutralizing the shunt capacitance of the line and thus controlling losses and voltage rises.
- For the control of basic insulation level (BIL) by reducing the maximum voltage obtainable due to surges at a reasonable level.

The following conditions were decisive in determining the location of reactors and their ratings in the Swedish network, and they would apply generally to other systems[5]:

- Normal service conditions with the whole network intact and a low power flow.
- To ensure voltage stability while energizing a line.
- Disturbances, for example, tripping of a transformer, resulting in an increase in the effective line length and consequently of greatly increased capacitive generation. Reactors in this case should be placed so that the extension of the effective line length does not accentuate the risk of a major breakdown.

9.5 DETERMINATION OF SHUNT REACTOR RATING

In EHV and HV systems, minimum value of shunt reactors must be used. Otherwise the question of tan ϕ of the load will be transferred to the distribution system to be solved there by the use of switched capacitor banks, underexcited alternators, etc. Reactive consumption requirements were determined so that the system should operate within prescribed voltage limits with all lines in service at low load levels. At very low load levels, one line would be taken out of service. The general principle was to strive to maintain a constant voltage (e.g., 400 kV in a 400-kV system) at the receiving and sending ends of the line.

Measures taken to maintain balance during light load periods are

- Connection of shunt reactors
- Underexcitation of alternators
- Disconnection of shunt capacitors in adjoining sections
- Disconnection of a trunk line if load falls to a very low limit

(Note that in 1967, in the British Grid, minimum summer loads occurred during summer nights on public holidays when they fell to 15% of winter maximum values and, in substantial parts of the system, to below 10%.[6])

A further requirement that must be fulfilled is that reactive balance be main tained even with the largest reactor out of service. In this case, the voltage limit may be exceeded, but under excitation, is used to the full. Reactors at the receiving end must be switched off during periods of heavy load; otherwise, they would need to be compensated for by extra ratings of shunt capacitors or synchronous condensers.

9.6 CHOICE OF VOLTAGE LEVEL FOR SHUNT REACTOR CONNECTION

It is technically possible to connect reactors at any voltage level. For this, the following factors should be considered.[5]

9.6.1 Effective Compensation

Reactors connected to the LV side of the transformers may have to work in parallel with one or more alternators or synchronous condensers. Even though the HV level is kept constant, the reactor terminal voltage will vary depending on the reactive power generation in the machine. The lowest reactor terminal voltage (and lowest effective reactive output) occurs when the machines are most underexcited; that is, during periods of lowest transmission when reactors are most needed. To compensate for this, LV reactors must be considerably overdimensioned (reduction in the reactor output can be in the range 10–20%).

Reactors connected to the HV side of transformers suffer little variation of voltage if the operating policy is to keep the HV level as constant as possible. Voltage drop in transformers has no influence on the reactor terminal voltage. Connections to a tertiary winding are also more effective than connection to the LV side. Direct connection of reactors to the HV lines is hence preferred. However, where some reactors are to be switched off under peak-load connections, such reactors can be connected to the tertiary windings through circuit breakers.

9.6.2 Influence on Transformers and Generators

The maximum permissible rating differs considerably for different levels of voltage connection.[5] They are primarily determined by the tolerable voltage variation. As reactor compensation is required mainly for EHV levels, connections other than direct connections to these lines involve transformer losses and hence affects the rating of relevant transformers. Thus, it becomes necessary to transform the leading MVArs generated on the line. This increase can be of the order of 3–10%.

Reactors connected on the LV side of the transformers in parallel with synchronous machines give rise to unfavorable operating conditions for such machines because of the extent of the voltage variations and the troublesome voltage surges that occur when reactors are connected and disconnected. It is particularly noticeable when the reactor rating is comparable with machine size. Alternators directly connected to transformers are seldom subjected to voltage variations greater than ±5% even under conditions of high-reactive generation to maintain system voltage regulation. However, with an LV-connected reactor, it is possible to have variations up to 15–20% if the reactor is large compared to the machine.

9.6.3 Switchgear Requirements

A faulty LV-connected reactor would draw fault current from both the alternator and transformer and hence impose heavier requirements on its circuit breaker than those imposed on either the alternator or transformer breakers.[5] Alternatively, both the alternator and transformer breakers must be arranged to trip if a reactor fault occurs. Both these conditions are undesirable because they result in the loss of generating or transforming capacity.

HV switchgear requirements vary widely, but it is not usual to provide a complete switchgear bay to connect a reactor. Sometimes a circuit breaker and an isolator connect the reactor directly to the line. Another method is to use a bus-transfer breaker.

9.6.4 Influence on Operation Reliability of the System

Satisfactory reactive balance can be obtained irrespective of where reactors are connected, that is, at the EHV, HV, or LV level.[5] However, if a reactor is connected to the LV side of a transformer, and a fault occurs such that the reactor becomes separated from the line and the line is still energized, line capacitance will no longer be compensated. This gives the directly connected HV or EHV reactor a definite advantage. Although the reactors are admittedly disconnected during high-transmission periods, they can easily be automatically reconnected simultaneously with a transformer tripping. A further advantage is obtained when reenergizing the system after a breakdown if the reactor is already connected to the line that it compensates.[5]

9.6.5 Influence on Insulation and Overvoltage Conditions

Shunt reactors reduce voltage levels under fault and normal switching conditions.[5] If the beneficial influence of reactors is to be exploited in the form of reduced insulation levels, reactors must be continuously connected. However, continuously connected reactors are disadvantageous in the case where the load exceeds the surge impedance loading for long periods, as the reactors must then be compensated by shunt capacitors or synchronous condensers. The magnitude of overvoltages increases with increasing line length and decreases with increasing short-circuit power.

9.7 SINGLE-POLE AUTORECLOSING OF TRANSMISSION LINES

In an EHV line, up to 70–80% of the faults occurring are single line to ground, and most of these are transitory due to lightning.[7,8] Some of the single-line-to-ground faults occurring on one circuit of a double-circuit line with vertical configuration are prone to spread to the second circuit. Hence, the reliability of both single-circuit and double-circuit lines in the transmission system can be improved if single-pole autoreclosing is successful for transient faults. In addition, the system is more stable with single-pole autoreclosing than with three-pole autoreclosing.

Effective single-pole switching would increase the reliability of a line approximately as much as the addition of overhead ground wires at a much lower cost would.[9] Further, with the deregulation of the electricity supply industry, there is pressure on the utilities to increase reliability and, at the same time, reduce operating costs with a minimum number of operating personnel. Single-pole auto-reclosing operation not only increases reliability but also reduces the work load on the operational staff, which in turn can result in lowering their number.

9.7.1 Arc Extinction with Single-Pole Switching

When one conductor of a three-phase line is opened at both ends to clear a ground fault, this faulted conductor is capacitively and inductively coupled to the two unfaulted conductors that are still energized at approximately normal circuit voltage and are carrying load current. This coupling has two effects:

- Before extinction of the fault arc, it feeds current to the fault and maintains the arc.
- After the arc current becomes zero (as it does twice per cycle), the coupling causes a recovery voltage across the arc path. If the rate of rise of the recovery voltage is too great, it will reignite the arc.

Of the two types of coupling, capacitive coupling is the more important.[9] Its importance increases with increase of circuit voltage, and it is the only type of coupling considered in detail in later analysis.

The arc on the faulted conductor after it has been switched off is called the *secondary arc*. Extinction of the secondary arc depends on its current, recovery voltage, length of the arc path, wind velocity, and perhaps on other factors such as the harmonic current contents of the arc, etc. Recovery voltage and length of the arc path both increase with circuit voltage, and thus, the effect of one factor may partially offset the other. This leaves the magnitude of the secondary arc current as the most significant index of whether the arc will be self-extinguishing.

For a given interphase capacitance, the secondary arc current is proportional to the circuit voltage and to the length of the line section that is switched out. Hence, the length of the section on which single-pole switching can be employed successfully is inversely proportional to the circuit voltage. Table 9.3 gives line lengths in the case of 230- and 765-kV lines for single-pole autoreclosure without supplemental arc-extinction devices such as shunt reactors.[8]

If the line is longer than that given in Table 9.3, and single-phase tripping and autoreclosing are to be applied, Knudsen[10] and Kimbark[9] have proposed a neutral reactor to ensure the extinction of the secondary arc current. Because coupling of the faulted conductor to the sound conductors through the shunt capacitive reactance between phases is the chief cause of the secondary arc current and recovery voltage, it is proposed that this capacitive reactance be neutralized by means of lumped shunt inductive reactance, equal and opposite to the capacitive reactance. The scheme proposed by Knudsen and Kimbark is analogous to the use of a Peterson coil, and both might well be called *ground-fault neutralizers*. However, the fourth reactor in single-pole autoreclosing projects would neutralize the capacitance between phases, amounting to $C_1 - C_0$ per phase (i.e., the difference between positive-sequence and zero-sequence capacitances); whereas the Peterson coil neutralizes the capacitances to ground, that is, $3\ C_0$. A further difference is that, whereas one Peterson coil can

TABLE 9.3
Line Lengths for Single-Phase Autoreclosure without Supplemental Arc-Extinction Devices (e.g., shunt reactors)

	Line Length (mi)	
Line-to-Line Voltage (kV)	**Successful Range**	**Doubtful Range**
765	0–50	50–80
500	0–60	60–100
345	0–140	140–260
230	0–300	300–500

Source: From IEEE Power System Relaying Committee Working Group (1992). Single Phase tripping and Auto Reclosing of Transmission lines—IEEE Committee Report, *IEEE Transactions On Power Delivery*, 7(1), 182–191. With permission from IEEE.

suppress ground faults anywhere on an entire transmission network, the reactors used with single-pole switching must be used on every transmission line that is too long for secondary arc extinction without them, and they must be switched with the line. Thus, a large number of reactors might be required for the latter. However, many EHV lines require shunt reactors for wholly or partly compensating the normal, positive-sequence charging current. By appropriate connections, these reactors can be made to serve the additional purpose of ground-fault suppression at a moderate additional expense.

According to Kimbark,[9] with a 0.4-s dead time, there is a high probability of successful reclosure if the secondary arc current does not exceed 18.0 A. Regarding the discussion on the maximum value of the secondary arc current, see References 11–13. If the neutral reactor is chosen to satisfy the theoretical condition for complete neutralization of the interphase charging current as derived by Knudsen[10] and Kimbark,[9] it will lead to an uneconomic system due to the higher basic insulation level required on all four reactors; Kimbark[11] and Carlsson et al.[13] have reported this point. Hence, the power-frequency reactance of the neutral reactor should be chosen only to partially compensate the interphase capacitance. A reduction in the secondary arc current to 9.0 A is considered satisfactory to ensure the extinction of the arc within the dead time of the auto-reclosing operation.

9.7.2 Laboratory Tests to Determine the Secondary Arc-Extinction Time

Experimental data are required to ascertain the dead time required for reliable extinction of secondary arcs on a given line.[11] References 14–17 present the results of tests made in laboratories on single-phase circuits that are predominantly capacitive, the only inductances being the leakage inductances of transformers. Kimbark is of the opinion that the results of such tests are valid for lines having no shunt reactors. They show that, for secondary arcs of about 20 A on 400- or 500-kV lines, prompt extinction is almost certain in time to permit successful 0.5-s reclosures, whereas at

30 A, the extinction time is likely to be too long. Then the next question to answer is whether the results of such tests are valid generally, which means, are they valid for application to lines having shunt reactors—either three reactors connected in Y-form with a solidly grounded neutral or the more special four-reactor bank adjusted for expediting extinction of the secondary arc.

Kimbark states that such test results are valid for Peterson's and Dravid's arc-extinction scheme[18] with no shunt reactors on the line because the added circuit elements are capacitors. Such test results are not valid for lines with shunt reactors, but they may give pessimistic results so that they may safely be used for reactored lines. We may be able to promptly extinguish greater secondary arc currents on reactored lines than those shown in the tests, perhaps up to 40 or 50 A.

9.7.3 Choice of Neutral Reactor

In this section,[7] we will try to derive an expression for the ratio X_n/X_1, where

X_n = neutral reactor value in ohms at power frequency
X_1 = shunt reactor positive-sequence value in ohms per phase at power frequency

The construction of a three-phase reactor will determine the ratio X_0/X_1, where

X_0 = shunt reactor zero-sequence value in ohms per phase at power frequency

For a three-limb core, this value could be as low as 0.4. If the $X_0/X_1 < 1$, then it is equivalent to connecting a capacitor between the neutral and ground of the shunt reactors, and this connection will increase the secondary arc current and recovery voltage. Hence, the ratio X_0/X_1 of the shunt reactors will also influence the neutral reactor size.

The requirement for 100% neutralization of the interphase capacitance is given by Knudsen[10] and Kimbark,[9] and is defined by the equation

$$1/\omega L_1 - 1/\omega (L_0 + 3L_n) = \omega (C_1 - C_0) \tag{9.1}$$

where
L_1 = positive-sequence inductance of the shunt reactor (in henrys)
L_0 = zero-sequence inductance of the shunt reactor (in henrys)
L_n = inductance of the neutral reactor (in henrys)
C_1 = positive-sequence capacitance of the line (in Farads)
C_0 = zero-sequence capacitance of the line (in Farads)
ω = angular frequency in radians per second at the power frequency 60 or 50 Hz

Let us define α, β, γ as

$$\alpha = X_0/X_1$$

$$\beta = 1/\omega^2 L_1 C_1 = \text{shunt compensation factor} \tag{9.2}$$

Hence

$$\omega^2 = 1/\beta L_1 C_1 \tag{9.3}$$

$$\gamma = C_0/C_1$$

Then from Equation 9.1, an expression for the ratio X_n/X_1 can be derived in terms of α, β, γ as follows.

Multiplying both sides of Equation 9.1 by $\omega(L_0 + 3L_n)$ yields

$$\omega(L_0 + 3L_n)/\omega L_1 - 1 = \omega^2(C_1 - C_0)(L_0 + 3L_n)$$

Substituting for ω^2 from Equation 9.3,

$$(X_0 + 3X_n)/X_1 - 1 = (1/\beta L_1 C_1)(C_1 - C_0)(L_0 + 3L_n)$$

$$\alpha + 3X_n/X_1 - 1 = (1/\beta)(1 - C_0/C_1)(L_0/L_1 + 3L_n/L_1)$$

$$= (1/\beta)(1 - \gamma)(X_0/X_1 + 3X_n/X_1)$$

$$= (1/\beta)(1 - \gamma)(\alpha + 3X_n/X_1)$$

$$= (\alpha/\beta)(1 - \gamma) + (1/\beta)(1 - \gamma)(3X_n/X_1)$$

$$(3X_n/X_1)(1 - (1 - \gamma)/\beta) = -\alpha + 1 + (\alpha/\beta)(1 - \gamma)$$

$$3(X_n/X_1)(\beta + \gamma - 1)/\beta = [-\alpha\beta + \beta + \alpha(1 - \gamma)]/\beta$$

Further algebraic simplification yields

$$X_n/X_1 = (1/3)[-\alpha + \{\beta/\beta + \gamma - 1\}] \tag{9.4}$$

If the shunt reactor details such as positive- and zero-sequence reactance values (X_1, X_0) and line data are given, then α, β, γ can be calculated. The preceding formula can then be used to compute X_n for 100% compensation of interphase capacitance.

9.7.4 Secondary Arc Current and Recovery Voltage

For lines or line sections of about 400 km in length, the voltage in the disconnected phase that is induced electromagnetically by the current in the sound phases must also be taken into consideration.[9,10]

Assume that a line is connected between two buses 1 and 2. Further assume that the fault on phase A is at bus 2.

Let Z_0 and Z_1 be the zero- and positive- (or negative) sequence no-load impedances of the line viewed from bus 1, Y_0 and Y_1 be the corresponding admittances, and E_a be the voltage in the disconnected phase before the fault. Examining the conditions at bus 1, it can easily be shown that the induced recovery voltage $E_{a\,ind}$ is given by the following equation:

$$E_{a\,ind} = -E_a(Z_0 - Z_1)/(2Z_0 + Z_1) = -E_a((Y_1 - Y_0)/(2Y_1 + Y_0)) \tag{9.5}$$

(Note that this equation is the same as Equation 22 in Kimbark's paper, Reference 9, with change in notation, that is, $Y_1 = B_1'$ and $Y_0 = B_0'$.) If the impedance of the feeding system cannot be neglected, it must be included in the denominator in the preceding equation and E_a be replaced by the internal voltage of the system.

Then, the secondary arc current $I_{a\,ind}$ in amperes, neglecting the arc resistance, is found to be

$$I_{a\,ind} = (-1/3)\,E_a(Y_1 - Y_0) \tag{9.6}$$

(Note that this equation is the same as Equation 24 in Kimbark's paper, Reference 9, with change in notation, that is, $Y_1 = B_1'$ and $Y_0 = B_0'$.) The driving-point admittance of the system viewed from the fault is

$$Y_{a\,ind} = (1/3)\,(2Y_1 + Y_0) \tag{9.7}$$

This is the mean value of the positive, negative, and zero admittances of the open line viewed from the fault. The latter expression can probably be used in all cases regardless of the way in which the line is connected at its ends because the impedance of the connected systems has very little influence in this respect. Consequently the secondary arc current is easily calculated when the induced recovery voltage is known.

The induced voltage (recovery voltage) may be regarded as composed of two components, that is,

$$E_{a\,ind} = Z_0\,I_0 + Z_1\,I_{12}$$

where

I_0 = zero-sequence current

I_{12} = sum of the positive- and negative-sequence currents flowing from bus 1 out in the line

It is found that

$$I_0 = -I_{12} = -E_a/(2Z_0 + Z_1) \tag{9.8}$$

The induced voltage (recovery voltage) at an arbitrary point on the line is given by

$$E_{a\,ind} = -E_a\,(Z_0 \cosh\theta_0'/(2Z_0 + Z_1)\,(\cosh\,\theta_0) - Z_1 \cosh\,\theta'/(2Z_0 + Z_1)\,(\cosh\,\theta))$$

where θ and θ_0 are the complex electric angles of the line for positive and zero sequences, whereas θ' and θ_0' are the complex electric angles of the part of the line that lies between the regarded point and the open end for positive and zero sequences.

9.7.5 Single-Pole Autoreclosing of EHV Lines—Field Test Results

Single-pole autoreclosing projects ranging from 220 to 765 kV lines have been completed in many countries. A few of the projects are listed as follows:

1. Single-pole switching on the Tennessee Valley Authority's (TVA) Paradise–Davidson 500-kV line—Design concepts and staged fault test results.[19] The following conclusions have been drawn from this project:
   a. Without special compensation, single-pole tripping and reclosing within one-half second is unsuccessful on a 500-kV radial line that is 93 mi (150 km) in length.
   b. Synchronous stability of a large generator is maintained during single-pole switching, provided reclosing is accomplished within one-half second.

   c. Switching surges and breaker transient recovery voltages for single-pole switching are no more severe than those for three-pole switching, provided closing and opening resistors are used.
   d. Solid-state line-protective relay systems properly identify the faulted phase and, with two-cycle breaker opening time, provide simultaneous three-cycle clearing at both ends of the protected line. Electromechanical relay systems require up to two additional cycles of clearing time at the far end of the protected line.
   e. Line-protective relaying schemes external to the protected line section operate properly, and erroneous relay operations do not occur because of the open-phase condition on the Paradise–Davidson connection.
   f. The surge-protective packages supplied with the solid-state line-protective relays are very effective during both staged fault and special arcing disconnecting switch tests.
2. Power generated by the Iron Gates Hydropower Plant (1050 MW) was transmitted over a 400-km, 400-kV line, in Romania. Single-pole autoreclosing was employed on this line using a four-reactor scheme with a neutral reactor.[13] For economical reasons, the neutral reactors will be purposely detuned on the side of low reactance to lower the basic insulation level on all four reactors. A ratio of $X_n / X_1 = 0.25$ was used in this project.
3. Thirteen years of experience with single-phase reclosing at 345 kV has been reported by the Quebec power system, Alcan Smelters and Chemicals Ltd. (Reference 20) in 1978. The following conclusions have been reached based on 13 years of experience:
   a. A dead time of the order of 50 cycles (0.83 s) is sufficient for the extinction of the residual current arc of single-phase-to-ground faults on the 147-km (91.2-mi) lines. It is planned to try shorter dead times later.
   b. Low tower-footing resistances (i.e., approximately 25 ohms maximum) are essential for reliable relay operation and successful reclosing performance.
   c. The addition of two shield wires reduced the lightning-induced flashover rate by a factor of 10, down to 0.86 flashovers per 100 circuit-miles per year.
   d. With low-footing resistances, nearly all lightning-induced flashovers will involve only one phase conductor on lines of horizontal configuration. When new lines are contemplated, the omission of shield wires should be considered.
4. Analysis of single-phase-switching field tests on the AEP 765-kV system has been reported in Reference 21. This is a new scheme in which neutral and high-side reactor switches are installed, because this 765-kV line is untransposed. See also Reference 8 for details of the other schemes in addition to the usual four-reactor scheme.
   a. Single-phase reclosing was successful in 23 tests where the secondary arc current was less than 40 A root-mean-square (rms) and the recovery voltage did not exceed the air gap withstand capability. The secondary arc extinction time in these tests did not exceed 160 ms and was independent of the secondary arc current value.

   b. Single-phase reclosing was unsuccessful in four tests where the secondary arc current was greater than 45 A rms and the recovery voltage exceeded the air gap withstand capability.
5. Staged fault tests with single-phase reclosing on the Winnipeg–twin cities 500-kV interconnection.[22] The following conclusions have been drawn from this project:
   a. The secondary arc time (time from the opening of the last breaker to the initial secondary arc extinction) was found to be dependent on the amount of dc offset in the primary fault current. This, in turn, is a function of the point on wave at which the fault occurs.
   b. The arc recovery voltages, with an initial rate of rise of recovery voltage (RRRV) of 3–4 kV/ms, did not result in the restriking of the secondary arc.
   c. Initial oscillations in the secondary arc are not due to transformer saturation but are likely due to natural frequencies of the circuit formed by the transmission line, fault location, and the shunt and neutral reactors.
6. Single-phase switching compensation schemes for transposed and untransposed lines were tested during staged fault tests on the 243-km, 765-kV Kammer–Marysville (*K–M*) line (USA) and the 417-km, 750-kV Vinnista–Dnieper (*V–D*) line (USSR).[23] Conventional four-legged reactors were used on both lines and, in addition, a switched four-legged reactor bank was utilized on the untransposed *K–M* line.

   In the USSR, 750-kV lines are transposed and, therefore, the electrical parameters of their individual phases are approximately equal. Thus, the direct use of four-legged reactors is applicable. In the United States, 765-kV lines are untransposed and therefore have considerably different interphase parameters. To effectively limit the secondary arc current on these lines, special switching schemes were developed to supplement the conventional four-legged reactors.[21,23]

   a. The secondary arc-extinction time when the calculated secondary arc current was less than 40 A rms did not exceed 0.15 s on the *K–M* line and 0.3 s on the *V–D* line. The rate of rise of recovery voltage after arc interruptions was less than 8 kV/ms in these tests.
   b. The secondary arc-extinction time in the *V–D* tests varied from 0.33 to 1.42 s in the schemes corresponding to the calculated secondary arc current (I_s) of 49 A rms.
   c. The secondary arc-extinction time in schemes with calculated secondary arc current (I_s) less than 30 A rms is determined mainly by the first zero crossing of the secondary arc current. A larger initial dc component in the secondary arc current results, as a rule, in a longer extinction time.
   d. Gapless zinc-oxide arresters were successfully used for the protection of neutral reactors against switching surges and temporary overvoltages during single-pole switching operations. Stabilized spark gaps can also be used for this purpose.

   e. Resonant overvoltages on an open phase of the *V–D* line were limited to 1.25 pu due to corona losses.
7. The European practice on single-pole switching is summarized in Reference 24. See Table 9.4 for a summary. Single-pole switching is not used in the United Kingdom; there, three-phase reclosing is generally applied. Sweden introduced single-pole switching in the early 1950s, but has since abandoned its use.
8. The results of the digital and transient network analyzer (TNA) studies performed at the planning stage of the Greek 400-kV network are reported in Reference 25. This includes the details of the choice of neutral reactor and discussion of its specifications such as BIL, current rating, and so on.

Some relaying practices in single-phase tripping and auto-reclosing of transmission lines are discussed in an IEEE committee report given in Reference 8.

9.7.6 Effect of X_0/X_1 Ratio and X_N on the Secondary Arc Current and Neutral Voltage

The effect of the ratio X_0/X_1 and X_n on the secondary arc current and neutral voltages will be illustrated using the Ross–Chalumbin 275-kV line in Queensland, Australia. The single-line diagram of the system studied is shown in Figure 9.1. At the Chalumbin end, 30 MVAr (275 kV) of shunt compensation is proposed for each circuit of the double-circuit line. However, one circuit is to be strung initially. The tower configuration details are shown in Figure 9.2. The formula for the calculation of X_n is given by Equation 9.4, which is reproduced below for easy reference.

$$X_n/X_1 = (1/3)\,[-\alpha + \{\beta/\beta + \gamma - 1\}] \tag{9.4}$$

In the present case, shunt compensation factor $\beta = (1/\omega^2\, L_1\, C_1) = 0.56$, $\gamma = C_0/C_1 = 0.62$, and shunt reactor $X_1 = 2600$ ohms.

For different values of α, values of X_n are tabulated in Table 9.5. Assume that the fault is on phase A at one-third the distance from the Ross end. The three sections of the transposed line are represented by their multi-π equivalents in the Electromagnetic Transients Program (EMTP) simulations. Table 9.6 shows the calculated secondary arc current (I_f) and neutral voltages for an isolated fault, an unisolated fault, and neutral voltage with only one phase energized, for different values of neutral reactance and values of α for the three-phase shunt reactor. A system voltage of 1.1 pu is assumed for these calculations.

From Table 9.6, it may be seen that $\alpha = 0.4$ yields high neutral voltages under unisolated fault conditions and, in general, higher secondary arc currents when the neutral reactor size is not sufficient to completely neutralize the interphase capacitance of the line. If three-phase line reactors with $\alpha = 0.4$ are used with lines designed for single-pole reclosing, a higher basic insulation level for all four reactors and large neutral reactors will be required. Hence, they are not recommended to be used on 275-kV lines designed for single-pole reclosing operation.

In the case of the Ross–Chalumbin line, if a shunt reactor is chosen with $\alpha = 0.4$, then a neutral reactor will be required. Without a neutral reactor, single-pole autoreclosing

TABLE 9.4
Answers to the Questionnaire "European Practice on Single-Pole Switching"[a]

	France	FR Germany VEW	FR Germany 2 EVS	Belgium	Italy	Norway	Switzerland
Voltage levels using SPS	225 kV 400 kV	220 kV 380 kV	220 kV 380 kV	155 kV 225 kV 400 kV	220 kV 380 kV	300 kV 420 kV	220 kV 400 kV
Overhead ground wires	Yes, after 1960	Yes	Yes		Yes		Yes
Average line length (km)	225 kV: 104 400 kV: 140	220 kV: 33 380 kV: 72	220 kV: 50 380 kV: 60	155 kV: 20 225 kV: 40 400 kV: 40	80–90	300 kV: 60 420 kV: 70	220 kV: 42 400 kV: 48
Maximum line length (km)	225 kV: 173 400 kV: 238	220 kV: 69 380 kV: 117	220 kV: 152 380 kV: 117	155 kV: 60 225 kV: 114 400 kV: 115	220 kV:180 380 kV: 250	300 kV:140 420 kV: 184	220 kV:187 400 kV:149
Fault duration (s)	225 kV: 0.12–0.20 400 kV: 0.12–0.15	0.075–0.10	0.09–0.12	0.06–0.20[b]	0.09–0.12	0.06–0.15	0.06–0.08
Duration of phase opening (s)	1.5	1.0	220 kV: 0.6 380 kV: 1.2	0.5[c] 1.0	220 kV: 1 380 kV: 2–5	0.7–1.0	0.5 1.0[d]
Aid to suppress secondary arc	None	None	None	None	380 kV: neutral reactor	None	None

SPS introduced in year	Before 1960	220 kV: 1963 380 kV: 1975	220 kV: 1951 380 kV: 1974	155 kV: 1950 225 kV: 1973 400 kV: 1971	380 kV: 1964	300 kV: 1960 420 kV: 1963	220 kV: 1954
Main activation for SPS	Security of supply, stability	Security of supply, stability	Security of supply, stability	Security of supply, stability	Security of supply, stability	Security of supply, stability	High-rate single-phase faults, stability
Frequency of single-phase faults in # per 100 km year	225 kV: 8.7	0.7	220 kV: 3.86 380 kV: 2.95	155 kV: 4.28 225 kV: 5.22 400 kV: 1.20	380 kV: 3.0	300 kV: 0.5 420 kV: 0.8	220 kV: 2.2 400 kV: 5.9
Percentage of successful reclosure	225 kV: 88.5 380 kV: 82.0	90%	220 kV: 100 380 kV: 80 380kV: 80	155 kV: 72 225 kV: 84 400 kV: 89	380 kV: 80	300 kV: 22 420 kV: 15	220 kV: 72 400 kV: 84

[a] Single-pole switching not used in the United Kingdom, three-phase reclosing generally applied; Sweden introduced single-pole switching in early 1950s; since then abandoned.

[b] Minimum fault duration 0.06 s; average 0.20 s.

[c] Initially 0.5 s; 1.0 s for the last 10 years.

[d] In some cases 1.0 s.

Source: From IEEE Discrete Supplementary Controls Working Group (1986)—Report of a Panel Discussion: "Single-pole switching for Stability and Reliability," *IEEE Trans. PAS*, PWRS-1 (2) 25–36. With permission from IEEE.

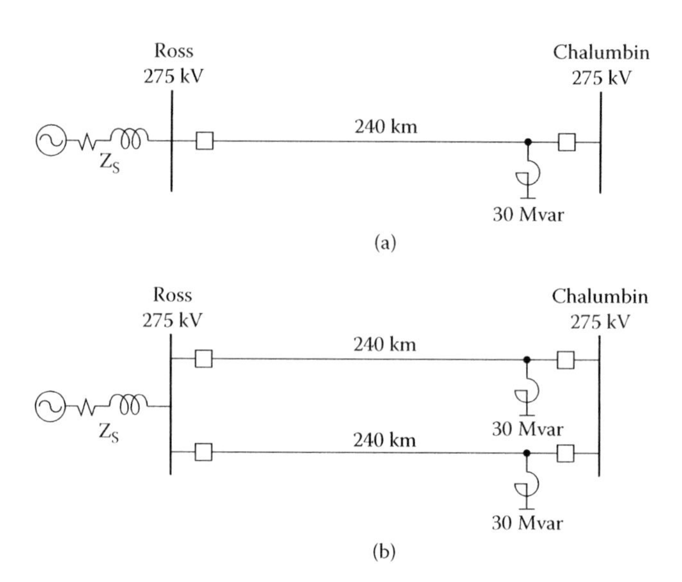

FIGURE 9.1 Ross–Chalumbin single-line diagram: (a) initial—single circuit strung and (b) ultimate—both circuits strung. (From Sastry, V.R. and Dunn, C.J. (1988). Factors affecting secondary arc current in EHV lines with single-pole auto reclosing, *Journal of Electrical and Electronic Engineering Australia,* 8 (2), 140–143. With permission from IE (Australia).

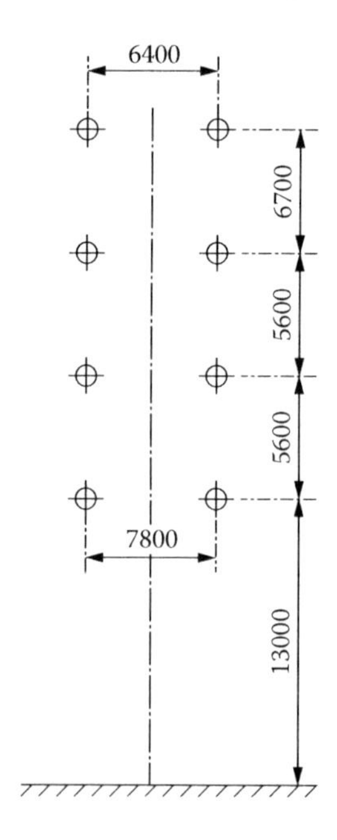

FIGURE 9.2 Tower configuration double-circuit steel tower. (From Sastry, V.R. and Dunn, C.J. (1988). Factors affecting secondary arc current in EHV lines with single-pole auto reclosing, *Journal of Electrical and Electronic Engineering Australia,* 8 (2), 140–143. With permission from IE (Australia).

TABLE 9.5
Effect of X_0/X_1 Ratio on the Neutral Reactor Size X_n

$\alpha = X_0/X_1$	X_n (ohms)
0.40	2.350
0.85	1.960
1.00	1.830

Source: From Sastry, V.R. and Dunn, C.J. (1988). Factors affecting secondary arc current in EHV lines with single-pole auto reclosing, *Journal of Electrical and Electronic Engineering Australia,* 8 (2), 140–143. With permission from IE (Australia).

will not be successful because the secondary arc current is 49.0 A. One of the proposals under consideration was to divert the existing reactors from one of the power stations for use as line reactors at Chalumbin. This is a three-phase reactor with an approximate value of $\alpha = 0.85$. If this reactor is used as a line reactor, the secondary arc current will be reduced to 19.3 A, increasing the probability of success of single-pole autoreclosing even without a neutral reactor.

TABLE 9.6
Effect of the Ratio X_0/X_1 and X_n on the Neutral Voltage and Secondary Arc Current

		Neutral Voltage, kV (rms)			
$\alpha = X_0/X_1$	X_n (ohms)	Isolated Fault	Unisolated Fault	One Phase Energized Only	I_f (A)
0.40	2350	42.5	91.3	82.9	0.17
0.85	1960	40.4	76.2	75.1	0.18
1.00	1830	39.5	71.1	72.1	0.18
0.40	1000	31.2	77.8	48.5	7.39
0.85	1000	31.2	60.4	48.5	4.15
1.00	1000	31.2	56.1	48.5	3.37
0.40	500	21.3	61.9	28.2	15.90
0.85	500	21.3	42.3	28.2	8.67
1.00	500	21.3	38.3	28.2	7.18
0.40*	—	—	—	—	49.0
0.85*	—	—	—	—	19.3
1.00*	—	—	—	—	15.4

Source: See Table 9.5.

9.7.7 Effect of Transposition Phasing of Double-Circuit Lines

In double-circuit lines with vertical configuration, the transposition sequences ABC (clockwise)/CBA (anticlockwise) yields minimum negative-sequence voltage at buses connecting these lines.[7] The transposition sequence ABC (clockwise)/CBA (anticlockwise) corresponds to ABC, CAB, BCA phasing on circuit 1, and CBA, BAC, ACB phasing on circuit 2, in the three sections of the transposed line. Table 9.7 shows the computed secondary arc currents, with two types of transpositions ABC (clockwise)/ABC (clockwise) and ABC (clockwise)/CBA (anticlockwise). The three sections of the transposed line are represented by their multi-π equivalents in the EMTP simulations.

At Chalumbin, 30-MVAr (275-kV) shunt reactors are connected to each circuit, with the neutrals solidly grounded. The faults are applied at the Ross end of the line. From Table 9.7, it may be seen that the secondary arc current is considerably increased (up to a maximum of 88%) for a fault on phase A or for faults on phase A of circuit 1, and phase C of circuit 2 with ABC (clockwise)/CBA (anticlockwise) arrangement.

If a 275-kV line is long (say 250 km) with heavy loading (200 MVA) and is likely to be tapped in the middle at a later date, it is preferable to have two barrels of transposition with opposite rotation, that is, one barrel ABC (clockwise) and the other BCA (anticlockwise). By having the middle one-third of the section at the same phasing with this arrangement, only one extra transposition tower is needed compared to a single barrel of transposition. Studies were conducted to check whether the secondary arc current will be affected significantly by a second barrel of transposition. Our studies indicate that there is no significant change in the secondary arc current and steady-state recovery voltage due to the additional transposition.

The phasing arrangement of transposed lines should be primarily chosen to reduce the negative-sequence voltages at buses because this will produce a continuous unbalance in the system. The secondary arc current can be reduced by suitable neutral reactors if they are found necessary. As some EHV lines (say 243-km,

TABLE 9.7
Effect of Transposition Pattern of Double-Circuit Steel Tower Line

Transposition Pattern	Faulted Phase		Secondary Arc Current, Amperes (rms)
	Circuit 1	Circuit 2	
ABC/ABC →/ →	A	—	8.9
ABC/ABC →/ →	A	C	8.0
ABC/ABC →/ →	A	A	19.6
ABC/CBA →/←	A	—	16.7
ABC/CBA →/←	A	C	14.6
ABC/CBA →/←	A	A	22.1

A system voltage of 1.1 pu is assumed.

Source: From Sastry, V.R. and Dunn, C.J. (1988). Factors affecting secondary arc current in EHV lines with single-pole auto reclosing, *Journal of Electrical and Electronic Engineering Australia*, 8 (2), 140–143. With permission from IE (Australia).

765-kV K–M line, USA) are untransposed (due to the cost of transposition towers, etc.), the preceding remarks are not applicable to them.

9.7.8 Selective-Pole Switching of Long Double-Circuit EHV Line

Kimbark[12] has developed the joint use of several new concepts for improving stability, reliability, and safety of a double-circuit three-phase EHV line[12,26]:

1. Use of selective-pole switching; that is, only the faulted conductors are opened and closed
2. Use of a bank of three shunt capacitors and nine shunt reactors for neutralizing the 15 interconductor capacitances and thereby eliminating the shunt capacitive coupling that tends to maintain the secondary arc (see Figure 9.3 for the arrangement of the nine shunt reactors)
3. Sectionalizing the faulted conductor or conductors into two or more longitudinal sections by remote-controlled switches to reduce the longitudinal resistive–inductive coupling that also tends to maintain the secondary arc
4. Exclusive use of the arrangement of conductors that gives the least current to ground from a large vehicle capacitively coupled to the line; like phases are in diagonally opposite positions

The optimum reactor compensation configurations were not practical for a 328-km, 500-kV double-circuit line in Electricity Generating Authority of Thailand (EGAT). Hence in Reference 26, Thomann et al. discuss nonoptimum compensation schemes for single-pole reclosing on this 500-kV double-circuit line.

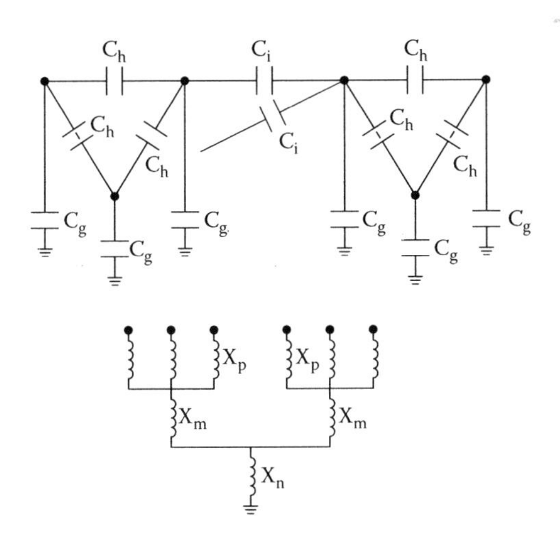

FIGURE 9.3 The intra- and intercapacitance of the double-circuit line and the reactor compensation configuration. (From Thomann, G.C., Lambert, S.R., and Phaloprakarn, S. (1993). "Non-Optimum Compensation Schemes for Single Pole Reclosing on EHV Double Circuit Transmission Lines," *IEEE Trans. on Power Delivery*, Vol. 8, No. 2, April 1993, pp. 651–659. With permission from IEEE.)

9.8 TYPES OF REACTORS BASED ON THEIR FUNCTION

Reactors are either series or shunt connected. Series reactors are generally used as current-limiting reactors. Shunt reactors are used to provide reactive compensation. Some of the applications are explained below.[27,29]

1. *Current-limiting reactors* are used to reduce the short-circuit current to a value within the rating of the equipment located after the reactor.
2. *Filter reactors* are used in series with a shunt capacitor bank to provide a tuned series resonant LC circuit for specific harmonic currents.
3. *Neutral grounding reactors* limit the line-to-ground fault current to the same value as the line-to-line fault current.

(Note that the purpose of the neutral reactor in single-pole autoreclosing applications is rather different. Its main purpose is to reduce the secondary arc current by neutralizing the interphase capacitance than to reduce single-line-to-ground fault current.)

4. *Capacitor reactors* are designed to be connected in series with a shunt-connected capacitor bank either to limit inrush current on energization or to control the resonant frequency of the system due to addition of the shunt capacitance.
5. *Smoothing reactors* are used to reduce the magnitude of the ripple current in a dc system. They are often required in HVDC schemes and may be required in series with rectifier-fed large dc motors used in the steel, mining, and process industries.
6. *Shunt reactors* are used to compensate for capacitive vars generated by long, lightly loaded transmission lines. Other applications are as var compensators where the reactive vars are continuously controlled by thyristors. 60-Hz ratings up to 25 MVA per coil are commercially available.
7. *Duplex reactors* provide a low reactance under normal conditions, and high reactance under fault conditions. These reactors can be applied to any system that has two separate feeder circuits that will remain isolated under all circumstances. The reactor is wound so that, under normal conditions, the magnetic field from each half of the coil is in opposition, giving low reactance between source and load. When a fault occurs in one feeder, the fault current flows only through one half of the coil. Because there is no fault current in the healthy feeder, there is no opposing flux from the other half of the coil. The fault current is therefore limited due to the increased reactance of the half of the coil in series with the faulted feeder.
8. *Motor-starting reactors* limit the starting current of ac motors.
9. *Load-balancing reactors* serve to control the current flowing in two or more parallel circuits.

9.9 CONSTRUCTION OF REACTORS

Consideration of various reactor designs is assisted by remembering that a reactor is an energy storage device and that, at a given frequency, its kVA rating is proportional to the stored energy in its magnetic circuit.[28]

From electromagnetic theory we know that[1]

$$\text{Energy stored per unit volume} = B^2/(2\mu) \tag{9.9}$$

where B = uniform flux density in an inductance coil and μ= permeability of the medium around which the coil is wound.

Further, we know that average energy stored in an inductance L is

$$W_{av} = LI^2/2 \tag{9.10}$$

where I = rms current in the coil.

Muliplying both sides of Equation 9.9 by 2ω, where ω = angular frequency in radians per second,

$$2\omega W_{av} = \omega LI^2 = X_L I^2$$

where X_L = reactance of the coil.

Hence, the kVA rating of a reactor $X_L I^2$ (X_L in kiloohms and I in amperes rms) is proportional to the volume of the magnetic circuit and $B^2/(2\mu)$.

$$\text{kVA rating} \propto W_{av} \propto \frac{1}{2}LI^2 \propto \text{Volume of the magnetic Circuit} \times (\text{B}^2)\frac{1}{2\mu}$$

In the above equation the symbol $\propto$ indicates proportionality. In other words kVA rating is proportional to W_{av}, which in turn is proportional to $\frac{1}{2}LI^2 \cdot \frac{1}{2}LI^2$ Is then proportional to Volume of the magnetic Circuit $\times (\text{B}^2)\frac{1}{2\mu}$.

It is evident from the foregoing discussion and Equation 9.9 why the designer uses a substantial volume of nonmagnetic (unity permeability) material in the reactor magnetic circuit. Consider an unloaded transformer; its core is of large volume and operates at high flux density, but its kVA as a shunt reactor is small because of the high permeability of steel.

Having chosen a low-permeability medium, the designer has the further choice of a magnetic field of large volume and low density, or conversely small volume and high density. Reference 29 describes a calculation method for voltage distribution in a large air-core power reactor to assist the designer. Good agreement was reported between the calculated results (using this method) and measured results for the impulse voltage distribution along the winding length at 0.5 μs after energization.

9.9.1 Types of Reactors Based on their Construction

Figure 9.4 shows the different types of reactor designs[1]:

a. Coreless
b. Barrel shielded
c. Magnetic shell
d. Gapped core

A coreless reactor consists of a coil with a substantial clearance between itself and the conducting shield in a cylindrical tank. The three-phase version is built by stacking the other phase coils coaxially, often using one coil in two pieces to obtain balanced mutual and self-inductances for phase balance. It can be made for Y or delta connection. This design has constant inductance, regardless of current, as there is no magnetic material to saturate. The 65-MVA, 230-kV, three-phase reactors have been in successful

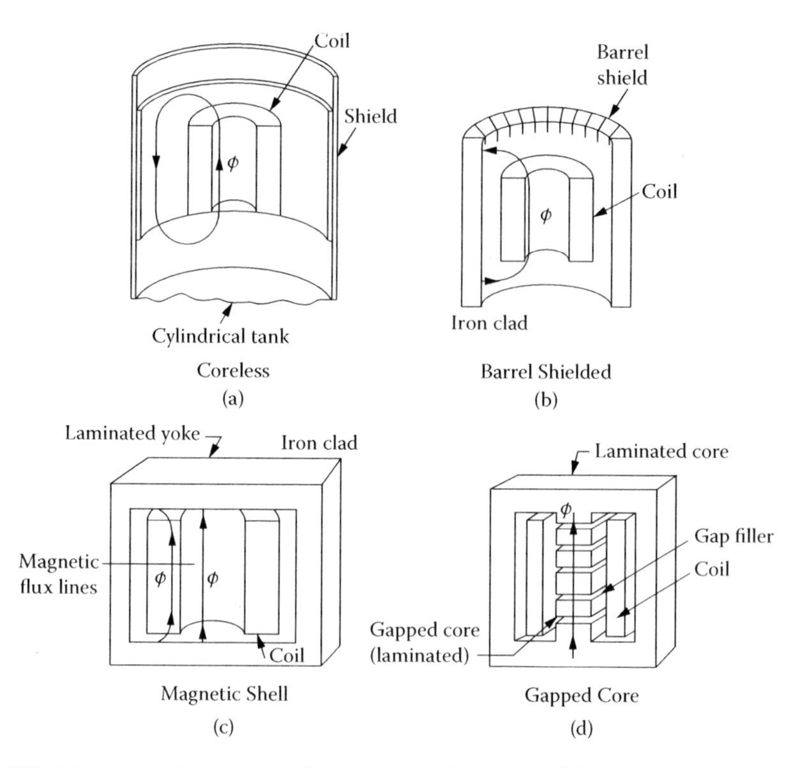

FIGURE 9.4 Types of reactor designs. (From Alexander, G.H., Hopkinson, R.H., and Welch, A.U. (1966). Design and Application of EHV Shunt Reactors, *IEEE Trans. on PAS,* 12, 1247–1258. With permission from IEEE.)

operation, and are characterized by sound levels several decibels below the NEMA (National Electrical Manufactures Association) standard for transformers of this rating. Size and voltage limitations of this design are determined by coil compressibility, that is, the ability to withstand compressive magnetic forces without undue vibration.

This design can be modified by using magnetic steel to form part of its external circuit. Hence, *barrel-shielded* and *magnetic shell* types can be combined, and are called *iron-clad* types. In the barrel-shielded type, the coil is enclosed in a long barrel made of narrow steel strips, the plane of lamination being radial. A large clearance to the coil is no longer needed, and the steel can be as close as permitted by the insulation. Because this adds to reactance rather than subtracting (as does the conducting shield), the coil is slightly reduced in size.

A more frequently used magnetic return circuit is the laminated picture-frame yoke in Figure 9.4c. Economics require that magnetic densities inside such coils be 3000–4000 G. The attractive forces on the yoke from this field over the large cross section of the coil become formidable when compared with the stiffness of the yoke structure. Vibration difficulties encountered with this construction can result in sound levels up to 100 dB.[29]

If a portion of the magnetic circuit contains steel, the usual copper content can be decreased. This generally results in lower losses because watts per pound in steel are less than in copper at practicable flux and current densities. Dollar evaluation of electrical loss for continuously operating reactors is high enough to influence the design in the direction of low losses. This incentive has led to design innovations for controlling vibration (i.e., keeping it within tolerable limits), yet obtaining loss improvement by using magnetic steel.

A magnetic shell design was used in 55-MVA, 735 kV, single-phase reactors that tested for a 73-dB sound level and 0.27% power factor. Figure 9.4d illustrates the principle of the *gapped-core design*. Flux density in the core is comparable to that in power transformers; thus the gap volume is only about one-tenth of that in the semicoreless designs (iron-clad). For reasons of linearity, the core flux density is a little lower than that in the power transformers. The effect on size is compensated by the lack of a low-voltage winding. Therefore, considerations of maximum size versus voltage or number of phases are the same as those of power transformers. The 70-MVA, 500-kV, three-phase reactors are constructed with the gapped-core design. Much larger three-phase reactors can be readily manufactured. The two following characteristics, i.e., linearity and phase balance affect the performance of the reactors installed in power systems. The design of the reactors i.e magnetic shell design and gapped core design shape these characteristics.

Linearity is important, since this characteristic determines the break point (knee point), i.e., the intersection point of the two portions of the magnetization curve, one representing the unsaturated portion of the curve, the other representing the saturation portion of the curve.

The phase unbalance will contribute to the current flowing between the neutral and ground causing relaying or telephone interference problems.

Linearity and phase unbalance are further discussed below.

Linearity. The coreless reactor has a constant reactance, regardless of voltage, that is, a linear characteristic. Any reactor using steel in its magnetic circuit will show some saturation when sufficient voltage is applied.

Gapped-core reactor excitation characteristics can be represented by two straight lines with a break point at 1.25 pu value, that is, the intersection of the magnetization curves of the two portions (See Figure 10 in Ref. 1). TNA studies with this gapped-reactor characteristic for typical systems have shown that overvoltages are not amplified during switching. Transformers usually saturate at voltages lower than this.

Saturation of the shell of iron-clad reactors results in reactances that are lower than that of the coil alone because of the screening effect of the current in the tank and in other metal structures.

Unbalance between duplicate single-phase reactors or between phases in a three-phase reactor. This is small because of the large size of reactors. Unbalance is caused by minor variations in dimensions; therefore, percentage variation will not be substantial in units of large physical size.

It seems likely that much larger ground currents will result from unbalanced ground capacitance and inductance of the rarely transposed EHV lines than from the phase unbalance of EHV reactors. A suggested phase unbalance criterion for EHV reactors is ±3% of the average of the phase currents of the bank. This is less than the unbalance of long untransposed lines and should not cause relaying or telephone interference problems.

Trench Electric in Canada designs and manufactures dry-type air-core reactors up to 25 MVA at 60 Hz (or 21 MVA at 50 Hz) per coil for both series and shunt applications.[27] The main features of these reactors are

- Epoxy-impregnated, continuous-filament fiberglass-encapsulated construction
- Epoxy resins compatible with the glass fibers under thermal-shock conditions
- Aluminum construction throughout, with all current-carrying connections being welded
- Aluminum conductors individually wrapped with polyester film
- Class B (150°C Hotspot) insulation with the hottest spot winding temperature by design not exceeding 130°C for series-connected coils and 120°C for shunt-connected coils
- Inherently weather-proof coil suitable for outdoor use without further protection
- Essentially zero radial voltage stress
- Noise level of less than 60 dB on the "A" scale at full-load current, maintained throughout the life of the reactor without mechanical adjustment
- No maintenance required
- High kVA-to-weight ratio

9.9.2 Testing of EHV Reactors

Full-load testing of EHV reactors has been difficult for most manufacturers because of the large power supplies needed.[1,30] In contrast to a new transformer that can be tested at rated load current with less than one-fifth of its rated kVA, a reactor requires full kVA and, because it has only one winding, it also requires full voltage. Very large capacitor banks are needed to supply the kVAr with generators of practical size. The required high voltage can be obtained from large test transformers or by using series resonance with capacitor banks. Measuring the power losses in a modern EHV reactor at full rating is difficult because the power factor is generally less than 0.3% and the losses are small in magnitude, even at full load. Power-factor errors of voltage and current transducers may be nearly as much as the powerfactor of the excitation current. Often, it is more accurate to test at low voltage and extrapolate from design knowledge to rating because the extrapolation errors will be less than the high-voltage measurement errors. When rated-voltage heat runs can be done, a reasonable loss approximation can be made by means of a dc heat run adjusted to cause the same oil temperature rise as an ac test.

Three-phase reactors require three-phase test power, regardless of design; no single-phase test can provide realistic measurement of losses, heating, or vibration. The applied potential insulation tests, sometimes termed the *induced tests,* are more easily obtained. General Electric Company in the United States routinely does this test by series resonance at 420 Hz with the moderate-size capacitor bank needed at this frequency. Also routinely, during this test, the radio noise measurement is made at full test voltage. Impulse testing is similar to that done on transformers.

9.10 CONCLUSIONS

In this chapter, we discussed the different planning aspects that deal with the choice of shunt reactors and their application in power systems. We have listed the other types of reactors, such as current-limiting reactors, filter reactors, neutral grounding reactors, capacitor reactors, smoothing reactors, duplex reactors, motor-starting reactors, load-balancing reactors, and their applications. We have also discussed the design and constructional details of the different types of EHV shunt reactors.

In an EHV line, up to 70–80% of the faults occurring are single-line-to-ground, and most of these are transitory due to lightning.[7,8] Hence, to improve reliability, several utilities use single-pole autoreclosing. We have discussed the four-reactor scheme with a neutral reactor to extinguish the secondary arc current by neutralizing the interphase capacitance. Further, we have derived an expression to facilitate the design of neutral reactors.

REFERENCES

1. Alexander, G.H., Hopkinson, R.H., and Welch, A.U. (1966). Design and application of EHV shunt reactors, *IEEE Trans on PAS,* 12, 1247–1258.
2. Croft, W.H., Bartley, R.H., Linden, R.L., and Wilson, D.D. (1968). Switching surge and dynamic voltage study of Arizona Public Service Company's proposed 345-kV transmission system utilizing miniature analyzer techniques, *IEEE Transactions on PAS*, 81, 302–312.
3. Johnson, W.R., Anderson, J.G., and Wilson, D.D. (1963). 500 kV line design, 1—Model Studies, *IEEE Transactions on PAS*, 82, 572–580, August 1963.
4. Wald, E.E. and Angland, D.W. (1964). Regional Integration of Electric Power Systems, *IEEE Spectrum*, 96–101.
5. Jancke, G., Jenkins, R., Nordstorm, B., and Norlin, L. (1962). The Choice of Shunt Reactors for the Swedish 400 kV system. *CIGRE Paper* 412, Vol 3.
6. Voltage Control on the British Grid (1968). *Electrical Review*, 276–277.
7. Sastry, V.R. and Dunn, C.J. (1988). Factors affecting secondary arc current in EHV lines with single-pole auto reclosing, *Journal of Electrical and Electronic Engineering Australia,* 8, (2) 140–143.
8. IEEE Power System Relaying Committee Working Group (1992). Single Phase tripping and Auto Reclosing of Transmission lines—IEEE Committee Report, *IEEE Transactions on Power Delivery*, 7(1) 182–191.
9. Kimbark, E.W. (1964). Suppression of Ground-Fault Arcs on single-pole switched EHV lines by Shunt Reactors, *IEEE Trans. PAS*, 83, 285–290.
10. Knudsen, N. (1962). Single-phase Switching of Transmission Lines using Reactors for Extinction of the Secondary Arc, *CIGRE Paper* 310.
11. Kimbark, E.W. (1975). Charts of three quantities associated with single-pole switching, *IEEE Trans. PAS,* 94, (2), 388–395. (See discussion by Kimbark on p. 394.)
12. Kimbark, E.W. (1976). Selective-Pole switching of Long Double-Circuit EHV Line, *IEEE Trans. PAS*, 95, (1) 219–230. (See discussion by Kimbark on p. 229).
13. Carlsson, L., Groza, L., Cristovici, A., Necsulescu, D.S., and Ionescu, A.I. (1974). Single-Pole Reclosing on EHV Lines, *CIGRE Paper* 31-03.
14. Haubrich, H., Hosemann, G., and Thomas, R. (1974). Single-Phase Auto-Reclosing in EHV Systems, *CIGRE paper* 31-09.

15. Ozaki, Y. (1965). Arc Characteristics of Insulator Strings for EHV Transmission Lines, *IEEE Paper* (31) CP65-18.
16. Balosi, A., Malaguti, M., and Ostano, P. (1966). Laboratory Full-Scale Tests for the Determination of the Secondary Arc Extinction Times in High Speed Reclosing, *IEEE Paper* (31), pp. 66–382.
17. Gary, C., Hesketh, S., and Moreau, M. (1971). Essais d'auto-extinction d'arcs greles en cas de declenchment monophase, *R.G.E.* 80 (5), 406–412.
18. Peterson, H.A. and Dravid, N.V. (1969). A Method for Reducing Dead Time for Single-Phase Reclosing in EHV Transmission, *IEEE Trans. PAS,* 88 (4) 286–292.
19. Edwards, L., Chadwick, J.W. Jr., Riesch, H.A., and Smith, L.E. (1971). Single-pole switching on TVA's Paradise-Davidson 500-kV line—Design concepts and staged fault test results, *IEEE Trans. PAS*, 91 (6), 2436–2450.
20. Haun, R.K. (1978). 13 years' Experience with Single-phase Reclosing at 345kV, *IEEE Trans. PAS*, 97 (2) 520–528.
21. Sherling, B.R., Fakheri, A.J., Shih, C.H., and Ware, B.J. (1981). Analysis of single phase switching field tests on the AEP 765kV system, *IEEE Trans. PAS*, 100 (4) 1729–1735.
22. Kappenman, J.G., Sweezy, G.A., Koschik, V., and Mustaphi, K.K. (1982). Staged fault tests with single phase reclosing on the Winnipeg-twin cities 500kV interconnection, *IEEE Trans. PAS*, 101 (3) 662–673.
23. Sherling, B.R., Fakheri, A.J., Shih, C.H., and Ware, B.J. (1978). Compensation scheme for single-pole switching on untransposed transmission lines, *IEEE Trans. PAS*, (97) 1421–1429.
24. IEEE Discrete Supplementary Controls Working Group (1986)—Report of a Panel Discussion: "Single-pole switching for Stability and Reliability" *IEEE Trans. PAS*, PWRS-1 (2) 25–36.
25. Cazzani, M., Clerici, A., Margaritidis, P., and Theloudis, J. (1974). Internal Overvoltages on the Greek 400 kV network, *CIGRE Report* 33.
26. Thomann, G.C., Lambert, S.R., and Phaloprakarn, S. (1993). "Non-Optimum Compensation Schemes for Single Pole Reclosing on EHV Double Circuit Transmission Lines," *IEEE Trans. on Power Delivery*, Vol. 8, No. 2, April 1993, pp. 651–659.
27. Dry-Type Air Core Reactors. (1978) *Trench Electric Bulletin* 100-05.
28. Thaler, G.J. and Wilcox, M.L. (1966). *Electric Machines: Dynamic and Steady State*, John Wiley, New York.
29. Salama, M.M.A. (1981) A Calculation Method for voltage distribution in a large Air Core Power Reactor, *IEEE Trans. PAS*, 100, 1752–1758.
30. Geijer, G., von. Jenkins, R.S., Sollergren, B., and Myklebust, R. (1964). Application design and testing problems in conjunction with large shunt reactors, *CIGRE Paper* (118).

10 Capacitors

10.1 INTRODUCTION

The installation of capacitors in power systems has several benefits. Some of them are as follows[1]:

1. Power-factor improvement: This will reduce the kVA demand, and hence, the tariff paid by customers for certain utilities, also reducing losses in the system. Further, in some cases, the installed capacity of the transformers and, in general, the kVA installed capacity of the utility are also reduced.
2. Reactive support: In the distribution and transmission systems, when the systems are heavily loaded, they require var support from the capacitors to obtain an acceptable voltage level at the different buses by compensating for lagging reactive loads in the system.
3. Capacitors for motors: One of the applications of capacitors is in induction motors. The power factor of an induction motor is between 0.25 and 0.90, depending on the size and speed of the motor. At lighter loads, the power factor is poor. Because many induction motors operate at less than full load for a substantial period of time, power-factor improvement can be obtained using shunt capacitors. However, care must be taken in the choice of capacitors because self-excitation of the motors can occur if switched off when using capacitors. Here, induction motors will act as induction generators, with the machine excitation being provided by shunt capacitors. With the motor slowing down, the voltage of self-excitation usually collapses in a few seconds. However, with high-inertia loads, the voltage due to self-excitation can be sustained for several minutes. These overvoltages can reach 1.40 pu and are not acceptable. Another problem that can arise with high-inertia loads such as large compressors and air conditioners is excessive torque when the motor is reconnected while still rotating. (See Reference 1 for a suitable choice of capacitors for induction motors.) Sometimes, capacitors in steps are used to avoid voltage dips when starting large induction motors.

Capacitors are also used to reduce voltage flicker with electric-arc furnaces and resistive spot-welding machines. Small capacitors are used with choke coils for fluorescent lamps and also for starting single-phase motors. To protect the electrical equipment from surges, surge capacitors are connected at the terminals of such equipment.

In extrahigh-voltage (EHV) systems, capacitors are used in the carrier communication circuits. The frequency spectrum of the carrier communication circuits is in the 30- to 500-kHz range. Capacitors are also used in a line trap, which is a parallel resonant circuit connected in series with the transmission line. This provides very high impedance for a specific carrier frequency and very low impedance for power frequency. A line-trap circuit is used to break the carrier current from one section to the other.

Other applications of the capacitors are

- Capacitive coupling voltage transformer (CCVT): In EHV systems, electromagnetic VTs are rather expensive. Hence, CCVTs are used.
- Coupling capacitors for potential-discharge measurement
- High-voltage (HV) potential dividers for 60- or 50-Hz measurement
- HV resistive-capacitive (RC) dividers for impulse measurement
- Power supplies for equipment testing

10.2 CAPACITOR BANKS

High-voltage capacitor units are constructed by connecting individual capacitor elements together in series and in parallel. These are generally manufactured as single-phase units, and are connected in wye or delta configuration for three-phase applications.

The element is the "heart" of the capacitor, and the solid dielectric in all-film capacitors uses polypropylene films with textured surfaces. The design offers easy impregnation and superior dielectric performance. To ensure high reliability, some manufacturers such as ABB test each element individually before it can be assembled into a capacitor unit.[2]

Polypropylene film was invented in the 1930s and was introduced in the capacitor industry in the 1960s. This material has a very low loss (loss factor = 0.0005) and relatively high permittivity, on the order of 2.25 in the range 50 Hz–1 MHz. The polypropylene film's dielectric strength is greater than 32 MV/m. The film thickness is very low, on the order of 10 μm, and hence, the overall volume of the capacitor for the given kVAr will be small.[1]

Elements are wound within a clean-room environment that has a submicronic filtering system, ensuring that dust particles as small as 1 μm are effectively removed.

The use of extended foils, folded foils, and textured films offers the following advantages[2]:

- Low dielectric losses—about 0.1 W/kVAr
- Little variation in the value of capacitance with temperature—ideal for harmonic filters
- Excellent partial-discharge properties
- Superior electrical connection to withstand high transient currents

The average loss/kVAr for other designs of the capacitors is given in Reference 1.

Paper, oil impregnated	2.0–2.5 W/kVAr
Paper, PCB impregnated[a]	3.0–3.5 W/kVAr
Plastic file/paper, PCB impregnated[a]	0.5–1.0 W/kVAr
Metallized film	under 0.5 W/kVAr

[a] These two types are not manufactured anymore due to the non-biodegradability of the polychlorinated biphenyls (PCBs).

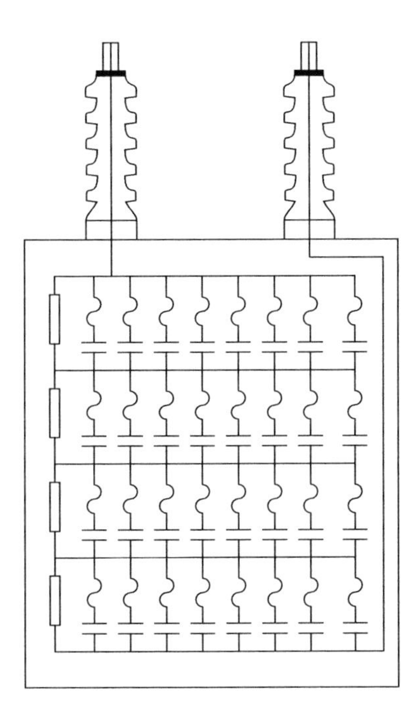

FIGURE 10.1 Capacitor with two bushings. (From Captech Web site http://www.captech.com.au. With permission from Captech P/L.) See permission from Bud Hartano dated 8 October 2007.

10.2.1 Fuses

Although some fuseless capacitor banks are manufactured, most have external or internal fuses.[1] All capacitor banks consist of series–parallel combinations. If one of the units fails, the bank can still continue operating even though the voltage across the capacitors in healthy phases increases slightly. These capacitor banks are used in distribution systems as pole-mounted capacitor banks. In externally fused banks, warning of a defective unit is achieved through fuse activation or neutral unbalance detection.

ABB pioneered internally fused capacitor banks in the 1940s. Failure of the fuse in internally fused capacitor banks can be identified usually through the current in the neutral for the grounded banks as discussed in later sections. However the fuses are not visible in the internally fused capacitor banks. Hence locating the faulty capacitor units requires significant effort, and hence maintenance time required is high.

Figure 10.1 shows a capacitor with two bushings.

10.3 CAPACITOR BANK CONNECTIONS

Large capacitor banks can be connected *wye ungrounded*, *wye grounded*, or *delta*. We will briefly discuss the relative merits of each of these connections.[3,4]

10.3.1 Ungrounded Wye-Connected Banks

This is the preferred connection from a protection point of view. Usually, fuses are used for both indoor and outdoor single-phase units. The fault current in any three-phase ungrounded wye connection is limited by the capacitors in the sound phases. Further, with this connection there is no ground path for harmonic currents to flow. However, if phase faults occur, large fault currents can result. The neutral should be insulated for full line voltage because it is momentarily at phase potential when the bank is initially switched.

10.3.2 Grounded Wye-Connected Banks

Grounded capacitor banks provide a low-impedance path to ground for lightning surge currents and afford some protection from surge voltages. They also provide a low-impedance path for triplen harmonics because most triplen harmonics are of zero-sequence nature. (Note that positive- and negative-sequence triplen harmonics will still flow in the lines instead of in the neutral-to-ground path.) Harmonic currents can cause interference with telephone lines if they are parallel to power lines and in close proximity to them. In some cases, they may cause harmonic resonance problems.

10.3.3 Delta-Connected Banks

These are used only in distribution systems.[3,4] As ungrounded wye-connected banks serve the same purpose, delta-connected banks are seldom used in HV or EHV systems. In the United States, delta-connected banks are frequently used at 2400 V because capacitor units for wye connection are not readily available.

10.4 PROTECTION OF CAPACITOR BANKS

The purpose of an unbalance protection scheme is the removal of a capacitor bank from the system in the event of a fuse operation in a fused bank.[5] Many methods are used in practice for detecting unbalances in capacitor banks. However, there is no practical method that will provide protection under all possible conditions.

In this section, we will discuss protection of grounded and ungrounded wye banks. As delta-connected banks are used only in distributions and not in HV or EHV systems, we will not discuss their protection.

10.4.1 Protection of Grounded Wye-Connected Capacitor Banks

Figure 10.2 depicts a healthy capacitor bank (no blown fuses) and balanced voltages. Assume that there are no significant manufacturing tolerances creating unbalances between different phases. Under these conditions, there will be negligible flow of neutral to-ground current. In real-life situations, however, due to manufacturing tolerances, there will be some unbalances between the different phase capacitances. Further, the system voltages may not be perfectly balanced. Due to these reasons, there will be some neutral-to-ground current even when all the elements in different units in the bank are healthy and no fuses are blown. A robust protection system should allow for this inherent unbalance due to manufacturing tolerances and normal unbalanced voltages in the system.

If one of the units in the bank becomes, an open circuit due to a blown fuse or a short circuit between elements of the same phase or between phases occurs. Then a neutral current will flow. This neutral to ground current will be larger than the earlier current due to normal unbalances, due to manufacturing tolerances, and so on.

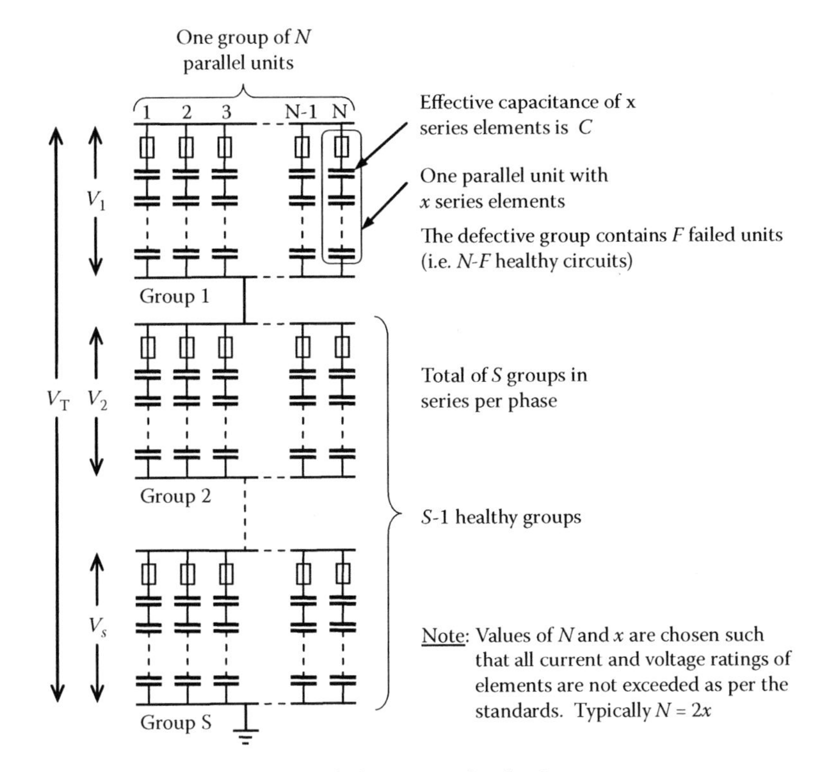

FIGURE 10.2 One phase of a grounded wye capacitor bank.

This neutral-to-ground current is made to pass through a relay with the help of the current transformers (CTs) or VTs, which in turn energize a trip coil of the circuit breaker controlling the capacitor bank.[5]

If one unit in each of the three phases fails (although unlikely), overvoltages will exist within the capacitor bank. However, all the three phases will still be balanced because all phases have the same reactance. Hence, there is no neutral current, and the protection system will not operate.

For a grounded wye bank or for each wye of a split grounded wye bank, the allowable number of units that can be removed from one series group, given a maximum % V_R on the remaining units, can be calculated using the following formula:

$$F = \left[\frac{NS}{S-1}\right]\left[1 - \frac{V_T}{SV} x \frac{100}{\%V_R}\right] \tag{10.1}$$

where

V_T applied line-to-neutral voltage
V rated voltage of capacitor units
V_R voltage on remaining units in group with F units removed
S number of series groups per phase
N number of parallel units in one series group
F number of units removed from one series group

If F is fractional, use the next lower number. The relay is set to signal the alarm upon failure of F units.

Let us assume that F units have failed and have been removed from one of the groups of capacitors connected in series to form a phase of a capacitor bank, as shown in Figure 10.2. We will try to derive expressions for the percentage of overvoltage on the remaining units in the failed group, the neutral-to-ground current flow, and the relay setting upon loss of F units.

Let

I_N = neutral-to-ground current flow

I = rated current in one unit

(Note that in one phase there are N units in parallel. Hence, total current in one phase is NI amperes.)

V_1 = voltage across a unit in the defective group

V_2 = voltage across a unit in the good group

Also, let

I_1 be the current in a unit in the capacitor in the defective group

I_2 be the current in a unit in the capacitor in the good group

x is the number of series elements in one parallel unit of the capacitor bank (see Figure 10.2)

C be the effective capacitance for x series elements

$$V_1 + V_2 (S - 1) = V_T \tag{10.2}$$

$$(N - F)I_1 = NI_2 \tag{10.3}$$

$$x\, I_1 (1/\omega C) = V_1 \tag{10.4}$$

$$x\, I_2 (1/\omega C) = V_2 \tag{10.5}$$

By Kirchhoff's law

$$(N - F)\, I_1 = NI_2 \tag{10.3}$$

From Equations 10.3, 10.4, and 10.5

$$I_1/I_2 = V_1/V_2 = N/(N - F) \tag{10.6}$$

$$V_2 = (N - F)\, V_1/N \tag{10.7}$$

From Equations 10.2 and 10.7

$$V_1 + \{(N - F)(S - 1)/N\}\, V_1 = V_T \tag{10.8}$$

$$V_1 [1 + \{(N - F)(S - 1)/N\}] = V_T$$

$$V_1 = \frac{V_T}{[1 + \{(N - F)(S - 1)/N\}]}$$

$$= \frac{NV_T}{[N + (N - F)(S - 1)]} \tag{10.9}$$

$$= \frac{NV_T}{[N + (NS - FS) - N + F]}$$

$$= \frac{NV_T}{[S(N - F) + F]} \tag{10.10}$$

Without any defectives capacitors,

$$V_1 \text{ normal} = \frac{V_T}{S} \tag{10.11}$$

$$\frac{V_1}{V_1 \text{ normal}} = \frac{NS}{[S(N-F)+F]} \tag{10.12}$$

Hence, the percentage of overvoltage of the remaining units is

$$= \frac{100\ NS}{[S(N-F)+F]} \tag{10.13}$$

If each capacitor unit is rated for V volts, and there are S series units, the maximum voltage that can be applied is SV volts. If the line-to-neutral voltage is V_T volts, then the ratio of V_T to the permitted rated voltage is V_T/SV.

Hence, the percentage of overvoltage for the remaining units with F failed units removed is

$$\%V_R = \left[\frac{V_T}{SV}\right] \times \left[\frac{100SN}{S(N-F)+F}\right] \tag{10.14}$$

Now we will derive an expression for the neutral current. Under slightly unbalanced currents, as can be seen from Figure 10.3, the currents in phase A will be nearly in phase opposition to the sum of the currents in phases B and C. Assuming that the fault occurs in phase A, the neutral current can be calculated as the difference between the earlier-mentioned currents.

$|I_B + I_C| = |I_A|$ because the phase angle difference between I_B and I_C is 120°, and numerically they are equal.

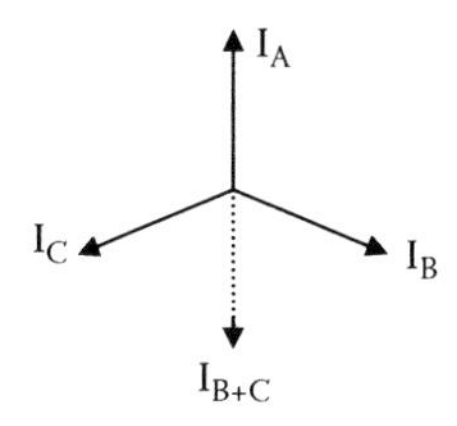

FIGURE 10.3 Phasor diagram of current in a capacitor bank.

$$|I_B| = \frac{N V_T I}{SV} \tag{10.15}$$

I is the rated current with V volts applied across a unit, V_T/S is the voltage across a unit under normal operating conditions, and there are N parallel capacitor elements in one unit.

$$|I_A| = \frac{(N-F)V_1 I}{V} \tag{10.16}$$

where I_A is the current in phase A. But from Equation 10.10

$$V_1 = \frac{N\,V_T}{[S(N-F)+F]} \tag{10.17}$$

$$|I_N| = |I_B| - |I_A|$$

$$= \frac{N\,V_T\,I}{SV} - \frac{(N-F)N\,V_T\,I}{V[S(N-F)+F]}$$

$$|I_N| = \frac{(I\,V_T)(NF)}{(SV)[S(N-F)+F]}\ \text{Amps} \tag{10.18}$$

The advantages of this scheme (in Figure 10.2 and 10.4) are[1]

1. In Figure 10.4, the capacitor bank contains twice as many parallel units per series group compared to the double wye bank for a given kVAr size, which reduces the overvoltage in the remaining units in a group in the event of a fuse operation.
2. May require less substation space and connections than a double wye bank.
3. Protection scheme is relatively inexpensive.

The disadvantages of this scheme are as follows:[5,1]

1. Sensitive to system unbalance. This may prevent its application on large banks.
2. Sensitive to triplen harmonics, and will generally require a filter circuit.
3. Will not provide protection when balance failures occur.
4. Not possible to identify the phase of the failed capacitor unit.

In the scheme shown in Figure 10.5, two CTs in the neutral-to-ground circuit are cross-connected. Hence, any current due to voltage unbalance in the system will circulate in the CT secondaries and not in the relay.

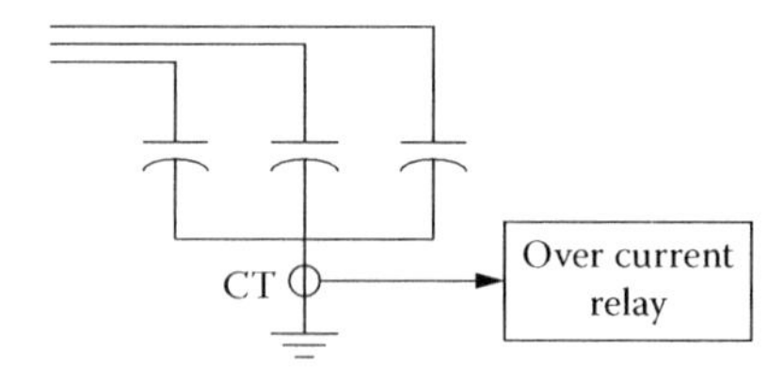

FIGURE 10.4 Unbalanced protection scheme using neutral-to-ground current sensing. (From Bishop, M., Day, T., and Chudhary, A. (2001). A primer on capacitance bank protection, *IEEE Transactions on Industry Applications*, 37(4), 1174–1179. With permission from IEEE.)

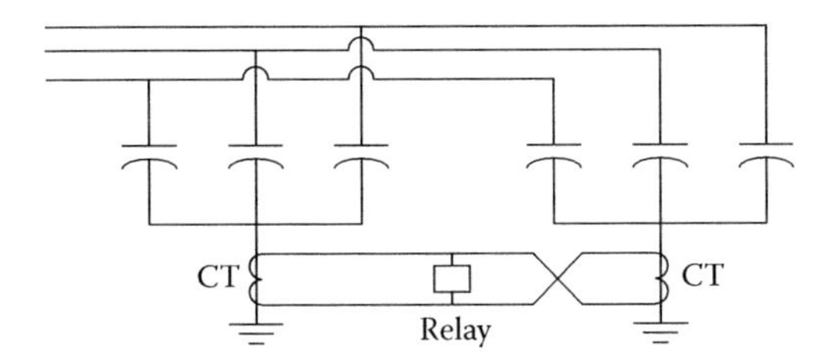

FIGURE 10.5 Neutral-to-ground sensing in a double-wye-connected bank. (From Bishop, M., Day, T., and Chudhary, A. (2001). A primer on capacitance bank protection, *IEEE Transactions on Industry Applications,* 37(4), 1174–1179. With permission from IEEE.)

The advantages of this scheme are[1,5]

1. Scheme not sensitive to system unbalance; thus, it is sensitive in detecting capacitor unit outages even on large banks.
2. It is not affected by harmonic currents.
3. It is possible to compensate for inherent capacitor tolerances.
4. Protection scheme is relatively inexpensive.

The disadvantages of this scheme are

1. The involved phase is not indicated.
2. It masks balance failures (such faults are very unlikely to occur).

Figure 10.6 shows the zero-sequence voltage-sensing method for a single grounded wye-connected bank. The relative merits of this scheme are similar to those of the scheme in Figure 10.4. In addition, if system unbalance changes, the detection circuit setting will have to be adjusted.

Figure 10.7 shows the differential voltage-sensing method for a single grounded wye-connected bank. The merits of this scheme are similar to those of the scheme in Figure 10.5. In addition,

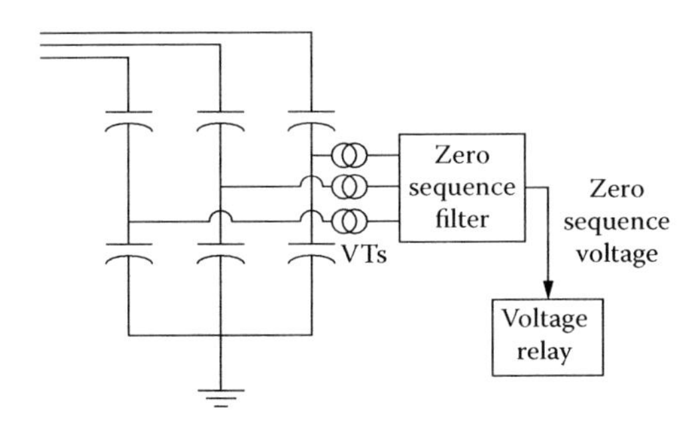

FIGURE 10.6 Zero-sequence voltage-sensing method for a single grounded wye-connected bank. (From Bishop, M., Day, T., and Chudhary, A. (2001). A primer on capacitance bank protection, *IEEE Transactions on Industry Applications,* 37(4), 1174–1179. With permission from IEEE.)

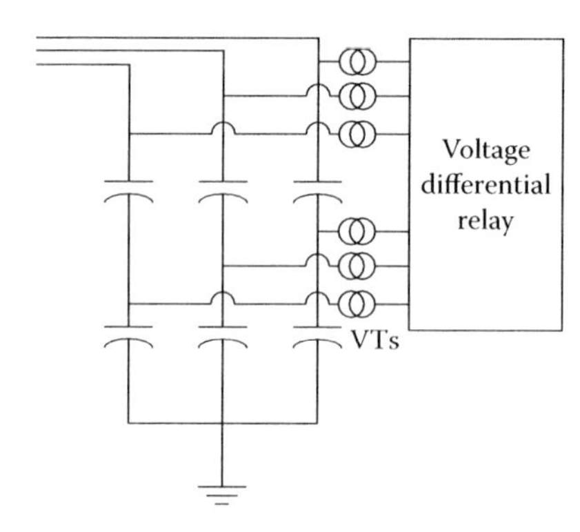

FIGURE 10.7 Differential-voltage sensing for a single grounded wye-connected bank. (From Bishop, M., Day, T., and Chudhary, A. (2001). A primer on capacitance bank protection, *IEEE Transactions on Industry Applications,* 37(4), 1174–1179. With permission from IEEE.)

1. It indicates the involved phase and, possibly, which portion of the phase is faulty
2. Protection scheme is relatively expensive due to the quantity and rating of the VTs.
3. It is subject to blocking in case of loss of potential from the bus VTs (although not likely).

Figure 10.8 shows an unbalance protection scheme for an ungrounded wye-connected capacitor bank using neutral voltage sensing.

Figure 10.9 shows an unbalance scheme for an ungrounded split-wye-connected capacitor bank using neutral CT.

Figure 10.10 shows a neutral voltage unbalance scheme for an ungrounded split-wye-connected capacitor bank.

The formulae to calculate the allowable number of units that can be removed from one series group, given the $\%V_R$ on the remaining units, the neutral shift voltage, and the percentage overvoltage on the remaining units after the faulty units are removed for the schemes in Figures 10.7 and 10.9, can be found in References 5 and 1. Also, these references give the formula for the neutral current instead of the neutral shift voltage for the scheme in Figure 10.8.

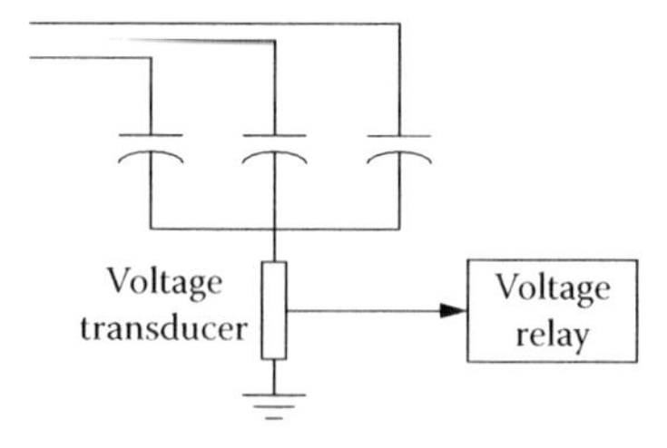

FIGURE 10.8 Unbalance protection scheme for ungrounded wye-connected capacitor bank using neutral voltage sensing. (From Bishop, M., Day, T., and Chudhary, A. (2001). A primer on capacitance bank protection, *IEEE Transactions on Industry Applications,* 37(4), 1174–1179. With permission from IEEE.)

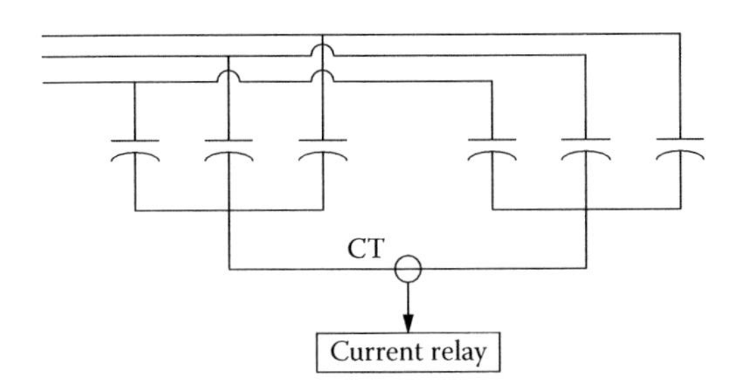

FIGURE 10.9 Unbalance protection scheme for an ungrounded split wye-connected capacitor using neutral current transformer (CT). (From Bishop, M., Day, T., and Chudhary, A. (2001). A primer on capacitance bank protection, *IEEE Transactions on Industry Applications,* 37(4), 1174–1179. With permission from IEEE.)

There are different IEEE, ANSI, NEMA, and IEC standards on capacitors, dealing with permissible overvoltages, currents, duration, tolerances, etc. Chapter 4 of the book (Reference 1), providing a good summary of the different standards, particularly IEEE and ANSI.

(Note that after the manuscript of this book was nearly completed, Dr. Arvind Chaudhary brought to the attention of the authors his recent work on the protection of multistring capacitors by measuring the impedance of the string units. Hence, the details of this method could not be incorporated into the main text. For more information, see the following reference:

1. Complete Protection of Multi-string Fuseless Capacitor Banks by Chaudhary, A., Day, T., Fender, K., Fendrik, L., and McCall, J., *IEEE Industry Applications Magazine*, November–December 2003, www.IEEE.org.IAS, pp. 34–39.
2. Complete Relay protection of Multi-string Fuseless Capacitor Banks by Fendrick, L., Day, T., Fender, K., Cooper Power Systems, Greenwood, SC, McCall, J., Chaudhary, A., Cooper Power Systems, South

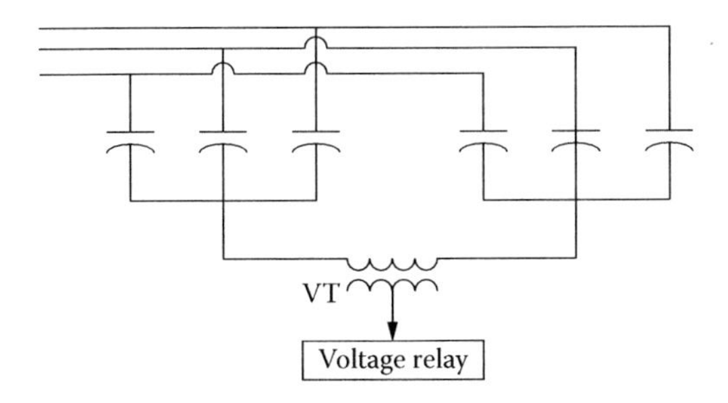

FIGURE 10.10 Neutral voltage unbalance scheme for an ungrounded split-wye capacitor bank. (From Bishop, M., Day, T., and Chudhary, A. (2001). A primer on capacitance bank protection, *IEEE Transactions on Industry Applications,* 37(4), 1174–1179. With permission from IEEE.)

Milwaukee, Wisconsin, Presented at the 55th Annual Conference for Protective Relay Engineers, Texas A&M University, April 9–11, 2002.
Affiliations:
Leo Fendrick, Lincoln Electric System, Lincoln, Nebraska
Tim Day, Cooper Power Systems, Franksville, Wisconsin
Karl Fender, Cooper Power Systems, Greenwood, SC
Jack McCall and Arvind Chaudhary, Cooper Power Systems, South Milwaukee, Wisconsin)

10.5 CAPACITOR BANK SWITCHING

10.5.1 Evaluation of Different Methods for Mitigating Remote Overvoltages due to Shunt Capacitor Energization

Energization of shunt capacitors on power systems can potentially create significant phase-to-ground and phase-to-phase overvoltages at the location of the switched bank and also at other locations in the system.[4] In the case of transformer-terminated lines, the phase-to-phase insulation of power transformers may be of special concern due to lack of specific standards for phase-to-phase switching surge-withstand level of power transformers.[7] Although arresters are applied at the transformer terminals, they offer very little protection for these phase-to-phase surges. This is because the surges comprise line-to-ground components that are approximately equal in magnitude but opposite in polarity. Thus, arresters that can limit the line-to-ground components to 2.2 per unit will allow phase-to-phase surges of 4.4 per unit (phase-to-ground base).

Preinsertion resistors, preinsertion inductors, synchronous closing of circuit breakers, and metal oxide varistors (MOVs) or surge arresters are used to reduce these transients. A parametric study to determine the effectiveness of preinsertion inductors for controlling remote overvoltages at the end of open-ended lines or transformer-terminated lines is described in Reference 6. We will briefly discuss the relative merits of the different methods used for mitigating remote overvoltages due to shunt capacitor energization.[6]

Preinsertion resistors have been used for years by the electric utility industry for controlling capacitor-bank energization overvoltages. However, their reliability is a growing concern for some U.S. utilities and circuit breaker/switcher manufacturers.[8]

Synchronous closing schemes use breaker control to synchronize the actual closing of contacts with the corresponding voltage zeros of the power frequency waves, thus reducing voltage transients.[9] In practice, synchronous closing schemes have closing tolerances that may impact their effectiveness scheme.

Inductors, which are primarily used for limiting inrush currents during back-to-back energization, can also provide voltage control. It has been suggested that preinsertion inductors be used as the preferred overvoltage control method over preinsertion resistors and synchronous closing.[10,11] (Note that preinsertion inductors are more economical than preinsertion resistors.)

MOVs or surge arresters can effectively limit the overvoltages to the arresters protective level at the point of application.[8] However, for arresters placed at the capacitor bank bus, the reduction of overvoltages at remote locations during capacitor energization may be inadequate.

Overvoltages at remote locations during restrike of capacitor switching devices are of particular concern due to the ineffectiveness of most of the overvoltage mitigation options discussed earlier. Restrike may occur for the breakers/switches not equipped with opening resistors when the magnitude or rate of the recovery voltage across the contacts exceeds specified limits after an opening operation. Although SF_6 breakers/switches are designated as restrike-free devices, restrike may occur if SF_6 gas pressure falls below manufacturer-specified limits. Restrike overvoltages cannot be protected by preinsertion devices because the preinsertion device is typically bypassed and disconnected at the time of breaker/switch opening.

The Southern California Edison (SCE) Company installed 48.6- and 78.6-MVAr shunt capacitors in their 115- and 220-kV systems, respectively. Because SCE's normal operating strategy is to leave unused lines open-ended for var support and quick energization in case of emergency, potential overvoltages at the end of the lines are of special concern during capacitor bank energization. To evaluate the impact of these capacitor installations, a switching study was performed on a transient network analyzer (TNA). Various overvoltage mitigation methods were investigated in the TNA study. The results of the study indicated that commercially available preinsertion inductors were not effective in reducing the remote overvoltages for some system contingencies.

The following discussion will assist in understanding why, under certain system conditions, preinsertion inductors are not effective. The power system with a simple radial line and a capacitor connected at the substation bus has two resonant frequencies. One, the damped resonant frequency of the system, is governed by the system equivalent impedance (i.e., short-circuit MVA), capacitor bank size, and the line surge impedance. The time-to-peak at the switched bus is equal to half of the period associated with this damped resonant frequency of the network. For open-ended lines, the first resonant frequency for line-end voltage magnification depends on the line length and corresponds to four travel times. The resonance repeats at integer multiples of the first resonance frequency. If the system damped resonance frequency is close to the resonance frequencies of the line, there is potential for high overvoltages at the open end. When a preinsertion inductor is utilized, the total inductance value is effectively increased, lowering the system resonance frequency. Therefore, the worst-case line-end overvoltages will occur for a line with lower resonant frequencies, which in turn corresponds to a greater length. It is also possible to have multiple peaks in the overvoltage performance curves because the damped resonant frequency may interact with the first resonance frequency of a line and may also interact with a second or higher resonant frequency of another line. Hence, preinsertion inductors should be carefully evaluated by means of detailed statistical system studies before being used for mitigating overvoltages due to capacitor energization.

Now we will discuss series reactors to reduce inrush currents and transients that occur due to the back-to-back switching of capacitors.

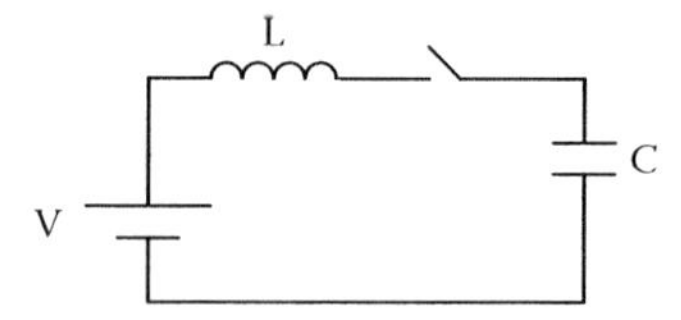

FIGURE 10.11 Capacitor energization.

10.5.2 Series Reactors for Capacitors

If a capacitor bank of C farads is energized from a constant voltage source V, with source inductance of L henrys, the inrush current is given by $V\sqrt{C/L}\ \sin \omega_n t$ amperes, where

$$\omega_n = 1/\sqrt{(LC)} = \text{natural frequency of the circuit} \tag{10.19}$$

and $\sqrt{(L/C)}$ is usually referred to as surge impedance of the capacitor (strictly that of the series LC circuit). This expression can be easily derived by considering the energization of the simple circuit shown in Figure 10.11. When the circuit is energized at time t = 0 with zero voltage on the capacitor C, we can write the following equation from Kirchhoff's voltage law:

$$L\, di/dt + 1/C \int i dt = V \tag{10.20}$$

If s is the Laplace operator, then

$$(Ls + 1/Cs)\, I\,(s) = V/s$$

Multiplying this equation by s/L yields

$$[s^2 + 1/(LC)]\, I\,(s) = V/L$$

$$\left[s^2 + \omega_n^2\right] I(s) = V/L$$

$$I(s) = V/(L\omega_n)\left[\omega_n/\left[s^2 + \omega_n^2\right]\right.$$

The inverse Laplace transform of the preceding equation yields

$$i(t) = V/(L\omega_n)\ \sin\omega_n t$$

$$\text{As } \omega_n = 1/\sqrt{(LC)},$$

$$i(t) = V\sqrt{C/L}\ \text{Sin}\,\omega_n t \tag{10.21}$$

If the system phase-to-neutral rms voltage is V_{ph}, the peak inrush current while energizing a single capacitor bank with no initial charge is calculated using the following formula[3]:

$$I_{peak} = (1.2)(\sqrt{2})\, V_{ph}\, \sqrt{C/L} \tag{10.22}$$

The 1.2 factor is used for system voltage variations and also possible current unbalance due to unequal pole operation of the breaker. If the capacitor is fully charged, with the terminal voltage equal to V_{ph}, the peak inrush current can be about twice this value.

In many cases, series reactors are used to reduce the inrush current during the energization of a capacitor bank. They also serve the purpose of reducing the phase-to-phase overvoltages of transformers located at the end of the radial feeders emanating from the substation, where the capacitor bank is energized. For a three-step bank with two steps energized and the third being energized,

$$C = 1/[\{(1/C_1 + C_2) + 1/C_3\}] \tag{10.23}$$

For delta-connected banks, the equivalent single-phase-to-ground capacitor kVAr must be used as though the bank is wye-connected. L in Eq. 10.19 is the inductance between the step being energized and that portion of the bank already energized.

10.5.3 Location of Series Reactors for Capacitors

When series reactors are used in each phase to reduce inrush currents, the question that arises is whether they should be installed at the neutral end or line end. The series reactors installed at the neutral end of the grounded wye capacitor banks need not be fully insulated to the line-voltage level, and hence, they are cheaper. However, they do not provide protection for the units near the line end in case of a ground fault because full line-to-ground voltage will be applied across those units. In such cases, series reactors are used near the line end to provide protection. However, they are more costly because they have to be insulated to the complete line-voltage level. Hence, in some power systems, series reactors are used on the neutral side till the 110- or 132-kV level, and above 275-kV levels they are provided on the line side. Also, most capacitor banks above the 275-kV level are usually grounded to reduce the cost of the capacitor-energizing circuit breakers.

10.5.4 Transient-Free Switching of Capacitors

As can be seen from Figure 2.13 (see Chapter 2), thyristor-switched capacitors (TSC) are switched in steps in parallel with a thyristor-switched reactor (TCR) to obtain the necessary variable susceptance in each phase of static var compensators (SVCs). If it is possible to have transient-free switching of capacitors, that is desirable. Even if transients are not severe from the power system point of view, still they have a bearing on the rating and cost of thyristor switches. This problem has been investigated by T.J.E. Miller and P. Chadwick.[12,13]

When switching an ideal capacitor, transient-free switching can be achieved when the capacitor voltage is equal to $\pm V_m$ before energization and the source voltage is $V_m \sin(\omega t + \alpha)$ by energizing when $\alpha = \pm\pi/2$. However, in real-life situations, there is always a series resistance and inductance (due to the source impedance of the power system) with the capacitor being energized, and further, the voltage on the capacitor can be any value, depending on its initial voltage and later rate of discharge. Most standards stipulate that capacitors rated above 600 V must discharge to a value of 50 V within 5 min for safety reasons (see Section 4.1e of the IEEE Standard C37.99 [2000][4]). In Reference 12, it has been proved by the analysis of a simple circuit that, in general, transient-free switching is not possible under all operating conditions. When we energize a discharged capacitor in series with an inductance, lower transients will be achieved if we energize at $v = 0$ rather than when $dV/dt = 0$; that is, when the source voltage is at its maximum $\pm V_m$.

10.6 SERIES CAPACITORS

Series capacitors are installed in transmission and distribution systems for the following purposes[1,3]:

1. To increase power-transfer ability: If there is a reactance X_L ohms between two buses with voltages V_1 and V_2, the power transfer P between two buses is given by

$$P = V_1V_2 \sin \delta/X_L \text{ watts} \tag{10.24}$$

   where δ is the phase angle difference between V_1 and V_2.
   If there is a series capacitor of X_C located on the transmission line between these two buses and if $X_L > X_C$ (i.e., the line is not overcompensated), the effective reactance will be reduced to $X_L - X_C$.
   Then, the power transfer (P) between these two buses will be increased to

$$P = V_1V_2 \sin \delta/(X_L - X_C) \text{ watts} \tag{10.25}$$

2. To improve voltage regulation on radial feeders: In a radial feeder, the voltage drop through the feeder is approximately

$$IR \cos \theta + IX_L \sin \theta \tag{10.26}$$

   where
   I = current through the feeder
   θ = power-factor angle
   R = resistance of the feeder
   X_L = reactance of the feeder
   In most situations, $X_L > R$, and if the power factor is poor, $IX_L \sin \theta$ will dominate the equation. Hence, if a series capacitor is used, the effective reactance of the feeder becomes $X_L - X_C$, and the voltage drop in the feeder decreases, improving regulation on the feeder. In most applications, $X_C < X_L$. If $X_C > X_L$, then overcompensation results. The disadvantages of overcompensation are as follows:
   - While starting a large motor, the lagging current may cause an excessive voltage rise, which introduces a light flicker.
   - With leading power factor loads, by the addition of a series capacitor the receiving end voltage will decrease instead of increasing.
3. To reduce light flicker on radial feeders: Series capacitors are used in radial circuits in which rapid and repetitive load fluctuations, such as frequent motor starting, varying motor loads, electric welders, and electric-arc furnaces, cause light flicker.[14] As the radial feeders from a substation

can be of different lengths, to reduce flicker series, capacitors must be installed in the supply circuit or circuits to the bus.

4. To control load sharing between parallel feeders: Suppose there are two parallel feeders of different lengths with different reactances. Then, load sharing will be different. By connecting a series capacitor in the feeder with higher reactance, the effective reactances of both the feeders can be made equal, thus ensuring equal distribution of the load.[14]

10.6.1 Protection of Series Capacitors

Figure 10.12 shows a thyristor-controlled series compensation (TCSC) module. A TCSC module consists of a series capacitor with a parallel path, including a thyristor switch with surge inductor. Also included are an MOV (metal oxide varistor) for overvoltage protection and a bypass breaker, typical of series capacitors. A complete TCSC system may be composed of several such modules in series and may accompany a conventional series capacitor bank as part of an overall project to aid system performance. Six TCSC modules, similar to that shown in Figure 10.12, are connected in series at the Slatt 500-kV substation in the United States.[15]

In some of the older schemes, a self-sparking spark gap or a triggered spark gap in parallel with the varistor was used for series capacitor protection. However, the spark gaps were rather unreliable and required too much maintenance in their cleaning and periodical calibration.[16]

The protective level of the varistor is selected to protect capacitors during power system faults.[16] Varistors are not affected by weather, unlike spark gaps. With this scheme, the varistor has a higher energy rating than that used in the triggered gap scheme. In the typical scheme, the varistor is rated to absorb the energy due to the worst fault current until the bypass switch closes.

When a power system fault occurs, the capacitor voltage rises and is limited by the condition of the varistor. Immediately at the start of conduction, closing of the

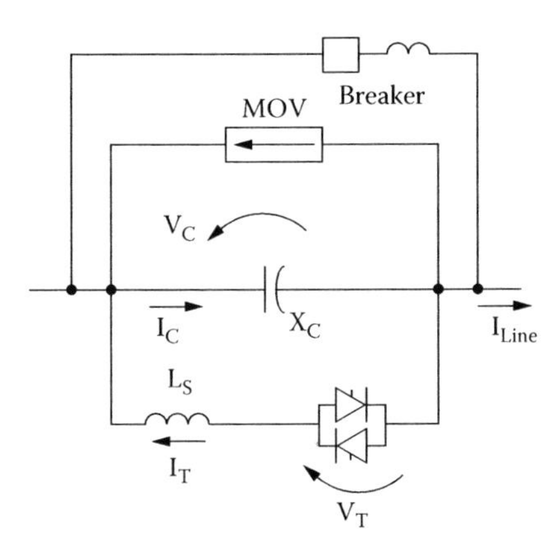

FIGURE 10.12 Thyristor-controlled series compensation (TCSC) module power circuit. (From Larsen, E.V. et al. (1994). Characteristics and rating considerations of Thyristor controlled series compensation, *IEEE Transactions on Power Delivery,* 9(2), 992–1000. With permission from IEEE.)

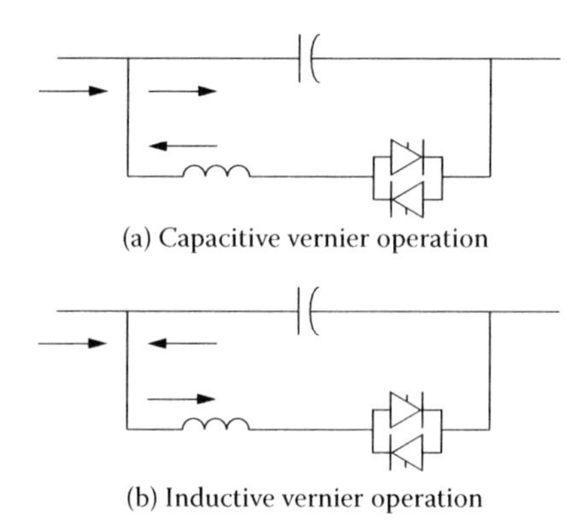

FIGURE 10.13 Vernier operating modes with partial thyristor conduction: (a) capacitive vernier operation and (b) inductive vernier operation. (From Larsen, E.V. et al. (1994). Characteristics and rating considerations of Thyristor controlled series compensation, *IEEE Transactions on Power Delivery*, 9(2), 992–1000. With permission from IEEE.)

bypass switch is initiated. The varistor continues to conduct until the bypass switch closes or the power system circuit breaker clears the fault.

In a few cases where the series capacitor has been applied to increase the fault current, the varistor's protective level has been raised, and the varistor is designed to withstand the full duration of the fault.

TCSC modules have three basic modes of operation[17,18,19]:

- Thyristors blocked (no gating and zero thyristor conduction).
- Thyristors bypassed (continuous gating and full thyristor conduction).
- Operating in vernier mode with phase control of gate signals and consequent partial thyristor conduction. In this vernier mode, we can have capacitive vernier operation or inductive vernier operation (see Figure 10.13a,b).

Figure 10.14 shows the fundamental frequency components of reactance, currents, and voltage with partial thyristor conduction. Note that, because we are evaluating a series capacitor in Figure 10.14, the capacitive impedance is defined to be positive, and inductive reactance is considered negative (just the opposite of the convention used for impedances).

Note that the gap in the control-range vernier operation can increase the apparent reactance in both directions but cannot produce smaller values of reactance. See also the unavailable portion in Figure 10.14. As can be seen from Figure 10.4 X_{net}, i.e, net reactance of TCSC in per unit of X_c cannot take values between −0.1 and +1.0. For voltage and current waveforms during capacitive and inductive vernier operations, see References 17 and 18.

Figure 10.15 illustrates the reactance capability versus the line current and duration for the Slatt 500-kV TCSC installation. The following control functions are provided:

Power-swing damping control (PSDC) is a modulation feature responsive to power swings in the frequency range of 0.3–1.5 Hz, which is intended to provide enhanced damping following major disturbances and also in ambient conditions.

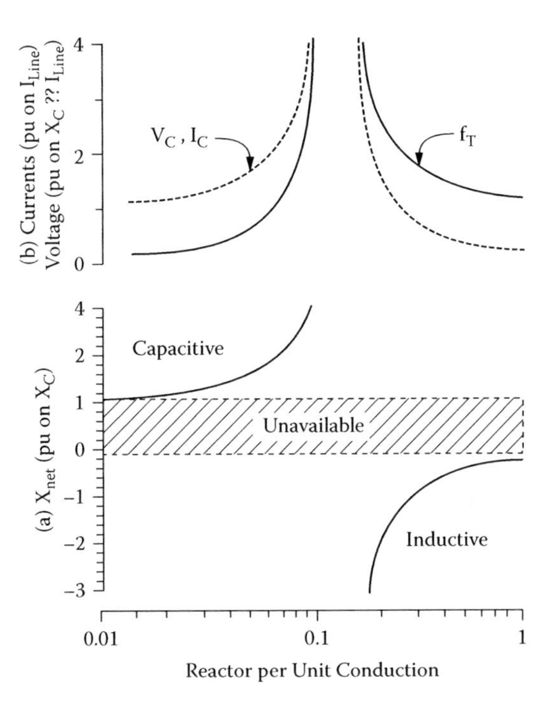

FIGURE 10.14 Fundamental frequency components of reactance, currents, and voltage with partial thyristor conduction. (From Larsen, E.V. et al. (1994). Characteristics and rating considerations of Thyristor controlled series compensation, *IEEE Transactions on Power Delivery,* 9(2), 992–1000. With permission from IEEE.)

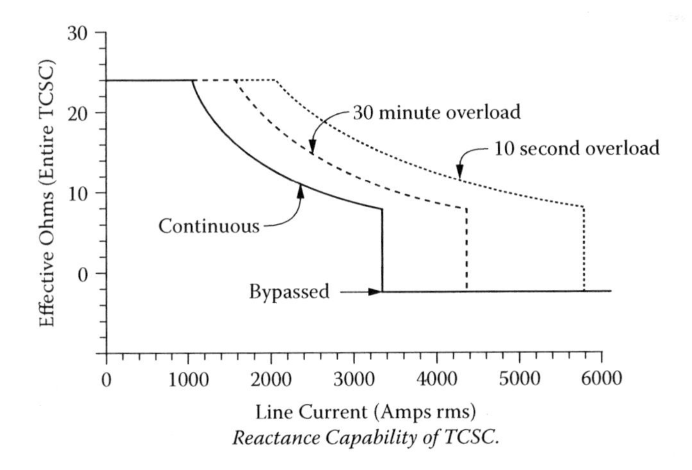

FIGURE 10.15 Reactance capability of Thyristor-controlled series compensation (TCSC). (From Urbanek, J. et al. (1993). Thyristor controlled series compensation protype installation at the Slatt 500 kV Substation, *IEEE Transactions on Power Delivery,* 8(3), 1460–1468. With permission from IEEE.)

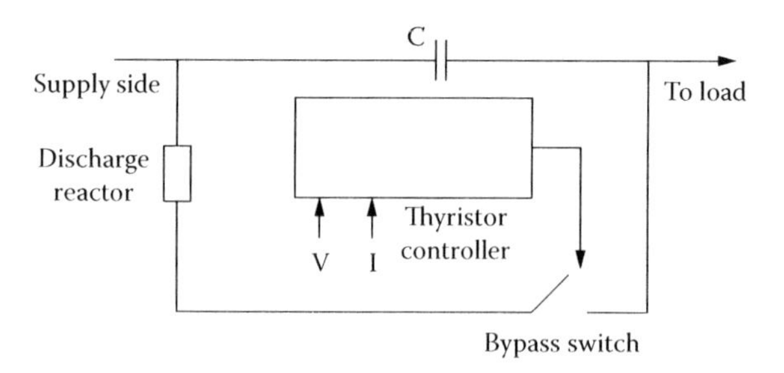

FIGURE 10.16 Series capacitor with thyristor-controller protection. (From Natarajan, R. (2005). *Power System Capacitors*, CRC Press, Boca Raton, FL. Exemption from permissions requested as this is a CRC publication.)

Transient stability control is an open-loop, preprogrammed response to a major disturbance. The transient stability control sequence is triggered by a single communicated signal, typically from a breaker contact at a remote location. This type of control is common in BPA (Bonneville Power Administration) power system.

Subsynchronous resonance (SSR) mitigation is an important function of TCSC and is provided in two ways: first, by ensuring a minimum level of reactance and second NGH scheme. In the next section, we will discuss the N.G. Hingorani scheme (NGH scheme) for damping subsynchronous resonance.

Figure 10.16 shows a thyristor-protected series capacitor scheme used at SCE's (Southern California Edison's) Vincent substation.[1,19] In this scheme, the thyristor can be triggered as soon as the overvoltage condition occurs as a result of a power-system fault. The firing angle of a thyristor can be controlled closely, and the protection is much smoother than in the other schemes. It should be noted that, in varistor protection, there may be differences in the phase voltages at which conduction occurs because of the mismatch in the devices.

10.6.2 NGH Scheme for Damping Subsynchronous Resonance

If there is a dc or subsynchronous frequency voltage, and the normal power frequency voltage 60 Hz is combined with it, in the resulting signal, some half-cycles will be longer than the normal half-cycle period (8.33 ms). Conversely, if there is no dc or subsynchronous component with 60 Hz frequency voltage, then each half-cycle will be 8.33 ms.[20,21]

The basic NGH scheme for one phase is shown in Figure 10.17. It comprises a linear resistor with back-to-back thyristors connected across the capacitor.

When a zero-voltage crossing point of the capacitor voltage is detected the succeeding half-cycle period is timed. If and when the half-cycle exceeds the set time (8.33 ms), the corresponding thyristor is fired to discharge the capacitor through the resistor and bring about its current zero sooner than it would otherwise. The thyristor stops conducting when the capacitor voltage, and therefore the thyristor current, reaches zero. Thereafter, the measuring of the half-cycle period restarts from a new voltage zero. No thyristor fires for half-cycles that are shorter than the set period.

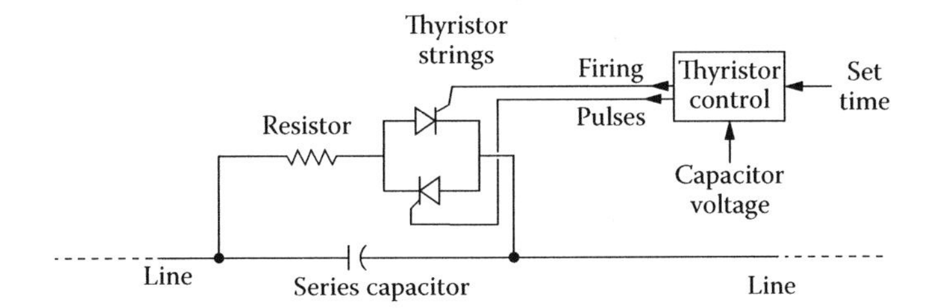

FIGURE 10.17 Basic NGH subsynchronous resonance (SSR) damping scheme. (From Hingorani, N. G. (1981). A new scheme for subsynchronous resonance damping of torsional oscillations and transient torque: Part 1, *IEEE Transactions on Power Apparatus and Systems,* 100(4), 1852–1855. With permission from IEEE.)

Timing starts afresh with each zero. The two thyristors are for the two polarities. For high voltages, the thyristor shown represents a series string of low-voltage thyristors. The resistor value is not critical. Generally, the lower its ohmic value, the more effective it is, except when too low. Some secondary effects associated with rapid changes in capacitor voltage may appear. The resistor's ohmic value in the range of 5–10% of the capacitor's 60-Hz ohmic value is satisfactory.

Principal Variations of the Basic Scheme

There are a number of possible variations and improvisations of this basic concept for meeting practical requirements. If the problem faced is only that of transient torque (no steady-state resonance), then the set period can be larger than 8.33 ms (say 8.5 ms), so that the thyristors will not fire at all during steady-state 60 Hz and small changes but will be equally effective in removing capacitor charge during large changes that lead to dc offset and significant subsynchronous component.

On the other hand, if there is a possibility of a steady-state resonance occurring, then the set period may be slightly less than 8.33 ms (say 8.1 ms); the thyristors will conduct during steady state at the tail end of each half-cycle of the capacitor voltage and provide a detuning effect against any gradual buildup of oscillations. Continuous power loss in this case will be very small and of little consequence to the thyristor rating or the cost of losses.

10.6.3 Limitations of Series Capacitor Applications

10.6.3.1 Ferroresonance

When an unloaded transformer is energized, the inrush current in the transformer is much greater than the rated current. The series capacitor can interact with the transformer inductance and cause oscillatory currents. This phenomenon is known as *ferroresonance.*

A fluctuating load also causes ferroresonance sometimes when a transformer is energized with a series capacitor. Further, in the case of series capacitors without the protection of varistors, ferroresonance problems may be caused by a single-line-to-ground fault. One such circuit is shown in Figure 10.18. Metal oxide arresters, bypass

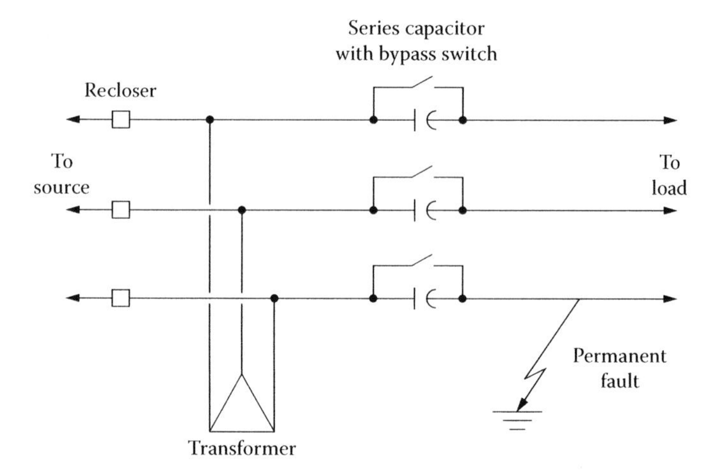

FIGURE 10.18 Fault resulting in ferroresonance. Following a permanent, single-line-to-ground fault, single-pole operation of closer and delayed reinsertion of the series capacitor resulted in a ferroresonance condition. (From Morgan, L., Barcus, J. M., and Ihara, S. (1993). Distribution series capacitors with high energy varistor protection, *IEEE Transactions on Power Delivery*, 8(3), 1413–1419. With permission from IEEE.)

switches, damping resistance, and suitable controls are employed to overcome the ferroresonance problem. Subharmonic ferroresonance oscillations can occur with a series capacitor installed in a long circuit that has a transformer of high no-load current. In many instances, a damping resistor can be used across the series capacitors.

10.6.3.2 Hunting of Synchronous Motors

Even without series capacitors, synchronous machines can hunt under lightly loaded conditions due to disturbances such as switching, changes in load, or excitation. Series capacitors reduce the effective reactance of the incoming line and cause violent hunting. A long line with overcompensation by a series capacitor will add to the hunting of a synchronous motor if the motor is started with a light load. Series capacitors should not be applied to circuits supplying either synchronous or induction motors driving reciprocating loads such as pumps and compressors. In addition to subsynchronous resonance, the motors once started may hunt, causing objectionable light flicker.

10.6.3.3 Subsynchronous Resonance

We have already discussed (Section 10.6.2) subsynchronous resonance associated with large synchronous machines and the NGH scheme for damping it. Even with induction machines, the rotor may lock and rotate at a speed below the rated speed, corresponding to 1/3 or 1/5 of its rated speed. Such a subsynchronous operation allows the motor to draw a large current and may damage it because of excessive heating and vibrations. In order to damp out such resonance, suitable damping resistance can be installed across the series capacitors. This resistance must be low enough to provide enough damping and, at the same time, as high as practicable to reduce continuous losses. It is advisable to check the possibility of subsynchronous resonance for all the larger motors with series capacitors.

10.6.3.4 Self-Excitation of Induction and Synchronous Machines

Occasionally, an induction motor with a series capacitor can act as an induction generator, producing current of lower than normal frequency. These low-frequency currents can become large if the impedance of the supply circuit is low, which in turn cause large pulsating torques and strong oscillations of the rotor. Such a self-excitation trend can also occur with synchronous machines.[22]

10.7 METAL OXIDE VARISTORS (MOVS)

Unlike old silicon carbide arresters with gaps, MOVs have no gaps.[1,23] They are made up of zinc oxide (ZnO) blocks and have a nonlinear volt-ampere characteristic. A typical volt-ampere characteristic of an MOV-type surge arrester is shown in Figure 10.19. MOVs are used for the surge protection of transformers, transmission lines, and series capacitors.[16]

Typical data provided by a manufacturer (ABB) are shown in Table 10.1. This consists of

1. Maximum system voltage
2. Rated voltage
3. Maximum continuous operating voltage
4. Temporary overvoltage
5. Maximum residual voltages usually with 8/20-μs waves and typical discharge currents from 1 kA to 40 kA

The energy and discharge current ratings for different applications are shown in Table 10.2.

Vicaud et al. have reported on the application of zinc oxide surge arresters on insulation coordination.[24]

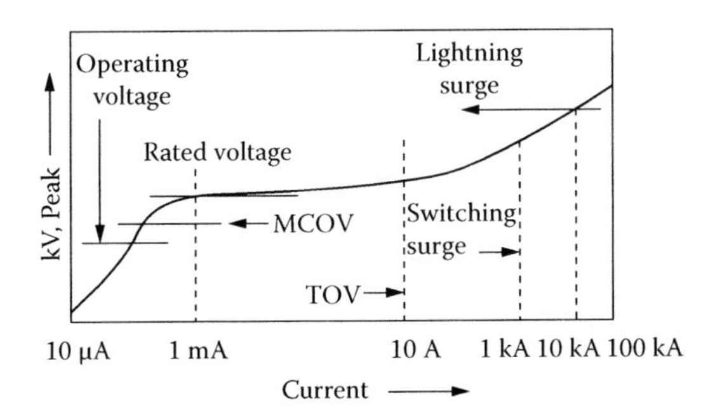

FIGURE 10.19 Typical volt-ampere characteristics of a metal oxide varistor (MOV) type. (From Natarajan, R. (2005). *Power System Capacitors*, CRC Press, Boca Raton, FL. Exemption from permissions requested as this is a CRC publication.)

TABLE 10.1
Guaranteed Protective Data of Surge Arresters

Maximum System Voltage (U_m)	Rated Voltage (U_r)	Maximum Continuous Operating Voltage[a]		TOV Capability[b]		Maximum Residual voltage (U_{res}) with Current Wave: 30/60 μs, 0.5 kA	30/60 μs, 1 kA	8/20 μs, 5 kA	8/20 μs, 10 kA	8/20 μs, 20 kA
kV_{rms}	kV_{rms}	kV_{rms}	kV_{rms}	kV_{rms}	kV_{rms}	kV_{crest}	kV_{crest}	kV_{crest}	kV_{crest}	kV_{crest}
145	108	86	86.0	126	119	230	239	273	292	232

[a] The continuous operating voltages Uc (as per IEC) and MCOV (as per ANSI) differ only because of deviations in type test procedures. Uc has to be considered only when the actual system voltage is higher than the tabulated one.

[b] With prior duty equal to the maximum single-impulse energy stress (2.5 kJ/kV U_r).

10.7.1 Modeling of MOVs in Computer Simulations

The nonlinear volt-ampere characteristic of an MOV arrester can be characterized by the following equation:

$$I = k\left(\frac{V}{V_{ref}}\right)^m \tag{10.27}$$

where k and m are constants.[23] On a log–log plane, the relationship between log I and log $\frac{V}{V_{ref}}$ will be a straight line. Using the principles of least-square fitting, values of the multiplier k and the exponent m are calculated.

TABLE 10.2
Energy and Discharge-Current Rating of MOV Arresters

Location	Discharge Current [kA (8/20 μs wave)]	Energy Rating (kJ/kV)
Major power stations and system voltages above 66 kV	10	7–14
Substations with nominal voltages below 66 kV	5	4–7
Small substations with nominal voltages around 22 kV	3	3
Rural distribution systems with voltages not exceeding 22 kV	1.5	3

Source: From Natarajan, R. (2005). *Power System Capacitors*, CRC Press, Boca Raton, FL.

TABLE 10.3
Zinc Oxide Arrester Parameters from the ATP Output

ZnO.pch

```
C <++++++> Cards punched by support routine on 28-Nov-00 09.20.03 <++++++>
C ZNO FITTER
C $ERASE
C            1          2          3          4          5          6          7          8
C     345678901234567890123456789012345678901234567890123456789012345678901234567890
C     BRANCH   TG1A       TG1B       TG1C
C     -1       3                     1                   480.0E3
C     240.0E3  240.0E3    1.0        1.0                                          2.0E-3
C     2.0E-3   240.0E3
C     0.5E3    461.0E3
C     1.0E3    476.0E3
C     2.0E3    495.0E3
C     5.0E3    536.0E3
C     10.0E3   564.0E3
C     20.0E3   621.0E3
C     40.0E3   694.0E3
C     BLANK CARD
C     Rating = 240000.0    v-mult = 1.00000E+00    I-mult = 1.00000e+00    Gapless
C            1          2          3          4          5          6          7          8
C     345678901234567890123456789012345678901234567890123456789012345678901234567890
92TG1A                                                   5555.
1
C                v-reference                             v-flashover
      4.8000000000000000E+05       -1.0000000000000000E+02
C                Multiplier                 Exponent                   v-min
      1.1217475433916030E+03   1.9099010704978640E+01   5.0003074254631690E-01
      1.3487395004260630E+03   1.2261268641014590E+01   1.0273172662519830E+00
      3.4704735510960010E+03   6.6721656473180250E+00   1.1842392530190250E+00
                       99999
92TG1B           TG1A                  5555.
92TG1C           TG1A                  5555.
```

The Alternative Transient Program (ATP), or the earlier Electromagnetic Transient Program (EMTP), has a subroutine to calculate these values of k and m. In system studies using the ATP, this method is used in computer simulations. Table 10.3 shows the output of the ATP, giving the values of the multipliers and exponents for different segments of an MOV arrester.

10.8 CONCLUSIONS

In this chapter, we have discussed the different types of capacitors used in power systems and, briefly, their construction. We have also discussed the different connections of shunt capacitor banks: grounded wye, ungrounded wye, and delta. Because most common connections were grounded wye and ungrounded wye, the protection of shunt capacitor banks with these types of connections were considered. Preinsertion resistors, preinsertion inductors, and series reactors were shown to reduce transients during capacitor switching. Location of the series reactors (line or neutral side and their relative merits) and the role of series reactors in reducing transients during back-to-back switching of capacitor banks and transient-free switching of capacitor banks associated with static var compensators were discussed.

It was shown that series capacitor banks are used in power systems for the following purposes:

- To increase power-transfer ability
- To improve voltage regulation on radial feeders
- To reduce light flicker on radial feeders
- To control load sharing between parallel feeders

We have discussed protection of capacitor banks with MOV arresters and with thyristor control, as in the Vincent substation. The NGH scheme for damping subsynchronous resonance was discussed. The MOV arresters are used for protection of series capacitors, transformers, transmission lines, power stations, substations. Their modeling for computer simulations, was reported.

Finally, the following limitations of series capacitors were pointed out:

- Ferroresonance
- Hunting of synchronous motors
- Subsynchronous resonance
- Self-excitation of induction and synchronous machines

REFERENCES

1. Natarajan, R. (2005). *Power System Capacitors*, CRC Press, Boca Raton, FL.
2. ABB leaflet on Capacitors.
3. Westinghouse Electric Corporation (1950). *Electrical Transmission and Distribution Reference Book, 4th Edition,* Oxford and IBH Publishing Company, New Delhi.
4. ANSI/IEEE Standard C37.99 (2000). IEEE Guide for Protection of Shunt Capacitor Banks.
5. Bishop., M., Day, T., and Chudhary, A. (2001). A primer on capacitance bank protection, *IEEE Transactions on Industry Applications,* 37(4), 1174–1179.
6. Bhargava, B., Khan, A. H., Imece, A. F., and DiPetro, J. (1993). Effectiveness of preinsertion inductors for mitigating overvoltages due to shunt capacitor bank energization, *IEEE Transactions on Power Delivery,* 8(4), 1174–1179.
7. Jones, R. A. and Fortson, H. S. (1986). Consideration of phase to phase surges in the application of Capacitor Banks, *IEEE Transactions on Power Delivery,* 1(3), 240–244.
8. Ribeiro, J. R. and McCallum, M.E. (1989). An application of metal oxide surge arresters in the elimination of need for closing resistors, *IEEE Transactions on Power Delivery*, 4(1), 282–291.
9. Alexander, R. W. (1985). Synchronous closing control for shunt capacitors, *IEEE Transactions on PAS*, 104(9), 2619–2626.

10. O'Leary, R. P. and Harner, R. H. (1988). Evaluation of methods for controlling the overvoltages produced by the energization of a shunt capacitor bank, *CIGRE Report*, Paper 13–05.
11. Bayless, R. S., Selman, J. D., Truax, D. E., and Reid, W. E. (1988). Capacitor switching and transformer transients, *IEEE Transactions on Power Delivery*, 3(1), 349–357.
12. Miller, T. J. E. and Chadwick, P. (1981). An analysis of switching transients in thyristor-switched-capacitor compensated systems, *IEE Conference Publication No. 205, Thyristor and Variable Static Equipment for AC and DC Transmission, London.*
13. Miller, T.J.E. (1982). *Reactive Power Control in Electric Systems,* John Wiley, New York.
14. Miske, S. A. (2001). Considerations for the application of series capacitors to radial power distribution circuits, *IEEE Transactions on Power Delivery,* 16(2), 306–318.
15. Urbanek, J. et al. (1993). Thyristor controlled series compensation protype installation at the Slatt 500 kV Substation, *IEEE Transactions on Power Delivery,* 8(3), 1460–1468.
16. Morgan, L., Barcus, J. M., and Ihara, S. (1993). Distribution series capacitors with high energy varistor protection, *IEEE Transactions on Power Delivery,* 8(3), 1413–1419.
17. Larsen, E. V. et al. (1994). Characteristics and rating considerations of Thyristor controlled series compensation, *IEEE Transactions on Power Delivery,* 9(2), 992–1000.
18. Noroozian, M., Angquist, L., and Ingerstorm, G. (1999). Series Compensation, Chapter 5, *Flexible ac Transmission Systems,* edited by Song, Y. H. and Johns, A. T., The Institution of Electrical Engineers, London.
19. Bhargava, B. and Haas, R. G., Thyristor Protected Series Capacitors Project, *IEEE Power Engineering Society*, Summer Meeting, Vol. 1, 2002, pp. 241–246.
20. Hingorani, N. G. (1981). A new scheme for subsynchronous resonance damping of torsional oscillations and transient torque: Part 1, *IEEE Transactions on Power Apparatus and Systems,* 100(4), 1852–1855.
21. Hedin, R. A., Stump, K. B., and Hingorani, N. G. (1981). A new scheme for subsynchronous resonance damping of torsional oscillations and transient torque: Part 2, *IEEE Transactions on Power Apparatus and Systems,* 100(4), 1856–1863.
22. Kimbark, E. W. (1966). Improvement of system stability by switched series capacitors, *IEEE Transactions on Power Apparatus and Systems,* 85(2), 180–188.
23. EMTP Rule Book M39 version (1984). Published by the Bonneville Power Administration.
24. Vicaud, A., Rousseau, A., and Hennebique, I. (1986). Knowledge and Application of Zinc Oxide Surge Arresters Effect on Insulation Coordination, *CIGRE Report,* 33–10.

11 Fast Fourier Transforms

In Chapter 4 we briefly introduced the Fourier series while discussing harmonics. Now we will discuss the Fourier series and fast Fourier transforms (FFTs) in more detail.

Most instruments used for harmonic analysis collect the data in digital form from the secondaries of the potential transformers (PTs) and current transformers (CTs). These data are later analyzed using FFT. This method is popular because much less computational effort is required compared with earlier methods. The FFT method was first reported by Cooley and Tukey[1] in 1965. In view of the application of this method in such diverse areas as acoustic wave propagation, speech transmission, linear network theory, transport phenomena, optics, and electromagnetic theory, considerable literature has been published in the technical press. For some of these applications, see the June 1967 issue of the IEEE Transactions on Audio and Electroacoustics.[2]

We will first discuss the Fourier series, Fourier transforms (FTs) and Discrete Fourier transforms (DFTs), before we develop the FFT algorithm.

11.1 FOURIER SERIES

If $f(\theta)$ is a periodic function of θ, the general trigonometric form of the Fourier series is[3–6]

$$f(\theta) = a_0 + \sum_{n=1}^{\infty} \cos n\theta + b_n \sin n\theta \tag{11.1}$$

We have derived the following formulas to calculate a_0, a_n, and b_n in Chapter 4.

$$a_0 = (1/2\pi)\int_0^{2\pi} f(\theta)d\theta \tag{11.2}$$

$$a_n = (1/\pi)\int_0^{2\pi} f(\theta)\cos(n\theta)d\theta, \quad \text{(except when } n = 0\text{)} \tag{11.3}$$

$$b_n = (1/\pi)\int_0^{2\pi} f(\theta)\sin(n\theta)d\theta \tag{11.4}$$

In Equations 11.2, 11.3, and 11.4, the limits on the definite integrals can be changed from 0 to 2π to $-\pi$ to $+\pi$, because both these cover one complete period.

Further, if one is interested in the magnitude A_n and phase angle ϕ_n of the n-th harmonic (say, voltage or current), they can be obtained from the following equations:

$$A_n \angle \phi_n = a_n + jb_n \tag{11.5}$$

where

$$A_n = \sqrt{\left(a_n^2 + b_n^2\right)} \quad (11.6)$$

$$\phi_n = \tan^{-1}(b_n/a_n) \quad (11.7)$$

The exponential form of the Fourier series can be derived using the following trigonometric identities:

$$\cos(n\theta) = (e^{jn\theta} + e^{-jn\theta})/2 \quad (11.8)$$

$$\sin(n\theta) = (e^{jn\theta} - e^{-jn\theta})/2j \quad (11.9)$$

When these identities are substituted in Equation 11.1, we can write

$$f(\theta) = a_0 + (1/2\sum_{n=1}^{\infty} a_n - jb_n)e^{jn\theta} + (1/2)\sum_{n=1}^{\infty}(a_n + jb_n)e^{-jn\theta} \quad (11.10)$$

To simplify this expression, if we introduce negative values of n in Equations 11.3 and 11.4, then

$$a_{-n} = a_n \quad \text{because} \quad \cos(n\theta) = \cos(-n\theta) \quad (11.11)$$

$$b_{-n} = -b_n \quad \text{because} \quad \sin(n\theta) = -\sin(-n\theta) \quad (11.12)$$

Hence, we can write

$$\sum_{n=1}^{\infty} a_n e^{-jn\theta} = \sum_{n=-1}^{n=-\infty} a_n e^{jn\theta} \quad (11.13)$$

$$\sum_{n=1}^{\infty} jb_n e^{-jn\theta} = -\sum_{n=-1}^{n=-\infty} jb_n e^{jn\theta} \quad (11.14)$$

Substituting Equations 11.13 and 11.14 in Equation 11.10 yields (except when $n = 0$)

$$f(\theta) = a_0 + (1/2)\sum_{n=-\infty}^{\infty}(a_n - jb_n)e^{jn\theta} \quad \text{(except when } n = 0) \quad (11.15)$$

and (except when $n = -0$)

$$f(\theta) = \sum_{n=-\infty}^{\infty} c_n e^{jn\theta} \quad (11.16)$$

where

$$c_n = (1/2)(a_n - jb_n), \quad n = \pm 1, \pm 2 + \quad , \quad (11.17)$$

and

$$c_n = c_{-n}$$

$$c_0 = a_0$$

The c_n terms can also be obtained by complex integration:

$$c_n = (1/\pi)\int_{-\pi}^{\pi} f(\theta)e^{-jn\theta}d\theta \tag{11.18}$$

$$c_0 = (1/2\pi)\int_{-\pi}^{\pi} f(\theta)d\theta \tag{11.19}$$

The expression of the Fourier series in exponential form in Equation 11.16 and the complex coefficients in Equations 11.18 and 11.19 are used mostly in analysis.

11.2 SYMMETRICAL PROPERTIES OF WAVEFORMS

Odd Symmetry:

$$\text{If } f(\theta) = -f(-\theta), \tag{11.20}$$

the waveform has odd symmetry and will contain only sine terms.[3-6]

Even Symmetry:

$$\text{If } f(\theta) = f(-\theta), \tag{11.21}$$

the waveform has even symmetry and will contain only cosine terms.

Half-wave Symmetry:

A function $f(\theta)$ has half-wave symmetry if

$$f(\theta) = -f(\theta + \pi) \tag{11.22}$$

That is the shape of the waveform over a period from $(\theta + \pi)$ to $(\theta + 2\pi)$ is the negative shape of the waveform over the period θ to $(\theta + \pi)$. Waveforms having half-wave symmetry can contain only odd-order harmonics.

In the rest of this chapter, we will reserve the symbol "f" for frequency. Hence, to avoid confusion, we will be representing the signal function by the symbols $x(\theta)$, $x(\omega t)$, $x(2\pi ft), x(2\pi t/T), \text{or} x(2\pi nt/T)$ as appropriate. The different symbols have the following meanings:

f—Frequency.
t—Time.
$\theta = \omega t = 2\pi ft$.
T—Period of the function ($1/f$).
n—Harmonic number; it also represents the time sample index in DFT, depending on the context.
k—In the DFT, $x(k)$ corresponds to a frequency sample,

and $x(n)$ corresponds to a time sample.

Though there are other definitions of Fourier transform in the literature,[6,7] we will define the FT as follows:

$$X(f) = \int_{-\infty}^{\infty} x(t)e^{-j2\pi ft}dt \tag{11.23}$$

$$x(t) = \int_{-\infty}^{\infty} X(f)e^{j2\pi ft}df \tag{11.24}$$

With this definition, Parseval's theorem becomes

$$\int_{-\infty}^{\infty} x^2(t)dt = \int_{-\infty}^{\infty} |X(f)|^2 df \tag{11.25}$$

11.3 SINC FUNCTION

Let us consider a rectangular function in Figure 11.1, defined by

$$x(t) = A \quad \text{for} \quad |t| \le T/2| \quad \text{and } 0 \quad \text{for} \quad |t| > T/2|.$$

That is, the function is continuous over all t but is zero outside the limits ($-T/2$, $T/2$).

Its FT is[6]

$$\begin{aligned} X(f) &= \int_{-T/2}^{T/2} x(t)e^{-j2\pi ft}dt \\ &= \int_{-T/2}^{T/2} Ae^{-j2\pi ft}dt \\ &= -(A/\pi f\ 2j)[e^{-j\pi fT} - e^{j\pi fT}] \end{aligned} \tag{11.26}$$

Using the identity

$$\sin\theta = (e^{j\theta} - e^{-j\theta})/2j \tag{11.27}$$

yields the following expression for the FT:

$$X(f) = (A/\pi f)\sin(\pi fT) = AT[\sin(\pi fT)/\pi fT] \tag{11.28}$$

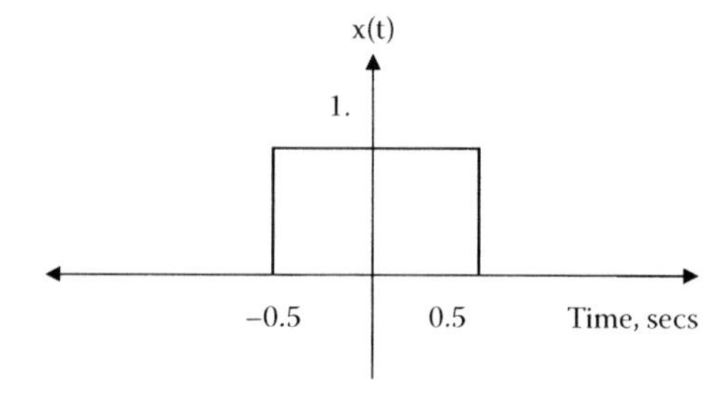

FIGURE 11.1 Rectangular function.

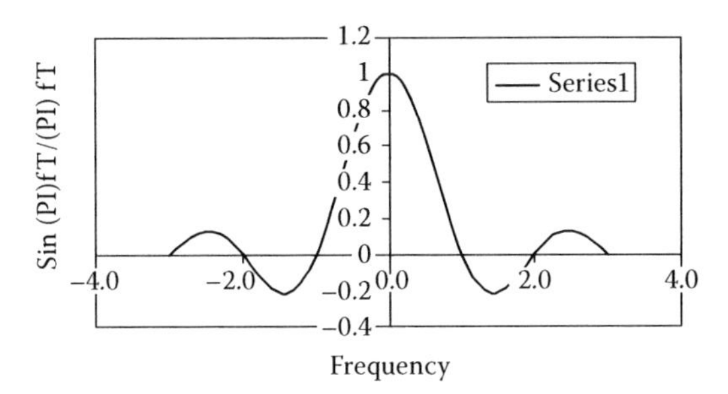

FIGURE 11.2 Sinc function.

The term in brackets, $(\sin \pi x)/\pi x$, or the sinc function, is shown in Figure 11.2.

Although the function is continuous, it has zero values at the points $f = n/T$ for $n = \pm 1, \pm 2, \ldots$, and the sidelobes decrease in magnitude as $1/T$ increases.

11.4 DISCRETE FOURIER TRANSFORM (DFT)

In most measurements these days, signals of interest (e.g., voltages or currents) are usually sampled at equal intervals of time and then recorded. For the analysis of these signals, DFT is used very extensively. The FFT is a highly efficient procedure for computing the DFT of a time series. Hence, we will study both DFT and FFT.

Analogous to the Fourier transform defined in Equations 11.23 and 11.24, we can define the DFT pair as follows:[7–9]

$$X(k) = (1/N)\sum_{n=0}^{N-1} x(n)e^{-j2\pi nk/N} \tag{11.29}$$

$$x(n) = \sum_{n=0}^{N-1} X(k)e^{j2\pi nk/N} \tag{11.30}$$

for $n = 0, 1, \ldots (N-1)$ and $k = 0, 1, \ldots (N-1)$. Both $X(k)$ and $x(n)$ are, in general, complex series. In quite a few situations, $x(n)$ may contain only real numbers.

[Note: The definition of the DFT is not uniform in the literature. If

$$X(k) = a1\sum_{n=0}^{N-1} x(n)e^{-j2\pi nk/N} \quad \text{and} \quad x(n) = a2\sum_{n=0}^{N-1} X(k)e^{j2\pi nk/N},$$

then $(a1)(a2) = 1/N$. Hence, some authors use either $a1$ or $a2$ to be $1/N$, whereas others use $a1 = a2 = 1/\sqrt{N}$. Still others use a positive exponent as in Equation 11.29 or a negative exponent as in Equation 11.30.

For the derivation of DFT from the continuous FT, see References 10 and 11.

When the expression $e^{-2\pi j/N}$ is replaced by the term W, the DFT pair takes the following form:

$$X(k) = (1/N)\sum_{n=0}^{N-1} x(n)\, W^{nk} \tag{11.31}$$

$$x(n) = \sum_{n=0}^{N-1} X(k)\, W^{-nk} \tag{11.32}$$

The DFTs of some of the signals, which are functions of time, are shown in Figure 11.3.

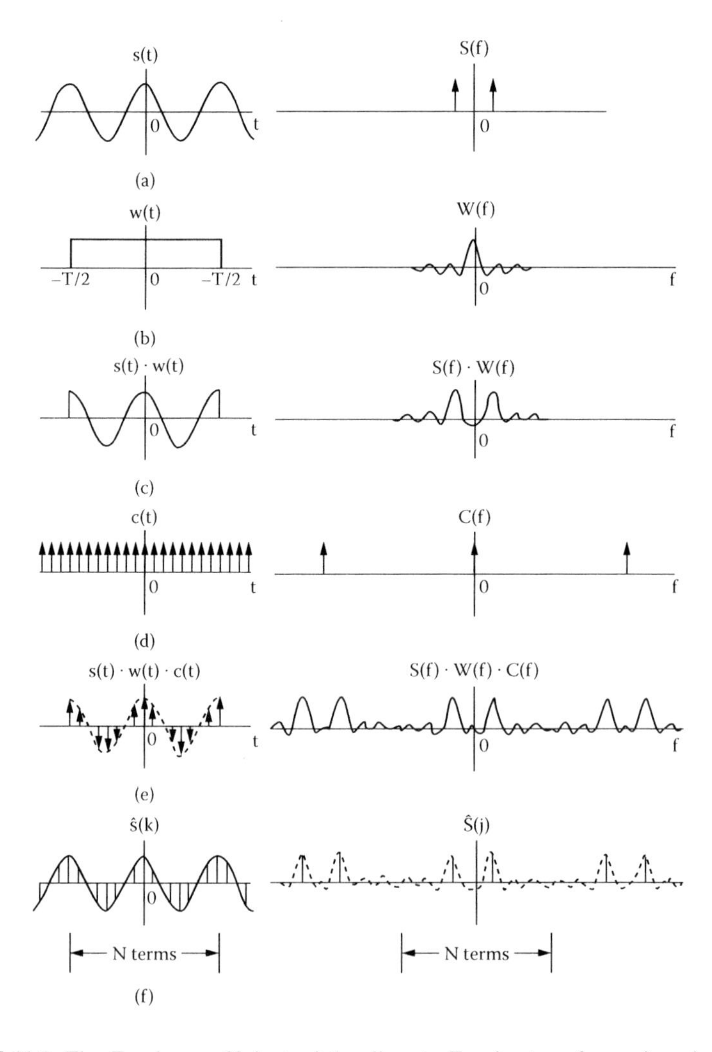

FIGURE 11.3 The Fourier coefficient of the discrete Fourier transform viewed as a corrupted estimate of the continuous Fourier transform. (From: Bergland, G.W., 1969. *A guided tour of the fast Fourier Transform. IEEE Spectrum*, July, 41–52. With permission of IEEE.)

11.5 FAST FOURIER TRANSFORM (FFT)

The two conventional forms of the FFT algorithms are known as *decimation in time* and *decimation in frequency*.[6–9]

We will first discuss the decimation in time form of the FFT algorithm.

11.5.1 Decimation in Time (DIT)

The series expression for the DFT series coefficient $X(k)$ may be separated into two series such that the first contains even sample numbers $B(r)$, and the other, odd sample numbers $C(r)$. Also, let $N = 2^m$, where $m > 0$ and is an integer. Then

$$X(k) = (1/N)\sum_{n=0}^{N-1} x(n)\, W^{nk} \tag{11.33}$$

Let $x(2r)$ and $x(2r+1)$ represent the even and odd sample numbers, respectively, in the original time series where r takes the values 0, 1, 2, …, $\{(N/2) - 1\}$.
Then, we can write

$$X(k) = (1/N)\sum_{n=0}^{N-1} x(n)\, W^{nk}$$

$$= (1/2)\{1/(N/2)\}\sum_{r=0}^{N/2-1} x(2r)\, W^{2rk} + W^k\left[(1/2)\{1/(N/2)\}\sum_{r=0}^{N/2-1} x(2r+1)\, W^{2rk}\right], \tag{11.34}$$

where k takes the values 0, 1, 2, …, $(N - 1)$ and r takes the values 0, 1, 2, …, $(N/2) - 1\}$

$$X(k) = (1/2)B(r) + W^k(1/2)C(r), \tag{11.35}$$

where $B(r)$ and $C(r)$ are the DFTs of the even and odd samples, respectively, each comprising the $N/2$ complex numbers, that is,

$$B(r) = \{1/(N/2)\}\sum_{r=0}^{N-1} x(2r)\, W^{2rk}$$

$$C(r) = \{1/(N/2)\}\sum_{r=0}^{N-1} x(2r+1)\, W^{2rk}$$

Because $B(r)$ and $C(r)$ are DFTs of $N/2$ samples, they obey the periodic property of the DFTs. Hence,

$$B(r + N/2) = B(r) \quad \text{and} \quad C(r + N/2) = C(r) \tag{11.36}$$

If we know the values of $B(r)$ and $C(r)$, all the DFT coefficients of the original sequence can be calculated using Equation 11.35 and giving k the values from 0 to $(N - 1)$. This gives the DIT algorithm

$$W^{(r+N/2)} = e^{-j2\pi(r+N/2)/N} = e^{-j2\pi r/N}\, e^{j\pi} = -e^{-j2\pi r/N} = -W^{r} \tag{11.37}$$

Hence,

$$X(r) = (1/2)B(r) + W^{r}(1/2)C(r) \quad 0 \le r < N/2 \tag{11.38}$$

$$X(r + N/2) = (1/2)B(r) - W^{r}(1/2)C(r) \quad 0 \le r < N/2 \tag{11.39}$$

Let us examine the reduction of time that we have achieved with this algorithm. (For the moment, let us ignore the factor 1/2, which will be just a scaling factor when the complete algorithm is implemented.) If the original sample with N terms required a computation time proportional to N^2, we can now calculate all the DFT coefficients in $2\,(N/2)^2$ time. As we can reduce our samples further to $N/4$, $N/8$, and so on till we finally reach a two-point function, we can show in general that $N \log_2 N$ complex additions and, at most, $(1/2)\, N \log_2 N$ complex multiplications are required for computing the DFT of an N point sequence where N is having a power of 2. For $N = 1024$, this represents a computational reduction from more than 200 to 1.

For $N = 8$, the signal flow graph of one of the DIT versions of the FFT algorithm is shown in Figure 11.4.

11.5.2 Decimation in Frequency (DIF)

This form was found independently by Sande,[12] and by Cooley and Stockham. In this algorithm, the roles of the time and frequency samples are interchanged. The time samples are divided into two sequences, are composed of the first $N/2$ points, and the second is composed of the last $N/2$ points:

$$\begin{aligned} X(k) &= (1/N)\sum_{n=0}^{N-1} x(n)\, W^{nk} \\ &= (1/N)\sum_{n=0}^{N/2-1} [x(n)\, W^{nk} + x(n + N/2)W^{k(n+N/2)}] \\ &= (1/N)\sum_{n=0}^{N/2-1} [x(n) + x(n + N/2)W^{kN/2}]W^{kn} \end{aligned} \tag{11.40}$$

The expansion $W^{kN/2}$ can take only two values, depending on whether k is even or odd. If k is even, say $2r$, where r is an integer, $W^{rN} = 1$, because $W^{N} = 1$. If k is odd and equal to $(2r + 1)$, then $W^{(2r+1)N/2} = W^{rN}W^{N/2} = e^{-(2\pi j/N).N/2)} = e^{-\pi j} = -1$.
When k is even, that is, $k = 2r$,

$$y(n) = x(n) + x(n + N/2) \tag{11.41}$$

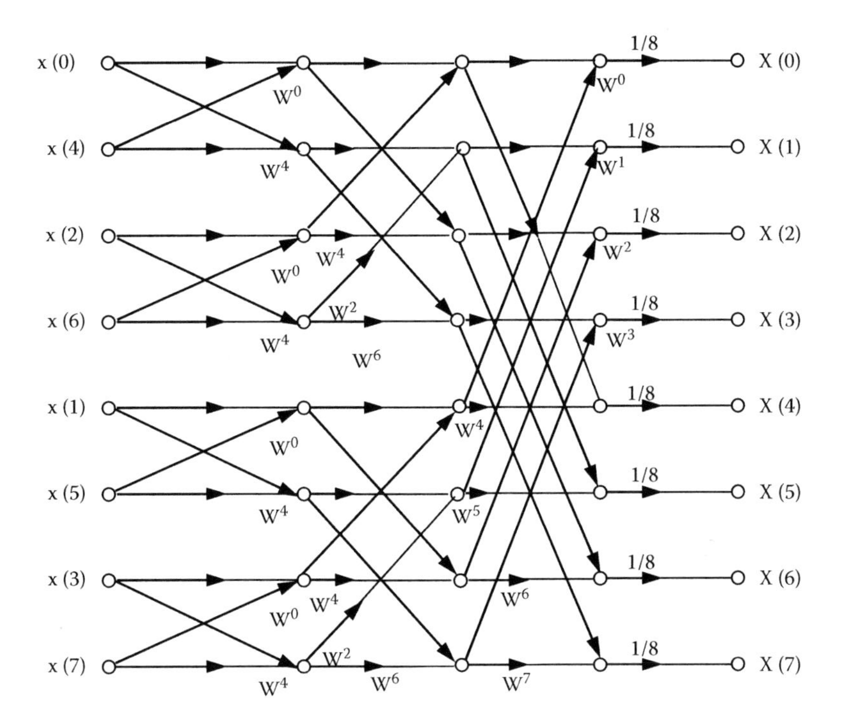

FIGURE 11.4 Flow diagram for eight-point DIT FFT.

When k is odd, that is, $k = 2r + 1$,

$$z(n) = x(n) - x(n + N/2) \tag{11.42}$$

Hence,

$$X(2r) = (1/N)\sum_{n=0}^{N/2-1}[y(n)]W^{2rn} = (1/2)\{1/(N/2)\}\sum_{n=0}^{N/2-1}[y(n)]W^{2rn} \tag{11.43}$$

The right-hand side of the equation is one-half times the $N/2$ point DFT of $y(n) = x(n) + x(n + N/2)$ because $W^2 = \exp[-2\pi j/(N/2)$ and the input sequence is $\{y\ (0), y\ (1)$, and $y\ (N/2-1)$.

When k is odd and $k = 2r + 1$,

$$X(2r+1) = (1/N)\sum_{n=0}^{N/2-1}[z(n)]W^{(2r+1)n} = (1/N)\sum_{n=0}^{N/2-1}[z(n)W^n]W^{2rn}$$

$$= (1/2\{1/(N/2)\}\sum_{n=0}^{N/2-1}[z(n)W^n]W^{2rn} \tag{11.44}$$

The right-hand side of the equation is one-half times an $N/2$ point DFT of $z(n)W^n$, that is, $[x(n) - x(n + N/2)]W^n$. The parameter W^n is sometimes called the *twiddle factor.*

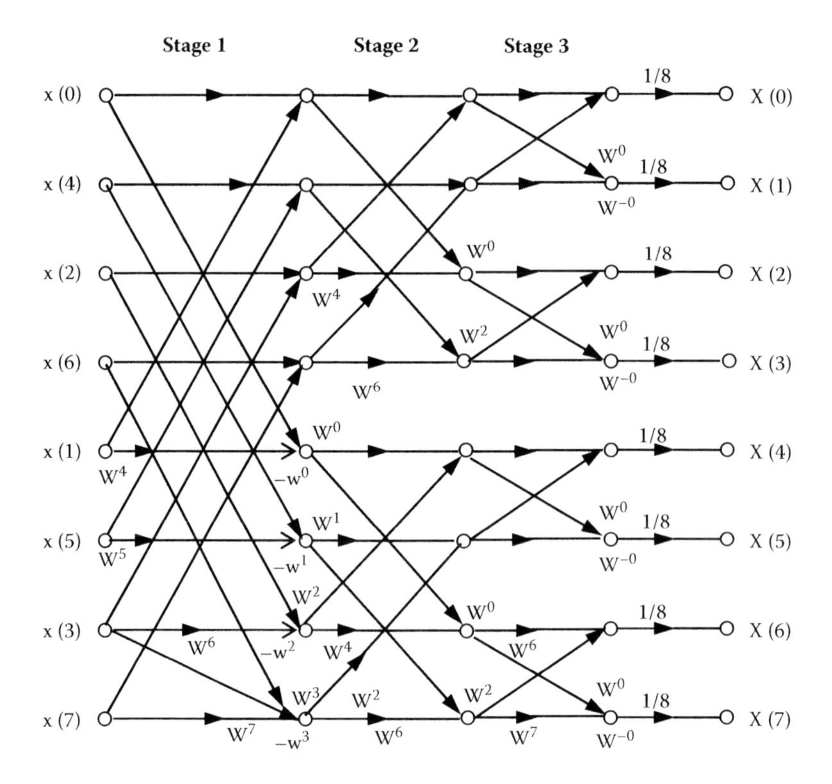

FIGURE 11.5 Flow diagram for eight-point DIF FFT.

As can be seen from Equations 11.43 and 11.44, we were able to obtain all the N coefficients of the DFT of the original sequence having N terms from two DFTs having $N/2$ terms. Hence, all factors regarding the reduction in computer time apply to the DIF algorithm also. Therefore, both methods require the $(N/2)\log_2 N$ complex multiplications and $N\log_2 N$ complex additions for computation of the DFT of an N point sequence, when N is having a power of 2.

For $N = 8$, the signal flow graph of one of the DIF versions of the FFT algorithm is shown in Figure 11.5.

11.5.3 Some Computational Details of FFT Algorithms

In-Place Computation

On careful observation, the signal flow graphs in Figures 11.4 and 11.5 show the values of two nodes in each stage, computed using the values of the earlier stage to its left and the twiddle factors. These values of the two nodes in the earlier stage are not needed in computing the values of any other nodes. Hence, we can write all the intermediate values on the original data sequence. Therefore, only two additional complex number storage locations are needed beyond that required for the original N complex numbers. In the second stage also, the computation of the quantities associated with the next vertical array of nodes to the right involves pairs of nodes, although these pairs are now two locations apart instead of one. This fact does not change the property of "in-place" computation because each pair of nodes affects only the pair immediately to its right. After a new pair is computed, this may be stored in the registers that held the old results, which are no longer needed. In the computation of the final array of nodes, corresponding to the values of the DFT, the computation involves pairs of nodes separated by four locations, but the "in-place" property still holds.

To obtain the version of the algorithm shown in Figure 11.4, initial shuffling of the data sequence $x(n)$ was necessary for "in-place" computation. The shuffling is due to the repeated movements of the odd-numbered members of a sequence to the end of the sequence during each stage of reduction in (see Figure 11.4). This shuffling has been called *bit reversal*[8] because the samples are stored in the bit-reversed order; i.e., $x(4) = x(100)$ is stored in binary position where 001 equals 1, or in $x(1)$ location. Note that this initial data shuffling can also be done in "in-place" computation.

Canonic Forms of the FFT

Reference 8 describes six canonical forms of the FFT along with the following properties:

1. "In-place" computation
2. Normally ordered inputs
3. Normally ordered outputs
4. Normally ordered coefficients, that is, twiddle factors

We may choose among the six canonical forms an algorithm with any one of the foregoing properties, but not all four at once. To achieve in-place computation, we must deal with bit reversal, and to eliminate bit reversal we must give up in-place computation. To obtain normally ordered outputs also, we must deal with bit reversal.

Space will not permit us to discuss all the six canonical forms of the FFT together with their signal flow graphs. The summary of the results reported in Reference 8 are shown in Table 11.1.

Twiddle Factor Determination[6]

Let us consider the DIT algorithm where inputs are normally ordered and outputs are scrambled. Also let us assume that the number of points is $2^5 = 2^m = 32$, and $m = 5$, that is, we will have five stages in the computation. Let us use the index k to denote all the points in each array at each stage, and p to denote the stage numbers. In other words, $x_3(12)$ refers to a node with $k = 12$ and $p = 3$. The expression k (12) in binary form with five bits is 01100. We scale this number $5 - 3 = 2$ places to the right and fill in with zeros; the result is 00011. We then reverse the order of the bits to yield the binary number 11000 or the integer $1.\ 2^4 + 1.\ 2^3 = 24$. The twiddle factor at this node is $W^{-24} = W^{-16}\,W^{-8} = {}^{-}W^{-8}$ because $W^{-16} = -1$. Similarly, we can determine twiddle factors at any node.

TABLE 11.1
Summary of the Properties of the FFT Algorithms

Input	Output	Coefficients (twiddle factors)	Bit Reversal
		Decimation in Time	
Scrambled	Normally ordered	Normally ordered	Needed only for inputs
Normally ordered	Scrambled	Need bit reversal	Needed for outputs and coefficients
		Decimation in Frequency	
Normally ordered	Scrambled	Normally ordered	Needed only for outputs
Scrambled	Normally ordered	Need bit reversal	Needed for inputs and coefficients

Note: All the algorithms have "in-place" computation facility.

There are more complicated algorithms that yield normally ordered inputs and outputs without the facility of "in-place" computation. See Reference 8.

Source: Cochran, W.T. et al., 1967. What is the Fast Fourier Transform? *Proceedings of the IEEE*, Vol. 55, October, 1664–1667.

Bit Reversal[6]

Let B be a binary number equal to $b_4\, b_3\, b_2\, b_1$. Divide B by 2, truncate, and multiply the truncated results by 2. Then compute $b_4\, b_3\, b_2\, b_1 - 2(b_4\, b_3\, b_2)$. If bit b_1 is zero, then this difference is also zero because the division by 2, truncation, and subsequent multiplication by 2 does not alter B. However, if the bit b_1 is one, then truncation changes the value of B, and the difference expression is nonzero. By this technique, we can determine whether bit b_1 is zero or one.

We can similarly identify the other bits, b_4, b_3, and b_2. Then, the bit-reversed number will be given by $b_1\, b_2\, b_3\, b_4$.

11.6 COOLEY–TUKEY ALGORITHM

For the development of computer programs, it is useful to have a set of recursive equations because then the coefficients of the DFT can be calculated iteratively.[6,7] First we will try to derive the Cooley–Tukey algorithm for $N = 8$, and then later generalize it for $N = m$, that is, any integer.

When $N = 8$, let us represent n and k as binary numbers, that is,

$$n = n_2.2^2 + n_1.2^1 + n_0.2^0 = n_2.4 + n_1.2 + n_0.1, \tag{11.45}$$

where $n_2.$, $n_1.$, n_0 can take on the values 0 and 1 only.

Similarly,

$$k = k_2.2^2 + k_1.2^1 + k_0.2^0 = k_2.4 + k_1.2 + k_0.1 \tag{11.46}$$

where $k_2.$, $k_1.$, k_0 can take on the values 0 and 1 only.

Using the earlier definition of DFT in Equation 11.31,

$X(k) = (1/N)\sum_{n=0}^{N-1} x(n)\ W^{nk}$ and letting $A(n) = (1/N)x(n)$, we can write

$$X(k) = \sum_{n=0}^{N-1} A(n)\ W^{nk} \qquad (11.47)$$

Using the representations of n and k in Equations 11.45 and 11.46, Equation 11.47 becomes

$$X(k_2, k_1, k_0) = \sum_{n_0=0}^{1} \sum_{n_1=0}^{1} \sum_{n_2=0}^{1} A(n_2, n_1, n_0) W^{(n_2.4 + n_1.2 + n_0.1)(k_2 4 + k_1 2 + k_0 1)} \qquad (11.48)$$

$W^{m+n} = W^m W^n$, we have

$$= W^{(n_2.4 + n_1.2 + n_0.1)(k_2 4 + k_1 2 + k_0 1)}$$

$$= W^{(k_2.4 + k_1.2 + k_0.1)4n_2} + W^{(k_2.4 + k_1.2 + k_0.1)2n_1} + W^{(k_2.4 + k_1.2 + k_0.1)n_0} \qquad (11.49)$$

If we look at these terms individually, it is clear that they can be written in the following forms,

$$W^{(k_2.4 + k_1.2 + k_0.1)4n_2} = W^{[8(2k_2 + k_1)n_2]}.W^{4k_0 n_2} \qquad (11.50)$$

$$W^{(k_2.4 + k_1.2 + k_0.1)2n_1} = W^{[8 k_2 n 1]}.W^{2(k_0 + 2k_1)n_1} \qquad (11.51)$$

$$W^{(k_2.4 + k_1.2 + k_0.1)n_0} = W^{(k_0 + 2k_1 + 4k_2)n_0} \qquad (11.52)$$

$W^8 = [e^{-2\pi j/8}]^8 = e^{-2\pi j} = 1.$

Therefore, the term W^8 in Equations 11.50 and 11.51 can be replaced by one. This means that Equation 11.48 can be written in the form

$$X(k_2, k_1, k_0) = \sum_{n_0=0}^{1} \sum_{n_1=0}^{1} \sum_{n_2=0}^{1} A(n_2, n_1, n_0) W^{4k_0 n_2}\ W^{2(k_0 + 2k_1)n_1}\ W^{(k_0 + 2k_1 + 4k_2)n_0} \qquad (11.53)$$

$A_1(k_0, n_1, n_0)$

$A_2(k_0, k_1, n_0)$

$A_3(k_0, k_1, k_2)$

In this form, it is possible to perform each of the summations separately and to label the intermediate results. Note that each set consists of only eight terms and that the latest set needs to be saved. Thus, Equation 11.53 can be rewritten in the following forms:

$$A_1(k_0,n_1,n_0)=\sum_{n_2=0}^{1} A(n_2,n_1,n_0)W^{4k_0n_2} \tag{11.54}$$

$$A_2(k_0,k_1,n_0)=\sum_{n_1=0}^{1} A_1(k_0,n_1,n_0).\,W^{2(k_0+2k_1)n_1} \tag{11.55}$$

$$A_3(k_0,k_1,k_2)=\sum_{n_0=0}^{1} A_2(k_0,k_1,n_0)\,W^{(k_0+2k_1+4k_2)n_0} \tag{11.56}$$

Finally, bit reversal is performed to obtain the DFT coefficients in the natural order.

$$X(k_2,k_1,k_0)=A_3(k_0,k_1,k_2) \tag{11.57}$$

Following similar reasoning, we can generalize these results for the case where $N=2^m$. We will state the results without proof. For details of the complete derivation, see Reference 6.

By interchanging the roles of k and n, another set of similar algorithms can be derived.

$$A_1(k_0,n_{(m-2)},\; n_1,n_0)=\sum_{n_{m-1}=0}^{1} A(n_{(m-1)},n_{(m-2)}\cdots n_2,n_1,n_0)W^{k_0|2^{(m-1)}\,n_{(m-1)}|}$$

$$A_2(k_0,k_1,n_{(m-3)},\ldots\; n_1,n_0)=\sum_{n_{m-2}=0}^{1} A_1(k_0,n_{(m-2)}\;\; n_1,n_0).\mathrm{W}^{(k_0+2k_1)|2^{(m-2)}\,n_{(m-2)}|}$$

....

....

$$A_m(k_0,k_1,k_2,\ldots,k_{(m-2)},k_{(m-1)})=$$

$$\sum_{n_0=0}^{1} A_{(m-1)}(k_0,k_1,\; k_{(m-2)},n_0)\times W^{|2^{(m-1)}k_{(m-1)}+\;2^{(m-2)}k_{(m-2)}+\;2^{(m-3)}k_{(m-3)}+\;+k_0|n_0} \tag{11.58}$$

Finally, to account for the bit reversal,

$$X_{(m)}(k_{(m-1)}\,k_{(m-2)},k_{(m-3)},\ldots,k_1,k_0)=A_{(m)}(k_0,k_1,k_2,\ldots,k_{(m-2)},k_{(m-1)}) \tag{11.59}$$

Let us estimate the number of complex multiplications and additions that we have to perform for calculating the DFT coefficients X using this algorithm. We have m equations, and in each equation N terms are multiplied by a complex number twice. However, one of these multiplications is unity because the multiplication is of the form $W^{an_{(m-1)}}$, where $n_{(m-1)} = 0$. Thus, only Nm or $N \log_2 N$ complex multiplications are required. Further, we know that $W^q = -W^{q+(N/2)}$ because $W^{(N/2)} = -1$. Hence, the number of multiplications is reduced by 2. The number of complex multiplications is $(N/2)\log_2 N$. The number of complex additions is Nm or $N \log_2 N$. As earlier explained, this will result in considerable savings in computation time compared to the direct evaluation of an N point DFT, which requires N^2 complex multiplications.

11.7 FFT OF TWO REAL FUNCTIONS SIMULTANEOUSLY

Using the symmetrical properties of DFT, we will develop an algorithm to transform two real-value sequences simultaneously by computing one complex fast fourier transform (CFFT).[13–16] From the definitions of DFT (Discrete Fourier Transform) and IDFT (Inverse Discrete Fourier Transform), we see that the following symmetrical properties exist:

1. $X(k)$ has complex conjugate symmetry (Hermian symmetry) when $x(n)$ is real. In other words, if we define $X(k) = X_r(k) + jX_i(k)$ with real and imaginary parts, $X_r(k)$ has even symmetry, and $X_i(k)$ has odd symmetry; that is,

$$X_r(k) = X_r(-k) \tag{11.60}$$

$$X_i(k) = -X_i(-k) \tag{11.61}$$

2. $x(n)$ is real if and only if $X(k) = X^*(-k) = X^*(N-k)$. (11.62)
3. $X(k)$ is real if and only if $x(n) = x^*(-n) = x^*(N-n)$. (11.63)
4. $x(n)$ is real and even if and only if $X(k)$ is real and even. (11.64)
5. $x(n)$ is real and odd if and only if $X(k)$ is purely imaginary and odd. (11.65)

Let us consider two real functions $x(n)$ and $y(n)$. Let us define a function $z(n)$ as

$$z(n) = x(n) + j\,y(n) \tag{11.66}$$

That is, $z(n)$ is constructed to be the sum of two real functions, where one of them is taken to be imaginary. The DFT has linearity property. Hence, the DFT of $z(n)$, $Z(k)$, is given by the following expression:

$$\begin{aligned} Z(k) &= Z_r(k) + j\,Z_i(k) = \text{DFT}\,[x(n) + j\,y(n)] \\ &= \{X_r(k) - Y_i(k)\} + j\,\{X_i(k) + Y_r(k)\} \end{aligned} \tag{11.67}$$

where subscripts r and i denote real and imaginary parts, respectively.

Further,

$$\begin{aligned} Z(N-k) = Z(-k) &= \{X_r(k) + Y_i(k)\} - j\,\{X_i(k) - Y_r(k)\} \\ &= Z_r(N-k) + j\,Z_i(N-k) \end{aligned} \tag{11.68}$$

From Equation 11.60, $X_r(k) = X_r(-k)$, and from Equation 11.61, $X_i(k) = -X_i(-k)$.

Hence,

$$\frac{1}{2}[Z_r(k) + Z_r(N-k)] = X_r(k) \tag{11.69}$$

$$\frac{1}{2}[Z_i(k) - Z_i(N-k)] = X_i(k) \tag{11.70}$$

Therefore,

$$\text{DFT}\,[x(n)] = X_r(k) + j\,X_i(k)$$

$$= \frac{1}{2}[Z_r(k) + Z_r(N-k)] + j\,\frac{1}{2}[Z_i(k) - Z_i(N-k)] \tag{11.71}$$

$$K = 0, 1, \ldots, \frac{N}{2}$$

Similarly,

$$\text{DFT}\,[y(n)] = Y_r(k) + j\,Y_i(k)$$

$$= \frac{1}{2}[Z_i(N-k) + Z_i(k)] + j\,\frac{1}{2}[Z_r(N-k) - Z_r(k)] \tag{11.72}$$

$$K = 0, 1, \ldots, \frac{N}{2}$$

Equations 11.71 and 11.72 constitute algorithms to compute the DFTs of two real sequences $x(n)$ and $y(n)$ from one complex FFT of $Z(n)$.

For reviews of different algorithms for the computation of the DFT of a real-value series, see Reference 13.

11.8 MIXED-RADIX FFT

Radix-2 FFT algorithms are well established and widely used in the industry. However, there is a significant speed increase from radix-2 to radix-4. The coding of radix-4 takes slightly longer. In general, the speed of calculation increases as the radix number increases; however, the coding becomes more complicated. Single-radix FFTs require points that are integer powers of the radices. The number of points required for the given radices are shown here:

Radix of the FFT	Number of Points Required
3	3, 9, 27, 81, 243
5	5, 25, 125, 625
10	10, 100, 1000

In practice, because of the large number of points required, it is not practicable to use a radix greater than ten.

TABLE 11.2
Typical Calculation Speeds of the Mixed-Radix FFTs and Other FFTs or DFT

Number of Data Points	Calculation Algorithm in 80486 Assembly Language	Time in Seconds
1024	DFT	7.14
1024	Packaged radix-2 FFT	0.06
1024	10 single radix-2 FFT	0.13
1024	$4 \times 4 \times 4 \times 4 \times 4$ FFT	0.09
1000	$10 \times 10 \times 10$ FFT	0.08
1000	$2 \times 2 \times 2 \times 5 \times 5 \times 5$ FFT	0.10
7200	$8 \times 9 \times 10 \times 10$ FFT	0.75
7200	$2 \times 2 \times 2 \times 2 \times 2 \times 3 \times 3 \times 5 \times 5$ FFT	0.99

The power system frequencies 60-Hz and 50-Hz, not being an exact power of two, present some problems when used for harmonic analysis and measurements. Although the sampling speed can be changed to reduce incompatibility, this works only to a limited extent.

We will discuss the windowing technique later to reduce the spectral leakage problem.

Mixed-radix FFTs provide increased flexibility in the choice of data size and sampling rate. For example, a 12-point FFT can be arranged as a series combination of $2 \times 3 \times 2$. These mixed-radix routines can be easily programmed in an Intel 80486 assembler and run on a commonly available 33-MHz clock speed personal computer. Typical calculation speeds of various FFTs and the DFT are listed in Table 11.2. No special attempt has been made to fine-tune the routines to improve their performance.

11.9 SPLIT-RADIX FFT

Lu and Lee in Reference 17, Duhamel and Hollman in References 18 and 19 have reported the basic ideas behind the split-radix algorithm. In this algorithm, they applied a radix-2 index map to the even-indexed terms, and a radix-4 map to the odd-indexed terms. The basic definition of the DFT

$$C_k = \sum_{n=0}^{N-1} x_n W^{nk} \tag{11.73}$$

with $W = e^{-j2\pi/N}$ gives

$$C_{2k} = \sum_{n=0}^{N/2-1} [x_n + x_{n+N/2}] W^{2nk} \tag{11.74}$$

for the even-indexed terms; for the odd-indexed terms, it gives

$$C_{4k+1} = \sum_{n=0}^{N/4-1} [(x_n - x_{n+N/2}) - j(x_{n+N/4} - x_{n+3N/4})]W^n W^{4nk} \quad (11.75)$$

and

$$C_{4k+3} = \sum_{n=0}^{N/4-1} [(x_n - x_{n+N/2}) + j(x_{n+N/4} - x_{n+3N/4})]W^{3n} W^{4nk} \quad (11.76)$$

This results in an L-shaped "butterfly" that relates a length N DFT to one length $N/2$ DFT, and two length $N/4$ DFTs with twiddle factors. Repeating this process for the half- and quarter-length DFTs until scalars result, SRFFT (Split Rapid Fast Fourier) algorithms are obtained in much the same way as the DIF radix-2 Cooley-FFT is derived.[20,21] It is the location of the twiddle factors that makes this algorithm different from a radix-2 FFT; otherwise the signal graph for the algorithm calculated in place looks like a radix-2 FFT.

According to Reference 22, the split-radix FFT algorithm has an efficiency exceeding that of a radix-8 FFT, a size comparable to that of a radix-4 FFT, and the flexibility of a radix-2 FFT. They included in the appendix of their paper example Fortran programs for the decimation in frequency and time algorithms.

11.10 FFT PRUNING

There are basically four modifications of the $N = 2^m$-point FFT algorithm developed by Cooley and Tukey, which give improved computational efficiency.[23] These are

1. Inner-loop nesting instead of recursion.
2. Change in radix.
3. Data shuffling and unscrambling (not to be confused with bit reversal) when input data are real.
4. Eliminating operations on zeros (commonly referred to as pruning) when the number of nonzero input data points is considerably smaller than the desired number of output points (or the number of transform points is considerably smaller than the number of input points).

FFT pruning has applications in speech analysis. This technique can be applied effectively for evaluating a narrow region of the frequency domain by pruning a decimation in the time algorithm. Reference 23 has some estimates of time saving that can be achieved by FFT pruning. Space will not permit a detailed discussion of this topic; interested readers are advised to see Reference 23.

11.11 THE CONVOLUTION INTEGRAL

In a linear system, the system response $h(t)$ to an impulse input $\delta(t)$ can be utilized to determine the response $c(t)$ to any input $r(t)$.[6,7,24] The response $c(t)$ is determined by using the convolution integral, usually written as $r(t) * h(t)$.

$$c(t) = \int_{-\infty}^{\infty} r(\tau)h(t-\tau)d\tau = r(t) * h(t) \quad (11.77)$$

Convolution in the time domain is equivalent to multiplication in the frequency domain. We will first establish this equation.

Let us assume the following Fourier transform pairs:

Time Function	Fourier Transform
$h(t)$	$H(f)$
$r(t)$	$R(f)$
$c(t)$	$C(f)$

Then

$$C(f) = H(f)\,R(f) \tag{11.78}$$

Taking the Fourier transforms of both sides of Equation 11.77,

$$\int_{-\infty}^{\infty} c(t)e^{-j2\pi ft}dt = \int_{-\infty}^{\infty}\left[\int_{-\infty}^{\infty} r(\tau)h(t-\tau)d\tau\right]e^{-j2\pi ft}dt \tag{11.79}$$

This is equivalent to (assuming the order of integration can be changed)

$$C(f) = \int_{-\infty}^{\infty} r(\tau)\left[\int_{-\infty}^{\infty} h(t-\tau)e^{-j2\pi ft}dt\right]d\tau \tag{11.80}$$

By substituting $u = t - \tau$, the term in the bracket becomes

$$\int_{-\infty}^{\infty} h(u)e^{-j2\pi f(u+\tau)}du = e^{-j2\pi f\tau}\int_{-\infty}^{\infty} h(u)e^{-j2\pi fu}du$$

$$= e^{-j2\pi f\tau}\,H(f) \tag{11.81}$$

Now Equation 11.80 can be written as

$$C(f) = \int_{-\infty}^{\infty} r(\tau)e^{-j2\pi f\tau}H(f)\,d\tau = R(f)\,H(f) \tag{11.82}$$

The converse can be proved in a similar manner. In the case of the sampled functions, the convolution integral in Equation 11.82 can be written as

$$c(n) = \frac{1}{N}\sum_{\tau=0}^{N-1} r(\tau)h(n-\tau) \tag{11.83}$$

Therefore, the convolution integral in Equation 11.77 can be computed by calculating the inverse DFT (IDFT) of Equation 11.82. However in the definition of the convolution integral, it was assumed that the impulse response $h(t) = 0$ for $t < 0$. This condition is not satisfied when we try to represent the discrete samples $r(n)$ and $h(n)$ by their Fourier coefficients. Generally, the Fourier coefficients are periodic functions having values even for negative frequencies. This problem can be solved by adding extra N zeros for both the $r(f)$ and $h(f)$ samples and then taking the IDFT of the $2N$ samples. This aspect has been discussed in some detail in References 6 and 7.

In general, when two series have different numbers of samples, say N_1 and N_2, and they have to be convolved, the total number of samples should be $N_1 + N_2 - 1$. In other words, add $N_1 - 1$ zeros to the sample with N_2 numbers, and $N_2 - 1$ zeros to the sample with N_1 numbers, and obtain the IDFT of the two samples $N_1 + N_2 - 1$.[6]

11.12 AUTO- AND CROSS-CORRELATION FUNCTIONS

FFT can be used to calculate the auto- and cross-correlation functions also.

If $x(t)$ and $y(t)$ are two different functions, the correlation integral is defined as

$$z(t) = \int_{-\infty}^{\infty} x(\tau)y(t+\tau)d\tau \tag{11.84}$$

As in the case of the convolution integral in the previous section, it can be shown that the Fourier transform of Equation 11.79 yields[6]

$$Z(f) = X(f)\, Y^*(f) \tag{11.85}$$

where $Y^*(f)$ represents the complex conjugate of $Y(f)$.

By taking the IDFT of Equation 11.84, the cross-correlation functions of $x(t)$ and $y(t)$ can be calculated. If $x(t) = y(t)$, then it is referred to as the auto-correlation function.

In special cases when either $x(t)$ or $y(t)$ is an even function, convolution and correlation are equivalent.

11.13 PITFALLS OF THE DISCRETE FOURIER TRANSFORM

We will now discuss the three common problems encountered while using the DFT, namely, aliasing, leakage, and the picket-fence effect.[5,7]

In practice, when field harmonic measurements are made (for that matter, when any electrical signal is digitally measured), attention has to be paid to the following aspects:

- Sampling frequency—This is determined by the highest frequency in the signal that has significant power.
- Length of time record—This is determined by the resolution required, that is, the spacing between two frequency components in the frequency domain.
- The choice of a window function to reduce spectral leakage errors.

The following discussion should assist in the intelligent choice of the preceding items in different situations. Let us assume that the waveform of a signal contains frequencies up to a maximum of f_{max}. Let us also assume that we sample this waveform with a sampling interval of T seconds. If $T < \frac{1}{2f_{max}}$ and FFT is used to analyze the signal, the results for various harmonics (amplitudes and phase angles) will be correct. Corresponding to this case in the frequency domain (that is, signal power versus frequency or power spectra), curves between zero and f_{max}, and between f_{max} and $2f_{max}$ do not cross each other. (See Figure 11.6a.) In other words, no spectral leakage occurs, that is, signal power from one frequency is not transferred

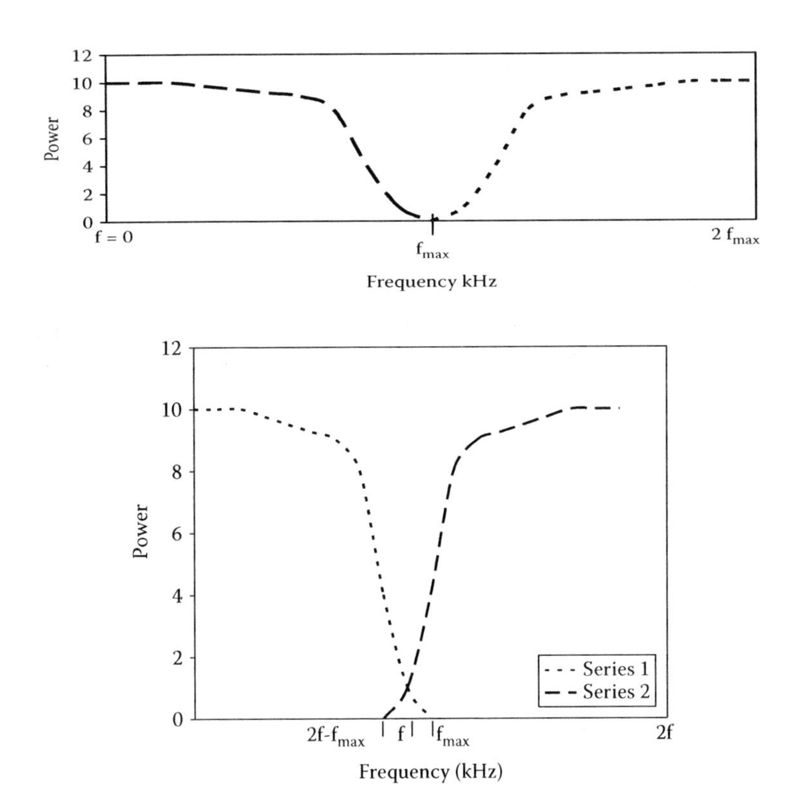

FIGURE 11.6 (a) Effect of sampling on the Discrete Fourier Transform; $T < \frac{1}{2f_{max}}$ (b) Effect of sampling on the Discrete Fourier Transform $T > \frac{1}{2f_{max}}$.

to another frequency due to inadequate sampling. Let the Nyquist frequency $f = \frac{1}{2T}$. If $T > \frac{1}{2f_{max}}$ in the frequency domain (i.e., signal power versus frequency), the curves cross each other in the region between $2f - f_{max}$ and f_{max}. This effect is usually known as *aliasing* (see Figure 11.6b).

11.13.1 Aliasing

The term *aliasing* refers to the fact that high-frequency components of a time function can impersonate low frequencies if the sampling rate is too low. This is demonstrated in Figure 11.7 by showing a relatively high frequency signal and a relatively low frequency that share identical sample points. This uncertainty can be removed by (a) anti-aliasing filters, or (b) increasing the sampling frequency to a value greater than twice the frequency to be evaluated. (This is necessary in view of Shannon's sampling theorem.)

11.13.2 Spectral Leakage

Spectral leakage will occur when the time record data used by the FFT algorithm does not contain an integral number of power frequency cycles. Power system frequency is subject to small deviations. For accurate measurements, people use a

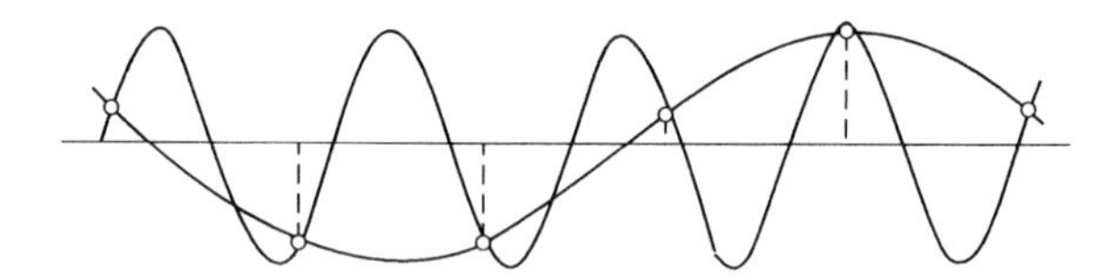

FIGURE 11.7 An example of high frequency "impersonating" a low frequency. (From: Bergland, G.W., 1969. A guided tour of the fast Fourier Transform. *IEEE Spectrum*, July, 41–52. With the permission of IEEE.)

phase-locked loop to vary the sampling rate of the data acquisition system so as to ensure that an integral number of cycles is always obtained. Personal computer-based systems generally operate with a fixed sampling interval, and phase-locked loop facility is usually not provided.

Whenever we analyze a finite record data, basically we are multiplying the finite data by a rectangular (or uniform) window, and then performing a DFT analysis. As one can see from Figures 11.1 and 11.2, the FT of a rectangular window has many sidelobes. Performing a multiplication in the time domain is equivalent to performing a convolution in the frequency domain. Thus, some of the power in the signal leaks into these frequencies represented by the sidelobes, which are spurious peaks. The objective is usually to localize the contribution of a given frequency by reducing the amount of "leakage" through these sidelobes.

Windows are used in harmonic analysis to reduce the undesirable effects caused by spectral leakage. The signal value is multiplied by the corresponding value of the window function, and the modified signal is analyzed using FFT. The ideal filter characteristic of a window function in the frequency domain is a flat characteristic having no sidelobes. This would reduce spectral leakage errors and give the correct magnitudes for frequency deviations up to the edge of the filter characteristic. Such an ideal characteristic is not really available, but a number of window functions used to preprocess the samples in the time record yield effective filter characteristics that have extended peak responses and rapid attenuation of the sidelobe amplitudes. A host of windows are discussed in the literature.[25,26] We will consider only the international standard window function, that is, the Hanning window. This is often incorporated into spectrum analyzers. It is defined as

$$w(n) = 0.5[1.0 - \cos(2\pi n/N)],\ n = 0, 1, \ldots, (N-1).$$

The Hanning window can be applied by convolving the DFT coefficients with the weights −1/4, 1/2, and −1/4.[7,25–27]

The Hanning (also known as von Hahn) window is compared with the unwindowed sinc-filter characteristic (also referred to as *uniform window*) in Figure 11.8 to show how spectral leakage can be reduced.[28] Figure 11.9 shows the Hanning window and its cepstrum. The FT of the log-magnitude spectrum is usually referred to as *Cepstrum*.

11.13.3 The Picket-Fence Effect

Spectral leakage also occurs when the time record data used by the FFT algorithm contains frequency components that do not correspond to one of the spectral lines.

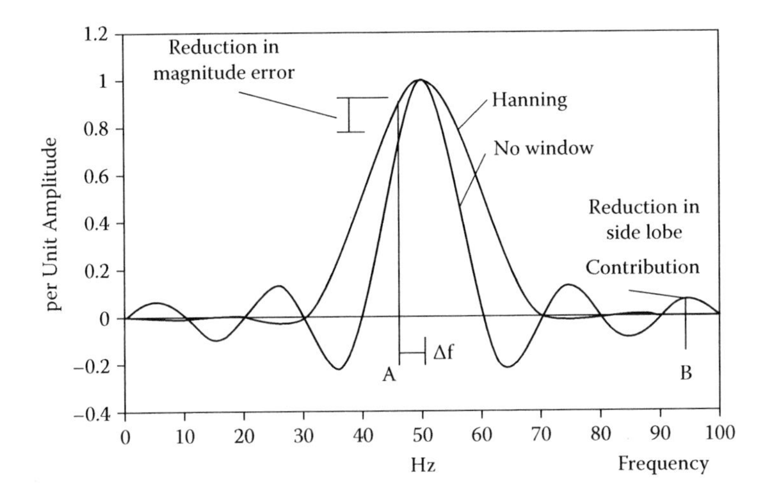

FIGURE 11.8 Reduction of errors using windowing. (From: George, T.A. and Bone, D., 1991. Harmonic power flow determination using the fast Fourier transform, *IEEE Trans. on Power Delivery,* 6(2), April, 530–535. With the permission of IEEE.)

(interharmonics or subharmonics). This effect is caused by interharmonics and is usually referred to as the picket-fence effect.[7]

This effect can be reduced by extending the record length by adding zeros, which is usually called *zero padding*. This increase in record length introduces extra DFT filters at points between the original filters. The bandwidth of the individual filters still depends on the original record length, that is, the width of the spectral window, and is therefore unchanged. This zero padding and the use of windows such as the Hanning window reduces the picket-fence effect substantially.

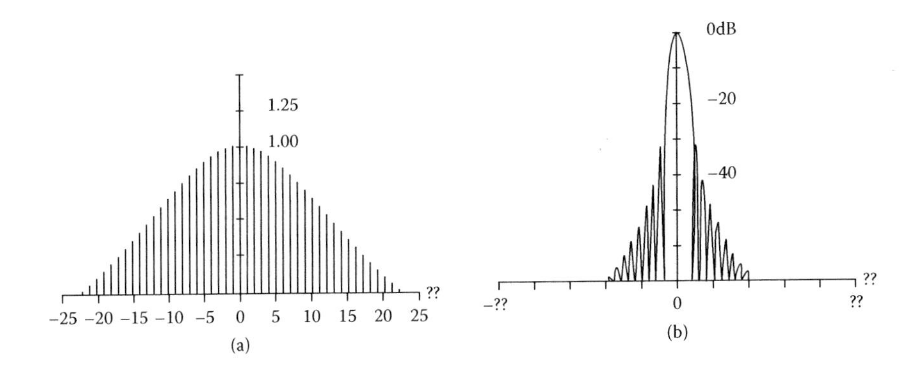

FIGURE 11.9 Hanning window: (a) Cos2 $\frac{n\pi}{N}$ window; (b) Log-magnitude of transform. (From: Harris, F.J., 1978. On the use of windows for harmonic analysis with discrete Fourier transform, *Proc of the IEEE,* 66(1), January, 51–83. As this research is funded by the U.S. federal government, no permission is necessary to reproduce this figure.)

11.14 GUIDELINES FOR USING FFT FOR HARMONIC ANALYSIS

The following guidelines are provided for FFT analysis of data.[28]

1. As the power system frequency, for most of the time, will have some deviation from the nominal frequency of 60 or 50 Hz, phase-locked loops must be used to vary the sampling rate of the data acquisition system to ensure that an integral number of cycles is always obtained in a record of measurement.
2. Windowing should always be used to reduce spectral leakage. Several windowing functions are available (e.g., Hanning, Hamming, and Kaiser) that will greatly reduce spectral leakage errors. Each has characteristics to suit different applications, but the Hanning window generally provides good results in the frequency ranges normally considered for harmonic investigations, that is, up to 3.0 kHz.
3. If one or both of the adjacent harmonics are much greater than the harmonic under consideration, it is possible that spectral leakage will cause significant magnitude and phase errors and results should be treated with caution. If the relative magnitudes differ by 10 (no-windowing) or 100 (windowing), excessive errors are likely.
4. Increasing the length of the time record increases the resolution in the frequency domain, that is, there will be more spectral lines between harmonics, and the leakage errors from adjacent harmonics will be further attenuated. This is one method of reducing errors caused by spectral leakage.
5. Increasing the sample frequency, but not the time length, will produce more spectral lines with the same frequency separation. Higher frequencies will thus be represented, but spectral leakage errors will be unaffected.
6. Where practical, anti-aliasing filters should be used on input channels. In cases where harmonics outside the system bandwidth are known to have low amplitudes, the aliasing error may be small enough to be ignored.
7. Frequency deviations in fundamentals cause proportionately larger deviations at harmonic frequencies. Without phase-locked loop facility and appropriate windowing, the calculated magnitudes and phase angles of higher-order harmonic components may be completely unreliable.
8. Power flow direction calculations are very sensitive to the phase angle when the voltage and current components are almost in quadrature. (This is the case with shunt reactors, capacitors, and tuned filters). Under these conditions, caution should be exercised when interpreting results.
9. A time record should contain either a static or pseudostatic signal if accurate power flows are to be obtained.
10. The frequency-response characteristics of the measuring potential and current transformers (PTs and CTs) can significantly affect the accuracy of harmonic measurements. In particular, the phase characteristics of these instruments are important for accurate power flow direction measurements. These are seldom available from the manufacturers for harmonic frequencies and also rather difficult to measure without sophisticated measurement facilities at higher voltages above 110 kV.

REFERENCES

1. Cooley, J.W. and Tukey, J.W., 1965. An algorithm for the machine calculation of complex Fourier series, *Math. Comput.*, Vol. 19, pp. 297–301.
2. Several papers in *IEEE Trans. Audio and Electroacoustics, June 1967.*
3. Kimbark, E.W., 1971. *Direct Current Transmission,* Vol. 1, John Wiley, New York.
4. Kreyszig, E. *Advanced Engineering Mathematics,* 2nd Edition, John Wiley, New York.
5. Arrillaga, J. and Watson, N.R., 2003. *Power System Harmonics*, John Wiley, Chichester, U.K.
6. Brigham, E.O., 1988. *The Fast Fourier Transform and its Applications,* Prentice Hall, Englewood Cliffs, NJ.
7. Bergland, G.W., 1969. A guided tour of the fast Fourier Transform. *IEEE Spectrum*, July 41–52.
8. Cochran, W.T. et al., 1967. What is the Fast Fourier Transform? *Proceedings of the IEEE,* Vol. 55, October, 1664–1667.
9. Elliott, D.F. and Ramamohan, Rao, K., 1982. *Fast Transforms, Algorithms, Analyses, Applications*, Academic Press, New York.
10. Blackman, R.B. and Tukey, J.W., 1958. *The measurement of Power Spectra,* Dover, New York.
11. Cooley, J.W., Lewis, P.A., and Welch, P.D., 1967. Application of the fast Fourier Transform to computation of Fourier Integrals, Fourier series and convolution integrals, *IEEE Trans. Audio and Electroacoustics,* June, 79–84.
12. Gentleman, W.M. and Sande, G., 1966. Fast Fourier Transforms—for fun and profit, *1966 Fall Joint Computer Conf., AFIPS Proc.*, Vol. 29, (Washington (D.C.), Sparton.)
13. Sorenson, H.V., Jones, D.L., Heideman, M.T., and Burrus, C.S., 1987. Real-valued Fast Fourier Transform Algorithms, *IEEE Trans. on Acoustics, Speech and Signal Processing,* Vol. ASSP-35 (6), June, 849–863.
14. Cooley, J.W., Lewis, P.A.W., and Welch, P.D., 1969. The finite Fourier Transform, *IEEE Trans. Audio and Electroacoustics,* Vol. AU-17, June, 77–85.
15. Ibid 1970. The fast Fourier transform algorithm: Programming considerations in the sine, cosine, and Laplace transforms, *J. Sound, Vib.,* Vol. 12(3), July, 315–337.
16. Rabiner, L.R. and Rader, C.M. (Eds.), 1972. *Digital Signal Processing*, IEEE Press, New York.
17. Lu, I.D. and Lee. P., 1994. Use of mixed radix FFT in Electric Power System Studies, *IEEE Trans. on Power Delivery,* Vol. 9(3), July, 1276–1280.
18. Duhamel, P. and Hollman, H., 1984. Split radix FFT algorithm. *Electronics Letters,* Vol. 20, Jan. 5, 14–16.
19. Duhamel, P., 1986. Implementation of split-radix FFT algorithms for complex, real and real symmetric data, *IEEE Trans. on Acoustics, Speech and Signal Processing,* Vol. ASSP-34 (2), April, 285–295.
20. Oppenheim, A.V. and Schafer, R.W., 1975. *Digital Signal Processing,* Prentice Hall, Englewood Cliffs, NJ.
21. Rabiner, L.R. and Gold, B., 1975. *Theory and Application of Digital Signal Processing,* Prentice Hall, Englewood Cliffs, NJ.
22. Sorenson, H.V., Heideman, M.T., and Burrus, C.S., 1986. On computing the split-radix FFT. *IEEE Trans. Acoustics, Speech, and Signal Processing,* Vol. ASSP-34 (1), February, 152–156.
23. Markel, J.D., 1971. FFT Pruning, *IEEE Trans. Audio and Electroacoustics,* Vol. AU-19, December, 305–311.
24. Kinariwala, B. et al., 1973. *Linear Circuits and Computation*, John Wiley, New York.

25. Harris, F.J., 1978. On the use of windows for harmonic analysis with discrete Fourier transform, *Proc. of the IEEE*, 66(1), January, 51–83.
26. Bingham, C., Godfrey, M.D., and Tukey, J.W., 1967. Modern techniques of power spectrum estimation, *IEEE Trans. Audio and Electroacoustics,* June, 56–66.
27. Probability Aspects Task Force of the Harmonics Working Group Subcommittee of the Transmission Distribution Subcommittee, "Time-Varying Harmonics: Part 1—Characterizing Measured Data" (1998), *IEEE Trans. Power Delivery* 13(3), July, 938–943.
28. George, T.A. and Bone, D., 1991. Harmonic power flow determination using the fast Fourier transform, *IEEE Trans. on Power Delivery* 6(2), April, 530–535.

Index

G

H

K

L

M

N

O

P

Q

R

S

T

U

V